Springer Series on

Wave Phenomena 1

Springer Series on

Wave Phenomena

Volume 1 **Mechanics of Continua and Wave Dynamics** 2nd Edition
By L. M. Brekhovskikh, V. Goncharov

Volume 2 **Rayleigh-Wave Theory and Application**
Editors: E. A. Ash, E. G. S. Paige

Volume 3 **Electromagnetic Surface Excitations**
Editors: R. F. Wallis, G. I. Stegeman

Volume 4 **Short-Wavelength Diffraction Theory**
Asymptotic Methods
By V. M. Babič, V. S. Buldyrev

Volume 5 **Acoustics of Layered Media I**
Plane and Quasi-Plane Waves
By L. M. Brekhovskikh, O. A. Godin

Volume 6 **Geometrical Optics of Inhomogeneous Media**
By Yu. A. Kravtsov, Yu. I. Orlov

Volume 7 **Recent Developments in Surface Acoustic Waves**
Editors: D. F. Parker, G. A. Maugin

Volume 8 **Fundamentals of Ocean Acoustics** 2nd Edition
By L. M. Brekhovskikh, Yu. P. Lysanov

Volume 9 **Nonlinear Optics in Solids**
Editor: O. Keller

Volume 10 **Acoustics of Layered Media II**
Point Sources and Bounded Beams
By L. M. Brekhovskikh, O. A. Godin

Volume 11 **Resonance Acoustic Spectroscopy**
By N. Veksler

Volume 12 **Scalar Wave Theory**
Green's Functions and Applications
By J. A. Desanto

Volume 13 **Modern Problems in Radar Target Imaging**
Editors: W.-M. Boerner, F. Molinet, H. Überall

Volume 14 **Random Media and Boundaries**
Unified Theory, Two-Scale Method, and Applications
By K. Furutsu

Volume 15 **Caustics, Catastrophes, and Wave Fields**
By Yu. A. Krautsov, Yu. I. Orlov

Volume 16 **Electromagnetic Pulse Propagation in Causal Dielectrics**
By K. E. Oughstun, G. C. Sherman

Volume 17 **Wave Scattering from Rough Surfaces**
By A. S. Voronovich

L. M. Brekhovskikh V. Goncharov

Mechanics of Continua and Wave Dynamics

Second Edition
With 99 Figures

Springer-Verlag
Berlin Heidelberg New York
London Paris Tokyo
Hong Kong Barcelona
Budapest

Academician Leonid M. Brekhovskikh
Dr. Valery Goncharov
Russian Academy of Sciences, P. P. Shirshov Institute of Oceanology,
23 Krasikowa St., 117218 Moscow, Russia

Title of the original Russian edition: *VVedenie v mekhaniku sploshnykh sred*

ISBN 3-540-57336-4 2. Auflage Springer-Verlag Berlin Heidelberg New York
ISBN 0-387-57336-4 2nd Edition Springer-Verlag New York Berlin Heidelberg

ISBN 3-540-13765-3 1. Auflage Springer-Verlag Berlin Heidelberg New York
ISBN 0-387-13765-3 1st Edition Springer-Verlag New York Berlin Heidelberg

Library of Congress Cataloging-in-Publication Data. Brekhovskikh, L. M. (Leonid Maksimovich) [Vvedenie v mekhaniku sploshnykh sred. English] Mechanics of continua and wave dynamics / L. Brekhovskikh, V. Goncharov. – 2nd ed. p. cm. – (Springer series on wave phenomena; 1) ISBN 3-540-57336-4 (Berlin: acid-free). – ISBN 0-387-57336-4 (New York: acid-free) 1. Continuum mechanics. 2. Wave mechanics. I. Goncharov, V. V. (Valeriĭ Vladimirovich) II. Title. III. Series. QA808.2B7413 1994 531'. 1–dc20 93-39737

Printed in Germany

Typesetting: Syntax International, Singapore 0513

54/3140-5 4 3 2 1 0 – Printed on acid-free paper

Preface to the Second Edition

The first edition of this text has sold steadily and received much praise from the scientific community. The new edition has given us the opportunity – apart from correcting a few misprints – to add some important material which was missing in the first edition. Following the suggestion of Dr. H. Lotsch, Springer-Verlag, we have presented the new material in the form of additional problems, which serve to guide the reader step by step towards a thorough comprehension of the subject.

November 1993 *L. M. Brekhovskikh · V. Goncharov*

Preface to the First Edition

This text is based on lectures given by the authors to students of the Physico-Technical Institute in Moscow in the course of many years. Teaching mechanics of continuous media to students in physics has some specific features. Unfortunately, the students rarely have enough time for this branch of science. Over a comparatively short period of time and without sophisticated mathematics, the lecturer has to set forth the principal facts and methods of this rather important aspect of theoretical physics. The goal can be achieved only if the knowledge and intuition obtained by students in other courses has been mobilized efficiently. These observations have extensively been taken into account when classical as well as more contemporary and actively developing branches of mechanics of continua are considered.

The theory of wave propagation is the most important topic in mechanics of continua for those who work in the field of physics and geophysics. That is why most attention is paid to this topic in the main text as well as in numerous exercises. The propagation of various kinds of hydrodynamic, magnetohydrodynamic, acoustic and seismic waves is considered in some detail.

We begin with simple questions and proceed to more complicated ones. Accordingly, we treat the elasticity theory first. In the limits this appears to be simpler than the mechanics of fluids. For further study of each topic the reader will find references in the bibliography.

The authors would like to thank very much G. I. Barenblat, A. G. Voronovich, G. S. Golitzin and L. A. Ostrovsky for reading the manuscipt and for many useful comments, and also V. Vavilova for translating the greater part of the book from Russian into English.

December 1984 *L. M. Brekhovskikh · V. Goncharov*

Contents

Part I Theory of Elasticity

Part II **Fluid Mechanics**

Part I **Theory of Elasticity**

All real bodies deform under the action of forces, that is, change their shape or volume or both. These changes are termed *strains*. The theory of elasticity deals with so-called *elastic solids*. Strains in such bodies vanish if forces are removed quasistatically, i.e. sufficiently slowly. When the applied forces are removed abruptly, oscillations can arise in an elastic solid.

Thus elasticity theory excludes such phenomena as plastic and residual strains, relaxation processes, etc. Moreover, we will consider below only weak strains where the distance between two neighbouring points of a solid changes little compared with that in the undeformed state. In this case terms of second and higher powers can be neglected in the equations and we obtain *linear theory of elasticity*.

Strain in an elastic body can also be caused or accompanied by temperature variations of the body. We confine ourselves, however, to the study of the strain at constant temperature or constant entropy, i.e., of *isothermic* or *adiabatic* strain. Quasistatic strain of elastic bodies is usually isothermal: a body has enough time to reach the temperature of the environment. On the contrary, in dynamical processes (in wave motion, for example) the heat transfer from one part of a body to another is frequently negligible, i.e. these processes are adiabatic. The laws of deformation for solids in isothermal and adiabatic processes, in principle, are the same and differ only in the empirically obtained numerical coefficients, called *elastic constants*. By virtue of this, as a rule we will usually not state what kind of process is considered in a concrete case.

1. The Main Types of Strain in Elastic Solids

In this chapter we will consider basic laws of the theory of elasticity such as Hooke's law and Poisson's relation. The main types of deformations of elastic bodies (tension of a rod, shearing, torsion, bending of a beam) will be treated by applying of these laws.

1.1 Equations of Linear Elasticity Theory

1.1.1 Hooke's Law

Elastic behaviour of solids is eventually determined by their molecular structure which is not considered in the mechanics of continua. Therefore, to establish the elasticity-theory laws, one requires experiments. In this case we have to distinguish between the statements based on experiment which are not derived in some logical way, and purely logical consequences from those laws.

The fundamental equations governing the elastic behaviour of *isotropic* solids—i.e. solids the properties of which are equal in every direction—can be deduced from two experimental facts:

i) the extension δl of an elastic bar of length l is proportional to the tensile force F (assumed to be small) applied to the bar's ends: $F \sim \delta l$;

ii) during extension the bar contracts in the transverse direction, the relative contraction being proportional to the relative longitudinal extension $\delta h/h \sim \delta l/l$, where h is the characteristic transverse dimension of the bar (Fig. 1.1).

The experimental fact (i) is Hooke's law expressing linearity of the stress-strain relation.

Let us establish now the relationship between the force F and the bar's length l. Connect the end faces of two bars, each of length l (Fig. 1.2) and again apply the force F to the ends of the composite bar. Each bar is apparently under the action of the same force F which leads to the extension δl. The extension of the

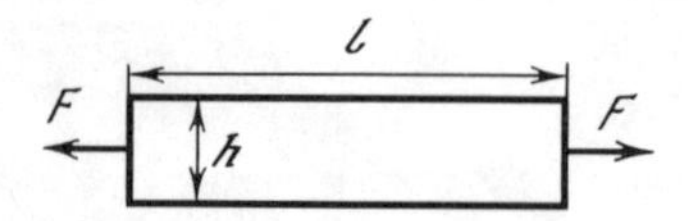

Fig. 1.1. Rectangular bar subjected to the tensile force F

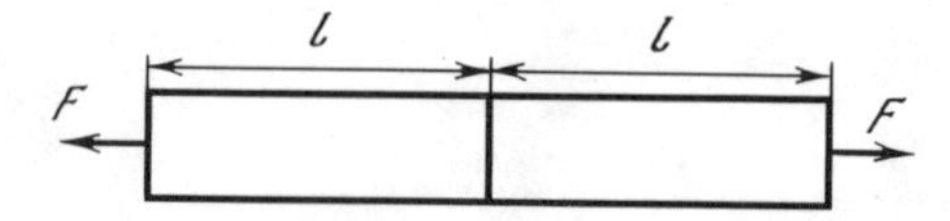

Fig. 1.2. A composite bar subjected to the tensile force F

composite bar is $2\,\delta l$. We see, as a result, that with the same F the extension is proportional to the bar's length.

We determine also the dependence of the force F on the bar's cross-sectional area S. To do this we consider two similar bars (their lateral faces are in contact) and apply the force F to the end of each bar. The extension is again δl as in the case of a single bar, but now the total force $2F$ is applied to the composite double bar. Hence, the cross section is doubled, an extension is the same as in the single bar but with the twice force. In other words, at constant extension the force is proportional to the cross-sectional area of the bar.

Now the relationship between the force F, the length l and the cross section S of the bar may be written in the form

$$F/S = E\,\delta l/l\,. \tag{1.1}$$

This is *Hooke's law*, and the constant E is termed *Young's modulus*. The quantity F/S is called *stress* and the relative extension $\varepsilon = \delta l/l$ *strain*. Thus, according to Hooke's law, stress is equal to the product of Young's modulus and strain.

In accordance with the experimental fact (ii), one more constant can be introduced—*Poisson's ratio* ν which relates longitudinal and transverse strains by

$$\delta h/h = -\nu\,\delta l/l\,. \tag{1.2}$$

Expressing $\delta l/l$ here in terms of F, according to (1.1), yields

$$\delta h/h = -(\nu/E)F/S\,. \tag{1.3}$$

The minus sign is taken in (1.2) because according to experiment, for an overwhelming majority of bodies, for positive δl we have negative δh.

Hooke's law and the relation (1.2) being experimental facts are the basis of the linear elasticity theory of isotropic bodies. Its laws follow from these facts as classic mechanics follows from Newton's laws and electrodynamics from Maxwell's equations, and thermodynamics from a few basic principles.

1.1.2 Differential Form of Hooke's Law. Principle of Superposition

It is often advisable to use Hooke's law (1.1) in differential form. To obtain this form consider the displacement $u(x)$ of the bar's points along its longitudinal axis (the x axis in Fig. 1.3) and calculate the length change of the bar's

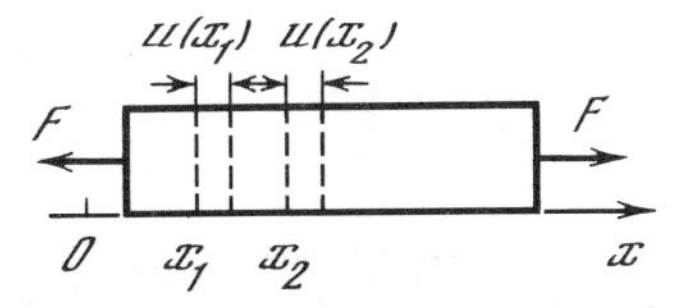

Fig. 1.3. The derivation of the differential form of Hooke's law

infinitesimal element whose length before deformation is $\Delta x = x_2 - x_1$. After deformation it is

$$x_2 + u(x_2) - [x_1 + u(x_1)] \simeq \Delta x + (\partial u/\partial x)_{x_1} \Delta x .$$

Thus, the length change for an arbitrary x value is: $\delta(\Delta x) = (\partial u/\partial x)\,\Delta x$. Hence, the longitudinal strain of the element Δx is $\varepsilon = \delta(\Delta x)/\Delta x \simeq \partial u/\partial x$. Substituting this for $\delta l/l$ in (1.1), we obtain *Hooke's law in differential form:*

$$F/S = E\,\partial u/\partial x . \tag{1.4}$$

An important consequence of the linearity of (1.1, 2) is the *principle of superposition.* Suppose we have a strain ε corresponding to some force F. Consider F as the sum $F = F_1 + F_2 + \cdots + F_n$, i.e. as a superposition of n forces. According to the principle of superposition, the strain can also be presented as a sum of n strains: $\varepsilon = \varepsilon_1 + \varepsilon_2 + \cdots + \varepsilon_n$. Each ε_i can be obtained from F_i, using (1.1, 2) as if no other forces would exist.

1.2 Homogeneous Strains

An elastic body is *homogeneously deformed* if the strain $\varepsilon = \partial u/\partial x$ is constant at all its points. Let us use the principle of superposition to consider some simple examples of this kind.

1.2.1 An Elastic Body Under the Action of Hydrostatic Pressure

Consider a body in the form of a rectangular bar with the dimensions l, h, w immersed in a fluid under pressure p (see Fig. 1.4). The forces acting on the bar's faces are normal to them and the stress [force per unit area] is also the same at each point and equal to p. Using the principle of superposition, we divide the problem into three:

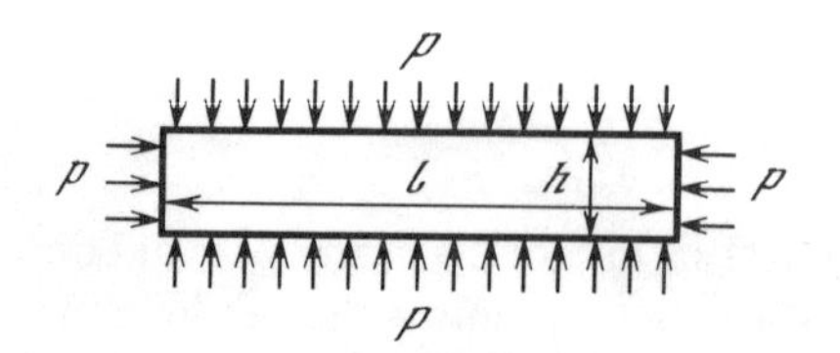

Fig. 1.4. Rectangular bar under hydrostatic pressure

i) A longitudinal contraction δl_1 in the l-direction caused by the action of forces applied to the ends of the bar. According to (1.1) we have $\delta l_1/l = -p/E$.

ii) A extension δl_2 in the same direction due to the forces applied to the horizontal lateral faces. By (1.3) we have

$$\delta l_2/l = \nu p/E .$$

iii) A similar extension as a result of the forces on the vertical lateral faces (parallel to the figure's plane): $\delta l_3/l = \nu p/E$.

The complete strain in the l-direction is

$$\delta l/l = \delta l_1/l + \delta l_2/l + \delta l_3/l = -p(1-2\nu)/E\,.$$

By virtue of the symmetry of the problem, we have similar expressions for the strains in the other two directions:

$$\delta h/h = \delta w/w = -p(1-2\nu)/E\,.$$

Prior to deformation, the bar's volume is $V = lhw$. Due to deformation it changes by δV which can be determined with an accuracy to linear (in strain) terms:

$$\begin{aligned}\delta V &= (l+\delta l)(h+\delta h)(w+\delta w) - V\\ &\simeq V\left(\frac{\delta l}{l}+\frac{\delta h}{h}+\frac{\delta w}{w}\right) = -3Vp\frac{1-2\nu}{E}\,.\end{aligned}$$

This relation can also be written as

$$p = -K\,\delta V/V\,, \qquad K = E/3(1-2\nu)\,. \tag{1.5}$$

K is called the *bulk modulus* or *modulus of compression.* It follows from (1.5) that $\nu < 1/2$, since in the opposite case $K < 0$ and the volume increases with increasing external pressure, which is impossible.

1.2.2 Longitudinal Strain with Lateral Displacements Forbidden

Let a bar extend or contract in the x-direction because of the force F_x. Suppose that a displacement in the y-direction is forbidden. In the case of contraction this can be easily implemented by putting the bar between two fixed, parallel walls (Fig. 1.5). The bar's faces are supposed to slip easily along the walls. Lateral faces perpendicular to the z-axis are supposed to be free. The force of reaction, F_y, is exerted by the walls on the faces normal to the y-axis which contributes additionally to the displacement δl_x. This contribution can be determined with

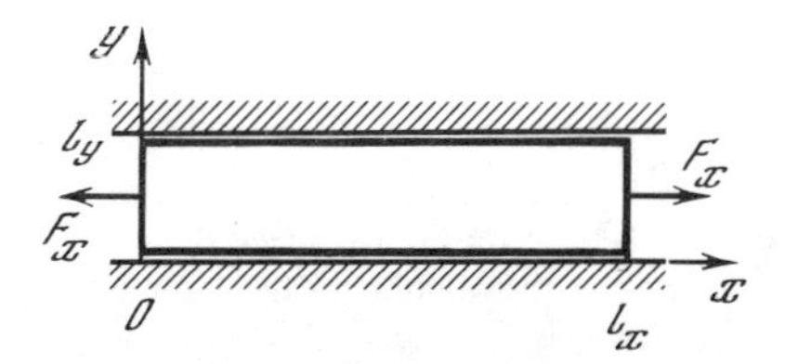

Fig. 1.5. Longitudinal deformation of a bar with lateral displacements forbidden

the help of an equation analogous to (1.3). As a result, we obtain for the uniaxial displacement in the x-direction:

$$\delta l_x/l_x = E^{-1}F_x/S_x - \nu E^{-1}F_y/S_y,$$

where S_x and S_y are the bar's cross-sectional area normal to the axes x and y, respectively; F_y is a yet unknown reaction force of the wall. The latter can be easily found if we write a similar expression for the strain in the y-direction:

$$\delta l_y/l_y = E^{-1}F_y/S_y - \nu E^{-1}F_x/S_x = 0,$$

which is zero in accordance with our assumption. Consequently, $F_y/S_y = \nu F_x/S_x$. Substituting this in the expression for $\delta l_x/l_x$, we obtain

$$\delta l_x/l_x = E^{-1}(1-\nu^2)F_x/S_x.$$

This can be written as

$$F_x/S_x = E_{\text{eff}}\,\delta l_x/l_x \quad \text{with} \quad E_{\text{eff}} = E/(1-\nu^2). \tag{1.6}$$

This relation is analogous to Hooke's law for the uniaxial deformation of a free rod. The difference is only that an *effective Young's modulus* E_{eff} takes the place of E.

1.2.3 Pure Shear

Consider another important kind of homogeneous strain. Figure 1.6a shows the sectional view of a square bar subjected to the shear forces G applied to its faces. These forces are supposed to be distributed uniformly over lateral faces of the bar. For this purpose, rigid slabs (such as in Fig. 1.6b for the side BC) can be used. We have apparently a two-dimensional problem. Consider the unit bar length adjacent to section ABCD.

Tangential strain arising in response to such forces is called shearing strain. We can show that the latter occurs at already considered extensions and contractions of a rectangular bar. Indeed, let us consider a bar of square section PQRT subjected to the action of extension and contraction forces applied to

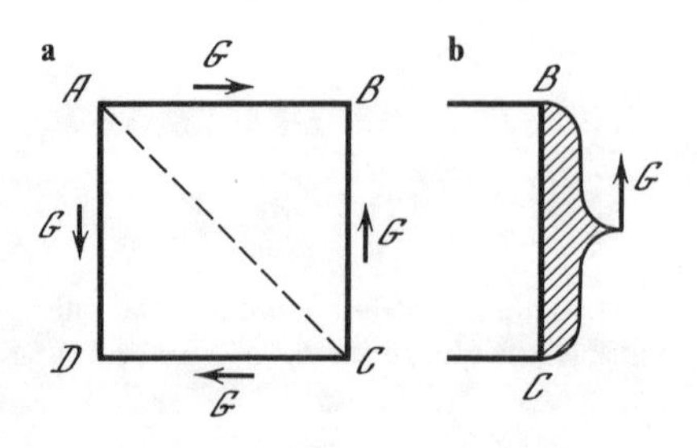

Fig. 1.6. **(a)** Rectangular bar subjected to shear forces. **(b)** The slab on the bar's face BC ensures a uniform distribution of the force G over this face

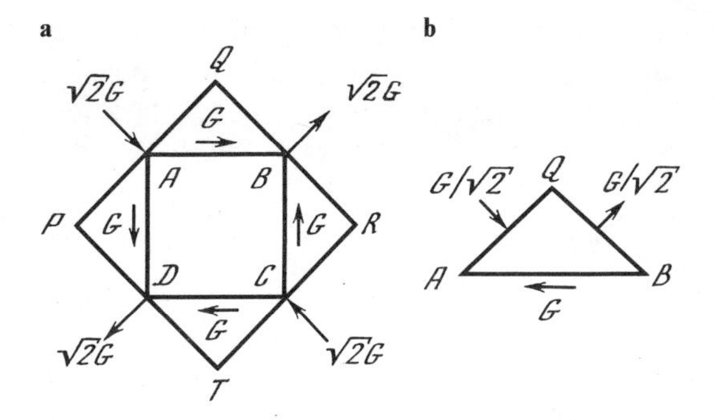

Fig. 1.7. (**a**) Shear stresses on the square ABCD as a result of the composition of a mutually perpendicular extension and contraction of the square PQRT. (**b**) Equilibrium conditions of an element AQB.

its sides (Fig. 1.7a) and containing the bar ABCD inside it. Now, considering the equilibrium conditions of the part AQB, for example, separately (Fig. 1.7b), we readily see that a tangential force $-G$ on the side AB is exerted by the square ABCD. Consequently, according to the law of action and reaction, the tangential force G acts on the side AB of the square ABCD. The same is true for the other sides of this square.

Using (1.1, 3) one can determine the extension and contraction of the faces of the bar PQRT, i.e., the deformations of the diagonals AC and BD of the bar ABCD. If d is the undisturbed length of the diagonal, then for its extension (or contraction) δd we have

$$\delta d/d = E^{-1}G/S + \nu E^{-1}G/S = E^{-1}(1+\nu)G/S\,, \tag{1.7}$$

where S is the area of the face AB. It has been taken into account that the area of the face PQ equals $\sqrt{2}S$.

The pure shearing strain is usually characterized by the changes of angles between the faces of the bar, which are directly related to the deformations of the diagonals. In practice, a simple pure shear is usually realized, as shown in Fig. 1.8. The side DC is assumed fixed. The shear force G is tangent to the side AB and distributed uniformly over it. In the process of deformation, the points

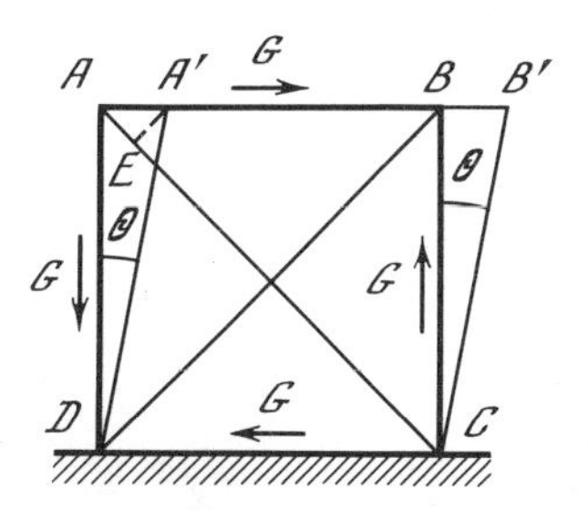

Fig. 1.8. Distortion of the square as a result of an applied shear force

A and B transfer to A′ and B′, respectively. The diagonal AC shortens, while the diagonal BD elongates. The strain shown in Fig. 1.8 is readily seen to be the same as in the case of Fig. 1.6. Hence, the system of forces is the same. The forces acting upon the sides AD, DC and CB indicated by arrows in Fig. 1.8 are due to the support reaction.

We relate the distortion angle θ to the force G. Evidently, $\theta = \mathrm{AA'/AD} = \sqrt{2}\mathrm{AE/AD}$, but $\mathrm{AE} = \delta d$, $\mathrm{AD} = \sqrt{2}d$, therefore,

$$\theta = 2\,\delta d/d = 2E^{-1}(1+\nu)G/S\,. \tag{1.8}$$

The material constant

$$\mu = E/2(1+\nu) \tag{1.9}$$

is called the *shear modulus*. If we also denote by $g = G/S$ the *shear stress*, then the relation holds

$$g = \mu\theta\,. \tag{1.10}$$

Obviously $\mu > 0$, otherwise the shear would be in the direction opposite to that of the applied force. Hence, it follows from (1.9) that $\nu > -1$. Taking also into account the restrictions on ν stated at the end of Sect. 1.2.1, we finally get the permissible range for ν: $-1 < \nu < 1/2$.

The case $\nu < 0$ (a transverse contraction of the bar accompanying its longitudinal contraction) does not occur apparently for real materials so, in fact, the following inequality holds

$$0 < \nu < 1/2\,. \tag{1.11}$$

1.3 Heterogeneous Strains

1.3.1 Torsion of a Rod

Now consider an important case of heterogeneous strain. Suppose that the end face on the left hand of the rod of circular cross section and finite length L is fixed (stuck to the wall, for example), and the torque (torsional moment) M is applied to the right one (Fig. 1.9). Twisting of the rod takes place as a result.

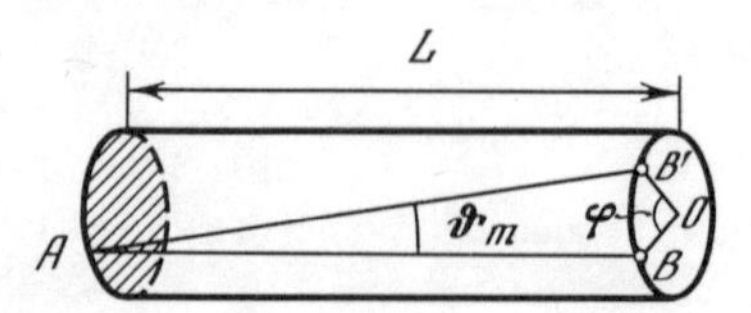

Fig. 1.9. Twist of a rod

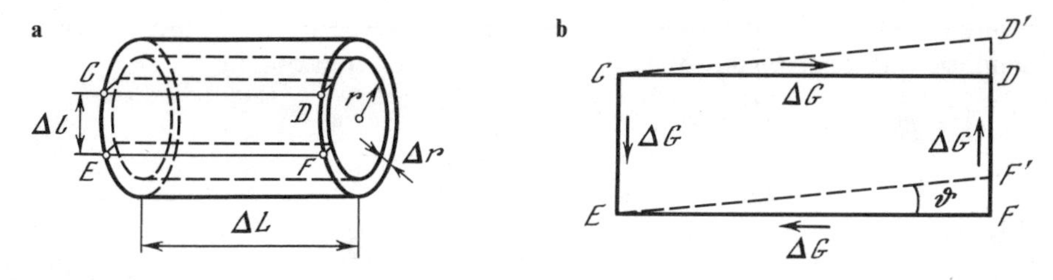

Fig. 1.10. (**a**) Infinitely small cylindrical tube separated from the rod. (**b**) Deformation of a small element of this tube

Points at its surface lying originally on the straight line AB are moved to AB′ while the radius OB on the right end face is displaced to OB′, turning by an angle φ. If the rod's radius is a, then the length of the arc BB′ equals $a\varphi$, and the angle between the lines AB and AB′ is $\vartheta_m = a\varphi/L$.

To relate the torque M and the angle φ, we consider a sufficiently small element of the rod, which may be regarded as homogeneously deformed, and use formulas derived in the preceding sections. First, consider a cylindrical layer (tube) of length ΔL in the rod sandwiched between two cylindrical surfaces with radii r and $r + \Delta r$ (of course, $r < a$). In Fig. 1.10a this cylindrical layer is shown on an enlarged scale. Then, separate a small element CDEF of volume $\Delta L\, \Delta l\, \Delta r$ from this layer. While twisting the rod, this element is displaced in the direction EC and becomes deformed, since the displacement of point E is somewhat less than that of point F. Displacement of the element as a whole does not influence the state of stress and hence is of no interest to us; the deformation is shown in Fig. 1.10b giving a side view (normal to the square CDEF) of the separated element. The distortion angle is readily seen to be $\vartheta = r\varphi/L$ (cf. an expression for ϑ_m). For the small separated element we have considered the strain of pure shear in Sect. 1.2.3. According to (1.10), the shear stress $g = \mu\vartheta = \mu r\varphi/L$ corresponds to this strain. Being applied to the area $\Delta l\, \Delta r$, it provides the force $\Delta G = g\, \Delta l\, \Delta r$ (Fig. 1.10b). The latter gives rise to the axial moment $\Delta M = \Delta G r = g r\, \Delta l\, \Delta r$. Now, if we consider all elements composing the section of tube (Fig. 1.10a), then in the last formula, Δl should be replaced by $2\pi r$. Therefore, the torque acting on the tube section is $M_{\Delta r} = 2\pi r^2 g\, \Delta r$ or, substituting the value of g, we obtain

$$M_{\Delta r} = f_{\Delta r}\varphi \quad \text{with} \quad f_{\Delta r} = 2\pi\mu r^3\, \Delta r/L\,. \tag{1.12}$$

The last formula gives the relationship between the torque $M_{\Delta r}$ and the angle of twist, φ, for a thin-wall cylindric tube with radius r, thickness $\Delta r \ll r$ and length L.

When a wall of a tube has finite thickness, the torque can be determined by integrating (1.12) over r, i.e.

$$M_{r_2 - r_1} = f_{r_2 - r_1}\varphi \quad \text{with} \quad f_{r_2 - r_1} = \pi\mu(r_2^4 - r_1^4)/2L\,, \tag{1.13}$$

where r_1 and r_2 are the inner and external radii, respectively. In the case of a cylindric rod, one should set $r_2 = a$, $r_1 = 0$ in (1.13) and obtain

$$M = f\varphi, \quad f = \pi\mu a^4/2L. \tag{1.14}$$

For a given φ and L, the torque is proportional to the fourth power of the radius. The quantity f is called the *torsion modulus* of a rod.

1.3.2 Bending of a Beam

A body whose length is much greater than its transverse dimensions is referred to as a *beam*. For the sake of simplicity, we consider a beam of uniform cross section.

First of all, consider the deformation of a beam by the action of transverse bending forces and relate the curvature of a beam at a fixed cross section to the moment of these forces. We will examine a small element (of length l) of the beam enclosed between sections AC and BD. The radius of curvature R due to the bending is assumed to be large compared with the transverse dimensions of the beam. Without going into particulars we assume, according to D. Bernouilli, that plane cross sections of the beam (for example, AC and BD) remain plane also after bending (A′C′ and B′D′ in Fig. 1.11). As a result, the part of the beam under consideration becomes deformed in such a way that the beam's fiber AB lengthens, CD shortens and some "*neutral*" fiber EF does not change its length. Such an unstrained fiber becomes a *neutral surface* in three dimensions. The further the fiber lies off the neutral surface, the greater its deformation. Consider a fiber at distance y from this surface. The fiber's length after bending is $(R + y)\chi$. Taking into account that its original length (the same as the one of the neutral fiber, Fig. 1.11) was $l = R\chi$, we have for the elongation $\delta l = y\chi$. However, $\chi = l/R$, hence $\delta l = yl/R$. According to Hooke's law (1.1), the stress acting in this layer is $\Delta F/\Delta S = E\,\delta l/l = Ey/R$, where ΔF is the force acting on the element of the sectional area ΔS. To obtain the resultant moment on the

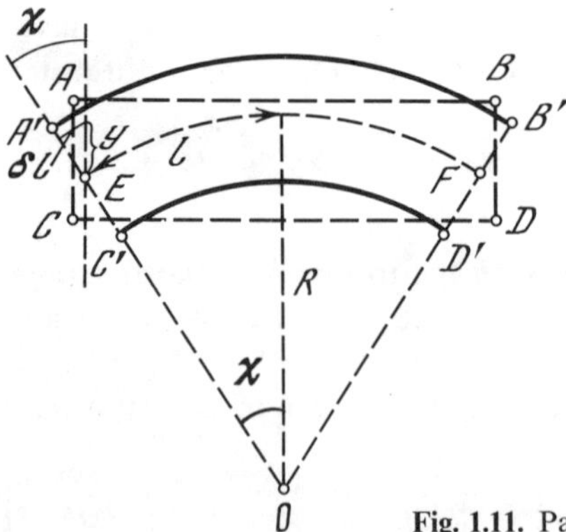

Fig. 1.11. Part of a beam under bending

cross section the product $\Delta F y = E y^2 \Delta S/R$ must be integrated over this section $M = \int_S y\, dF$ which yields

$$M = EI/R\,, \quad I = \int_S y^2\, dS\,, \tag{1.15}$$

where I is the *moment of inertia* of the beam's cross section about the axis normal to Fig. 1.11 and lying in the neutral surface. In the case of pure bending (extensions or contractions which are the same for all fibers of a beam are excluded), this axis passes through the centre of mass of the cross section. In fact, the resultant tensile force on the section is zero in this case ($\int dF = ER^{-1} \times \int y\, dS = 0$). This is satisfied only if the straight line $y = 0$ passes through the centre of mass of the cross section.

Equation (1.15) relating the bending radius of curvature to the moment of forces is basic to the beam-bending theory. It follows from this equation, in particular, that the beam is the more rigid the greater the moment of inertia I, i.e. the more the beam's material is concentrated off the neutral surface, as, for example, in the case of a flange beam.

1.3.3 Shape of a Beam Under Load

Suppose the x-axis lies in the neutral surface and is directed along the undeformed beam. The z-axis is perpendicular to it and lies in the bending plane. The equation for the neutral fiber which coincides originally with the x-axis after bending will be: $z = z(x)$. For the curvature we have, according to the known formula,

$$R^{-1} = d^2z/dx^2[1 + (dz/dx)^2]^{-3/2} \simeq d^2z/dx^2 \equiv z''\,. \tag{1.16}$$

Strains are assumed to be small so that we can neglect $(z')^2$ compared with unity.

Let us use (1.15, 16) to determine a deflection curve of a cantilever beam, one end of which is fixed in a wall, while the other is loaded (Fig. 1.12). Considering some section of the beam at distance x from the wall, we see that, according to (1.15, 16), the moment of forces, produced by stresses, acting in

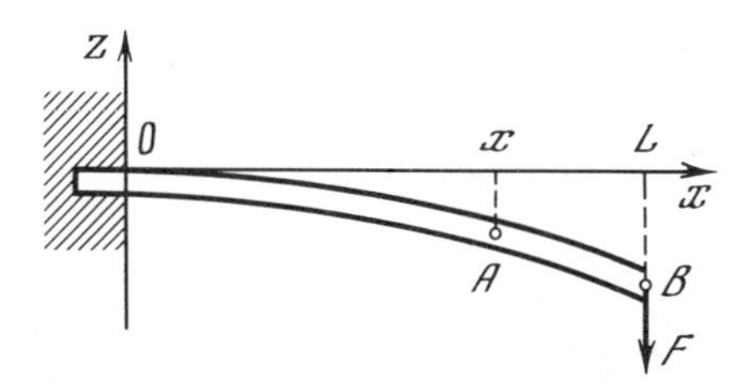

Fig. 1.12. Cantilever beam bending under load F acting at its end

this section is $M = EIz''$. Since the beam is in equilibrium, it must be compensated by the moment produced by the force F, namely, $(L - x)F$. This yields

$$EIz'' + (L - x)F = 0. \tag{1.17}$$

Integrating this equation twice we obtain

$$z(x) = -(F/2EI)x^2(L - x/3) + C_1 x + C_2.$$

According to the boundary condition $z(0) = 0$, $z'(0) = 0$ the constants C_1 and C_2 are equal to zero. Thus,

$$z(x) = -(F/2EI)x^2(L - x/3). \tag{1.18}$$

The bending deflection z_{max} at the end of the beam ($x = L$) is proportional to the cube of the beam length

$$z_{max} = FL^3/3EI. \tag{1.19}$$

If the bending moment varies along the beam, $M = M(x)$ as in the example just considered, then a *shear* force F_S perpendicular to the neutral surface arises at each beam section. To obtain an expression for this force we consider the equilibrium of a small element of the beam of axial length Δx. It follows from the balancing of the moment for this element that

$$M(x + \Delta x) - M(x) + F_S \Delta x \simeq (\partial M/\partial x)\Delta x + F_S \Delta x = 0 \quad \text{or}$$

$$F_S(x) = -\partial M/\partial x. \tag{1.20}$$

The shear force is due to tangential (shear) stresses at each section of the beam produced by bending, their sum being F_S. Tangential stresses cause shear strain which in turn give rise to warping of an originally plane beam section.

Thus, strictly speaking, the Bernouilli hypothesis used for the derivation of the basic equation (1.15) is not valid. On the other hand, all results obtained on the basis of (1.15) [for example, the cantilever beam shape (1.18)] are well confirmed by experiment. The explanation of this apparent contradiction is that in deriving the basic formula it is not necessary to assume that plane sections remain plane. It is sufficient just to accept the hypothesis of proportionality of longitudinal stresses to the distance of a fiber from the neutral surface of the beam. More sophisticated calculations based on general equations of the theory of elasticity confirm this hypothesis and also (1.15).

1.4 Exercises

1.4.1. Determine the distribution of stresses in a rod freely hanging in the gravity field and fixed at one end (Fig. 1.13). Find the total elongation of the rod

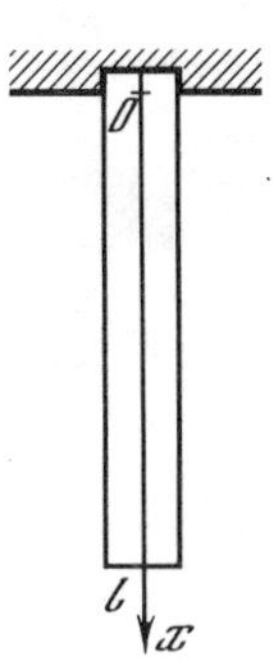

Fig. 1.13. The rod fixed at the upper end and freely hanging in the gravity field too. Denote as l the length of the rod, S its cross-sectional area, E Young's modulus, and ρ the density of the material.

Solution: Consider the rod's cross section at a distance x from the upper end of the rod O. The force $F = g\rho S(l - x)$, i.e., the weight of the lower part of the rod, acts on this cross section producing the stress $F/S = g\rho(l - x)$. Using the differential form of Hooke's law (1.4), write

$$du/dx = E^{-1}F/S = g\rho E^{-1}(l - x)\,.$$

Integrating this equation under the boundary condition $u(0) = 0$ we obtain $u(x) = \rho g E^{-1}(l - x/2)x$. The total elongation of the rod is $u(l) = g\rho l^2/2E$.

1.4.2. The force F_0 acts on the upper end of a vertical column of variable cross section standing on a fixed support (Fig. 1.14). What must be the sectional area distribution $S(x)$ for the stresses to be the same at all sections. Determine the total decrease of the column height.

Solution: Denote by S_m the area of an upper end face of the column, l its height, E and ρ Young's modulus and the density of material, respectively. At arbitrary x the stress

$$\frac{F_0}{S(x)} + \frac{\rho g}{S(x)} \int_x^l S(x)\,dx = \frac{F_0}{S_m}$$

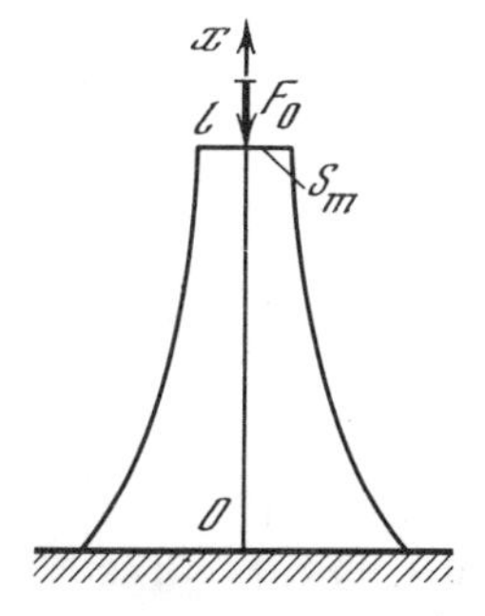

Fig. 1.14. A column of variable cross section under the action of a force F_0 at the upper end

is constant, according to our assumption. After multiplying by $S(x)$ and differentiating, we obtain $S'(x) = -\rho g S_m F_0^{-1} S(x)$, which is easily integrated:

$$S(x) = S_m \exp[\rho g S_m F_0^{-1}(l - x)] .$$

Since the strain is homogeneous in this case, the decrease in the column's height is: $\delta l = F_0 l / E S_m$.

1.4.3. Show that the effective Young's modulus in the case of uniaxial strain of a rectangular bar with displacements forbidden in both transverse directions is $E_{\text{eff}} = E(l - \nu)/[(1 + \nu)(1 - 2\nu)]$.

Hint: Write for the strains along each axis:

$$\frac{\delta l_x}{l_x} = E^{-1} \frac{F_x}{S_x} - \frac{\nu}{E} \frac{F_y}{S_y} - \frac{\nu}{E} \frac{F_z}{S_z}, \quad \frac{\delta l_y}{l_y} = 0 = E^{-1} \frac{F_y}{S_y} - \frac{\nu}{E} \frac{F_x}{S_x} - \frac{\nu}{E} \frac{F_z}{S_z},$$

$$\frac{\delta l_z}{l_z} = 0 = E^{-1} \frac{F_z}{S_z} - \frac{\nu}{E} \frac{F_x}{S_x} - \frac{\nu}{E} \frac{F_y}{S_y} .$$

An expression for E_{eff} can now be obtained analogously to (1.6).

1.4.4. For transmission of the torque M_0, a hollow shaft with the ratio α of inner radius to external one is used. Assuming that the maximum admissible stress for the shaft's material is $g_{\max}$, determine the minimum permissible outer radius R and the mass of a unit length of the shaft.

Solution: Torsion of the shaft generates shearing strain. One has $g_{\max} = \mu \nu_{\max} = \mu R \varphi / L$, where μ is the shear modulus, L the shaft length, and φ the twist angle. The latter can be found by using (1.13):

$$\varphi = \frac{2LM_0}{\pi \mu (R^4 - r^4)} = \frac{2LM_0}{\pi \mu R^4} (1 - \alpha^4)^{-1} .$$

R is determined by the condition $g_{\max} = \mu R \varphi / L = 2M_0 / \pi R^3 (1 - \alpha^4)$, which yields $R^3 = (2M_0 / \pi g_{\max})(1 - \alpha^4)^{-1}$. The mass of a unit length of the shaft is

$$m = \rho \pi R^2 (1 - \alpha^2) = \rho \pi \left(\frac{2M_0}{\pi g_{\max}} \right)^{2/3} \frac{1 - \alpha^2}{(1 - \alpha^2)^{2/3} (1 + \alpha^2)^{2/3}}$$

$$= \rho \pi^{1/3} \left(\frac{2M_0}{g_{\max}} \right)^{2/3} \frac{(1 - \alpha^2)^{1/3}}{(1 + \alpha^2)^{2/3}} .$$

Thus a hollow shaft is preferable compared with a solid one ($\alpha = 0$) if one is interested in economy of material.

1.4.5. The beam ends rest on two supports a distance L apart (Fig. 1.15). Determine the deflection curve of the beam if its weight per unit length is p.

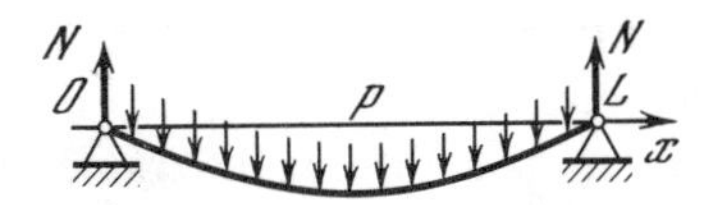

Fig. 1.15. Deformation of a beam under the action of its own weight

Solution: The reaction force on each support is $N = Lp/2$. The bending moment in the section x is the sum of the moment of the reaction force of the support $-pLx/2$ and of the moment of that part of the beam's weight $px^2/2$. Taking (1.15, 16) into account, we obtain

$$z'' = -\frac{p}{2EI}\,x(x - L)$$

whose solution, satisfying the boundary conditions $z(0) = z(L) = 0$, is

$$z = -\frac{px}{24EI}\,(L^3 - 2Lx^2 + x^3),$$

the maximum deflection being $z(L/2) = -(5/384)pL^4/EI$.

1.4.6. The same as Exercise 1.4.5 but the beam is weightless and the force F is applied at its middle.

Solution: The reaction force is $F/2$. The equation for the moments for $0 \le x \le L/2$ is $EIz'' = Fx/2$. Its solution satisfying the conditions $z(0) = 0$, $z'(L/2) = 0$ (the last due to symmetry with respect to the point $x = L/2$) is

$$z = \frac{F}{48EI}\,x(4x^2 - 3L^2).$$

The maximum deflection is $z(L/2) = -FL^3/48EI$ in this case.

1.4.7. A thin rod of length L is contracted by the uniaxial forces F applied at each end. Show that at sufficiently small $F < F_1$ the rod does not bend (state of stability). Calculate the critical force F_1 (Eulerian force).

Solution: When the rod is bent (Fig. 1.16), the moment $M(x) = Fz(x)$ arises in an arbitrary section x. According to (1.15, 16), the equilibrium equation $z'' + \alpha^2 z = 0$ with $\alpha^2 = F/EI$ holds, whose general solution is the function $z = A\sin\alpha x + B\cos\alpha x$. Displacement z must be zero at $x = 0$, L. Hence, for the possible forms of equilibrium of the rod, $B = 0$, A is arbitrary and $\alpha L = n\pi$,

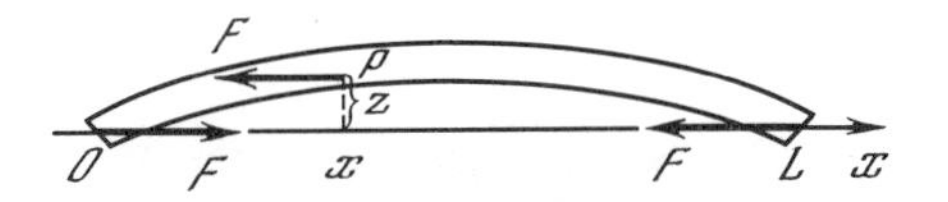

Fig. 1.16. Bending of a rod under the action of a longitudinal force

$n = 1, 2, \ldots$ For each n we have the force $F_n = n^2\pi^2 EI/L^2$. The smallest of them is $F_1 = \pi^2 EI/L^2$. If the applied force $F < F_1$, the rod remains straight. For $F = F_1$ equilibrium is possible at an arbitrary displacement within the framework of the linear theory. This theory becomes incorrect, however, when $(z')^2$ in (1.16) becomes comparable with unity. Calculations based on an accurate expression for the curvature, (1.16), gives the deflection curve of a rod at any $F > F_1$.

1.4.8. Determine the energy stored by the rod of length l while contracted by the force F_0.

Solution: This energy is equal to the work produced by the force during the process of quasistatic contraction of the rod. Therefore, we assume that Hooke's law (1.1) ($F/S = Eu/l$) holds in the process of contraction, u being the displacement of the rod's end to which force is applied. The work increment is $dA = F\,du = (ES/l)u\,du$. After integration from 0 to u, we obtain

$$A = ESu^2/2l = F_0 u/2.$$

The energy density (per unit volume) is equal to half of the product of stress and strain: $\varepsilon = A/Sl = (F_0/S)(u/l)/2$.

1.4.9. Determine the kinetic energy of the rod when abruptly contracted by the uniaxial force F_0.

Solution: Assuming that the force is constant during the process of deformation, we obtain $A = Fu$ for the work done. The latter is converted into the sum of elastic energy obtained in the previous exercise $U = F_0 u/2$ and kinetic energy T, i.e. $A = T + U$. Hence we easily obtain that $T = U = F_0 u/2$.

1.4.10. Determine the change of radius R of the thin-wall cylinder rotating about its axis.

Solution: For the cylinder element $\Delta l = R\,\Delta\varphi$ (Fig. 1.17) we have a centripetal force $F_c = \rho SR\,\Delta\varphi\omega^2 R$, where S is the sectional area of the cylinder by the

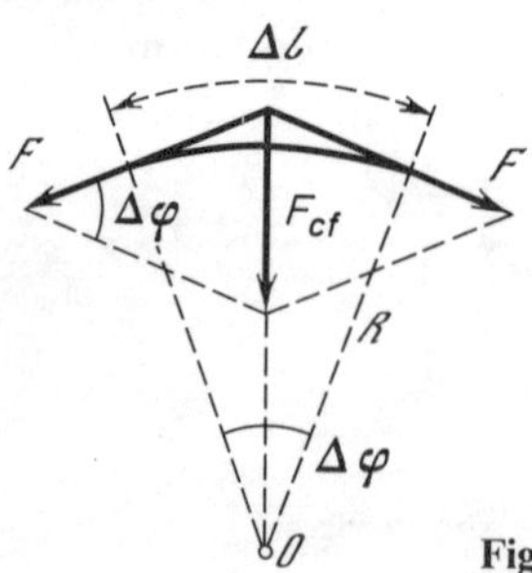

Fig. 1.17. Forces acting on the rotating cylinder segment

plane containing the axis of rotation and a generator of the cylinder; ρ is the material density. This force is the resultant of forces extending the given element $F_c = 2F\sin(\Delta\varphi/2) \simeq F\,\Delta\varphi$. Hence, we have for the stresses $F/S = \rho R^2\omega^2$. Now by Hooke's law,

$$\frac{\delta(2\pi R)}{2\pi R} = \frac{\delta R}{R} = E^{-1}\frac{F}{S} = \frac{\rho\omega^2 R^2}{E} \quad \text{or} \quad \delta R = \frac{\rho\omega^2 R^3}{E}.$$

1.4.11. Determine the relative displacement of points of a rod moving in the direction of its length with a constant acceleration a.

Solution: Let ρ be the density of a rod, E Young's modulus, l the rod's length, and S the area of its cross section. In an arbitrary section x, the force $F(x) = \rho Sxa$ acts causing the rod's motion with acceleration a. Using the differential form of Hooke's law (1.4), we obtain $F(x)/S = \rho xa = E\,\partial u/\partial x$. Integrating the last equation under the condition $u(0) = 0$, we find $u = (\rho a/E)x^2/2$.

1.4.12. Find the displacement of points of a rod rotating about its end with frequency ω.

Solution: The centripetal force of the rod's part $x > x_0$ is $F_c = \int_{x_0}^{l} S\rho\omega^2 x\,dx = S\rho\omega^2(l^2 - x_0^2)/2$. It is equal to the resultant of internal stresses in the x_0 section ($F_c = SE\,\partial u/\partial x$). Integrating the obtained equation, we find

$$u(x) = (\rho\omega^2/2E)x(l^2 - x^2/3).$$

1.4.13. Calculate the critical force (Eulerian force) when the end of the rod at point L (Fig. 1.16) is fixed and rotation is prevented, while the other at point 0: a) is free, b) can move along the X-axis only and rotations are not prevented.

1.4.14. A beam of triangular cross section rests on two supports. How must it be placed, on the side or on the edge if the material is not strong against tension.

2. Waves in Rods, Vibrations of Rods

Static forces and static deformations have been considered in the previous chapter. If, however, an elastic body is under the action of a force varying in time, different types of waves can arise in this body. Each type of wave can transfer disturbances from one part of the body to the other at a finite speed.

In this chapter we will consider the main types of the waves in rods. Equations will be obtained which govern the wave propagation in rods as well as vibrations of rods. Some general properties of wave propagation will also be discussed.

2.1 Longitudinal Waves

2.1.1 Wave Equation

Striking an end of a rod in the axial direction generates longitudinal strain propagating along the rod: *a longitudinal wave*. We proceed to derive an equation which governs this propagation. Let the displacement of the rod section at time t be $u(x, t)$, x being the section's position in an undisturbed state. When disturbed, this section has the position $x + u(x, t)$. Consider a small rod element between sections x and $x + \Delta x$. The force $F(x, t)$ acting on the x section is, according to the differential form of Hooke's law (1.4),

$$F/S = E\,\partial u/\partial x\,, \tag{2.1}$$

where S is the rod's sectional area and E is Young's modulus. A similar expression can also be written for the force acting on the section $x + \Delta x$. The difference between these forces $[F(x + \Delta x) - F(x) \simeq \Delta x\,\partial F/\partial x]$ gives rise to acceleration of the element under consideration. Since the mass of the latter is $\rho S\,\Delta x$ (ρ being the rod's density) and the acceleration is $\partial^2 u/\partial t^2$, we obtain, according to Newton's second law,

$$\rho S\,\Delta x\,\partial^2 u/\partial t^2 = \Delta x\,\partial F/\partial x\,.$$

Dividing this relation by Δx and substituting $F(x)$ from (2.1), we obtain the required wave equation for longitudinal waves in the rod:

$$\partial^2 u/\partial t^2 = c^2 \partial^2 u/\partial x^2\,, \qquad c = \sqrt{E/\rho}\,. \tag{2.2}$$

One can easily confirm by simple substitution that an expression

$$u(x,t) = f(x - ct) + g(x + ct) \tag{2.3}$$

is the solution of the wave equation, f and g being arbitrary functions. The first term in (2.3) represents a wave propagating in the positive x-direction with velocity c without changing its shape, whereas the second one is a wave propagating in the opposite direction with the same velocity.

2.1.2 Harmonic Waves

In the important case of *harmonic* waves where f and g are cosine functions, we have

$$U(x,t) = a\cos(kx - \omega t + \psi_1) + b\cos(kx + \omega t + \psi_2). \tag{2.4}$$

The constants a and b are called *the wave amplitudes*, ψ_1 and ψ_2 are also the constant initial *phases* of waves, ω is the *circular frequency*, and $k = \omega/c$ the *wave number*. The latter can also be written as $k = 2\pi/\lambda$, where $\lambda = 2\pi c/\omega = Tc$ is *the wavelength* and T is the *period* of wave. As in (2.3), the first term in (2.4) describes a wave propagating in the positive x-direction, and the second term describes one propagating in the opposite direction.

Instead of (2.4), it is often convenient to write the waves in complex form:

$$u(x,t) = A\exp[i(kx - \omega t)] + B\exp[i(kx + \omega t)], \tag{2.5}$$

where the constants $A = a\exp(i\psi_1)$, $B = b\exp(i\psi_2)$ are termed *complex wave amplitudes*. We will use this complex form in what follows, but taking into account only the real part of (2.5) which coincides with (2.4). In all linear operations with complex waves (summation of waves, differentiations with respect to time and coordinates, etc.), the real part of the final result will be equal to the result of the same operations applied to the real part of the original waves. In the case of nonlinear operations, however (e.g., estimation of the wave energy quadratic with respect to u), it is necessary to deal with the real notation from the very beginning.

2.2 Reflection of Longitudinal Waves

2.2.1 Boundary Conditions

In an infinite rod a wave propagates in a given direction without changing its form. But as a matter of fact the rod always has a finite length and a wave is reflected from its end. If a rod is in contact with another one, a wave partly penetrates the boundary and is partly reflected from it. Reflection and transmission of waves depend on the properties of the boundary and are governed by the boundary conditions. We consider some of them, assuming that the boundary is at $x = x_0$.

i) Absolutely rigid boundary. Particle displacement is forbidden:

$$u(x_0, t) = 0 . \tag{2.6}$$

In practice, this case can be realized if an end face of the rod is stuck to a massive wall of a material with very large Young's modulus.

ii) Absolutely soft (free) boundary, say, a rod in a rather rarefied medium or vacuum. The forces (stresses) vanish at such a boundary:

$$F/S|_{x0} = E\, \partial u/\partial x|_{x0} = 0 , \quad \text{or} \quad \partial u/\partial x|_{x0} = 0 . \tag{2.7}$$

An interface with air is a good approximation to an absolutely soft boundary.

iii) Contact between two rods with equal cross sections glued together with different material constants ρ_1, E_1 and ρ_2, E_2, respectively. The displacements and forces (stresses) on both sides of the boundary are the same in this case:

$$(u_1 - u_2)_{x0} = 0 , \quad (E_1\, \partial u_1/\partial x - E_2\, \partial u_2/\partial x)_{x0} = 0 . \tag{2.8}$$

2.2.2 Wave Reflection

We confine ourselves to the simplest case of harmonic waves, since conclusions concerning the law of reflection in the general case follow by the principle of superposition in the linear case under investigation. Assume that the pulse $f(x - ct)$ can be represented by the Fourier integral:

$$f(x - ct) = \int_{-\infty}^{\infty} A(\omega) \exp[\mathrm{i}(kx - \omega t)]\, d\omega .$$

We consider a longitudinal harmonic wave $u_+ = A \exp[\mathrm{i}(kx - \omega t)]$ propagating in a semi-infinite rod $-\infty < x \leq 0$ toward $x = 0$ (*incident wave*). Suppose that we have an absolutely rigid boundary at $x = 0$. Obviously, u_+ alone cannot satisfy the boundary condition (2.6). Hence, a wave propagating in the opposite direction (*reflected wave*) of frequency, say, ω', must arise: $u_- = AV \exp[-\mathrm{i}(k'x + \omega' t)]$, $k' = \omega'/c$. The constant V is termed the *reflection coefficient*. The displacement of the rod's particles is the sum of both waves:

$$u(x, t) = A\{\exp[\mathrm{i}(kx - \omega t)] + V \exp[-\mathrm{i}(k'x + \omega' t)]\} .$$

Applying the boundary condition (2.6) at $x_0 = 0$ yields

$$\exp(-\mathrm{i}\omega t) + V \exp(-\mathrm{i}\omega' t) = 0 \quad \text{or}$$

$$\exp[\mathrm{i}(\omega' - \omega)t] = -V. \tag{2.9}$$

Since the right-hand side of (2.9) is constant, the left-hand side must be constant, too. Therefore, $\omega' = \omega$, i.e., the frequencies of an incident and reflected waves are equal; hence $k' = k$, too. For the reflection coefficient we now have $V = -1$. Thus, the total wave field in the rod is

$$\begin{aligned} u(x,t) &= A[\exp(\mathrm{i}kx) - \exp(-\mathrm{i}kx)]\exp(-\mathrm{i}\omega t) \\ &= 2\mathrm{i}A\sin kx\exp(-\mathrm{i}\omega t). \end{aligned} \tag{2.10}$$

The reflection from a free boundary can be considered in a similar manner, leading to the reflection coefficient $V = 1$.

We noted that the frequencies of incident and reflected waves must be equal to satisfy the boundary condition at each moment t. This holds in every case where the boundary condition are independent of time, and simplifies the solution of wave-reflection problems.

Consider now two semi-infinite rods of different materials but of the same cross section glued together. As above, assume that the harmonic wave $A\exp[\mathrm{i}(k_1x - \omega t)]$, $k_1 = \omega/c_1$, $c_1 = \sqrt{E_1/\rho_1}$ is incident from the left. Since the frequency does not change, the solution of the corresponding wave equations, u_1, to the left ($x < 0$) and, u_2, to the right ($x > 0$) of the boundary can be written in the form

$$u_1(x,t) = \varphi_1(x)\exp(-\mathrm{i}\omega t), \qquad u_2(x,t) = \varphi_2(x)\exp(-\mathrm{i}\omega t).$$

For φ_1 and φ_2 we obtain the ordinary differential equations

$$\varphi_j'' + k_j^2\varphi_j = 0, \quad k_j = \omega/c_j, \quad c_j = \sqrt{E_j/\rho_j}, \quad j = 1, 2. \tag{2.11}$$

Their general solutions can be written as

$$\varphi_1 = A\mathrm{e}^{\mathrm{i}k_1x} + VA\mathrm{e}^{-\mathrm{i}k_1x}, \qquad \varphi_2 = WA\mathrm{e}^{\mathrm{i}k_2x} + A_2\mathrm{e}^{-\mathrm{i}k_2x}. \tag{2.12}$$

Here the terms $\exp(\mathrm{i}k_jx)$ correspond to waves propagating to the right, and $\exp(-\mathrm{i}k_jx)$ to those propagating to the left. Since at $x > 0$ there are no sources, only a wave leaving the boundary and propagating to the right can occur in the second rod, therefore $A_2 = 0$. Substituting (2.12) into (2.8) for $x_0 = 0$, we obtain a set of two equations to determine the reflection V and transmission W coefficients.

$$1 + V = W, \quad \mathrm{i}k_1E_1(1 - V) = \mathrm{i}k_2E_2W.$$

They lead to the Fresnel formulae

$$V = \frac{\sqrt{\rho_1E_1} - \sqrt{\rho_2E_2}}{\sqrt{\rho_1E_1} + \sqrt{\rho_2E_2}} = \frac{n - m}{n + m}, \qquad W = \frac{2n}{m + n}, \tag{2.13}$$

where $n = c_1/c_2$ is the *index of refraction* and $m = \rho_2/\rho_1$ the ratio of densities.

2.3 Longitudinal Oscillations of Rods

After the action of an external force ceases, a rod of finite length continues to oscillate with some *resonant frequencies.* The frequencies and the form of oscillations [described by the function $u(x, t)$] can be found by solving the wave equation (2.2) for the corresponding boundary conditions. Consider, for example, a rod with fixed ends at $x = 0$ and $x = l$, i.e., $u(0, t) = u(l, t) = 0$. Let us look for a solution in the form $u(x, t) = \varphi(x)\exp(-\mathrm{i}\omega t)$. Substituting it into (2.2), we obtain for the function $\varphi(x)$ the ordinary differential equation $\varphi'' + k^2\varphi = 0$, $k = \omega/c$. If we take $\varphi(x) = A\sin kx$, then the condition for $x = 0$ is satisfied automatically. The boundary condition for $x = l$ gives $\sin kl = 0$, whence we get a set of permissible wave numbers k_n:

$$k_n = n\pi/l\,, \qquad n = 1, 2, \ldots . \tag{2.14}$$

Since $k_n = \omega_n/c$, we obtain the *resonant frequencies*

$$\omega_n = n\pi c/l \tag{2.15}$$

and the corresponding *normal modes of vibration*

$$u_n(x, t) = A_n \sin k_n x \exp(-\mathrm{i}\omega t)\,. \tag{2.16}$$

Presenting the complex amplitude as $A_n = a_n\exp(\mathrm{i}\psi_n)$ and separating the real part, we have

$$u_n(x, t) = a_n \sin(n\pi x/l)\cos(\omega_n t - \psi_n)\,; \tag{2.17}$$

a_n and ψ_n remain arbitrary, of course.

Figure 2.1 portrays the distribution of the vibration amplitudes $\sin(n\pi x/l)$ over the rod length for the first three integers n. We see that the quantity $n - 1$ gives the number of *nodes* (points where $u_n = 0$) not counting its ends. Note that only those vibrations can occur in the rod for which the length of the bar is equal to an integer number of half waves. This can also be seen from (2.14), if, taking into account that $k_n = 2\pi/\lambda_n$, we write it as $l = n\lambda_n/2$.

In the case of a free rod of finite length [the boundary conditions are $u'(0, t) = u'(l, t) = 0$], the only difference from the case under consideration is a choice of

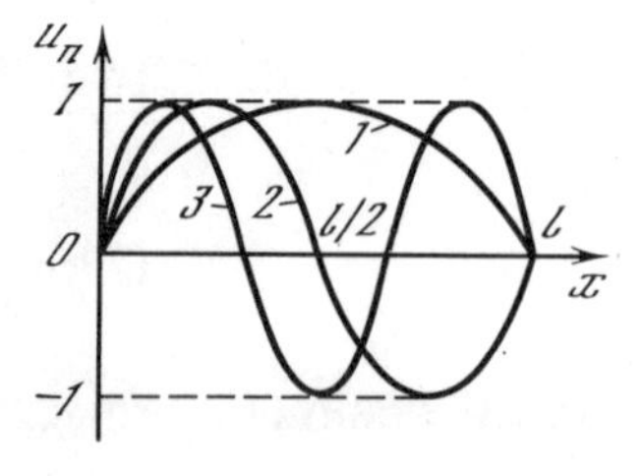

Fig. 2.1. Amplitude distribution of vibration along a rod with fixed ends for the first three modes of vibrations

solution in the form of $\varphi(x) = A\cos kx$ whose derivative vanishes at $x = 0$. The equation for the resonant frequencies remains the same ($\sin kl = 0$) so (2.14, 15) hold. For the normal modes of vibrations we have

$$u_n(x,t) = a_n \cos(n\pi x/l)\cos(\omega_n t - \psi_n). \quad (2.18)$$

The amplitude distributions of the first three modes along the rod are depicted in Fig. 2.2. The length of the rod must again be equal to an integral number of halfwaves. Due to this fact, the frequencies of vibration are the same for the fixed and free rods. Here, however, it is the quantity n (rather than $n-1$) that gives the number of nodes along the rod.

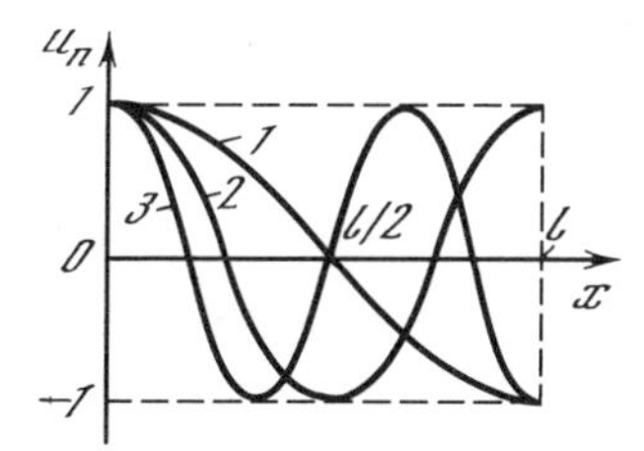

Fig. 2.2. The same as in Fig. 2.1 but for a rod with free ends

2.4 Torsional Waves in a Rod. Torsional Vibrations

Suppose that a torquc is suddenly applied to the end of the rod with circular cross section. Then a torsional wave propagates along the rod. The strain in each section x is specified by the twist angle $\varphi = \varphi(x,t)$ about the axis. The stress in the same section is specified by the torque $M(x,t)$. With (1.14) we can easily relate these two quantities. In fact, consider an element of the rod bounded by two sections a very small distance Δx apart. Its twist angle is

$$\Delta\varphi = \varphi(x + \Delta x) - \varphi(x) \simeq \Delta x\, \partial\varphi/\partial x.$$

Setting $L = \Delta x$ and $\varphi = \Delta\varphi$ in (1.14) we obtain

$$M(x) = f\Delta\varphi = \mu(\pi a^4/2)\Delta\varphi/\Delta x,$$

or in the limit $\Delta x \to 0$,

$$M(x) = \mu\frac{\pi a^4}{2}\frac{\partial\varphi}{\partial x}. \quad (2.19)$$

The net torque acting on the element Δx is equal to the difference of the torques at the two sections:

$$M(x + \Delta x) - M(x) \simeq \Delta x\, \partial M/\partial x = \mu(\pi a^4/2)\Delta x\, \partial^2\varphi/\partial x^2.$$

By Newton's second law, this torque leads to the angular acceleration of the rod's element considered about the x axis. Taking into account that the moment of inertia of the element is $\pi\rho a^4 \Delta x/2$ and the angular acceleration is $\partial^2\varphi/\partial t^2$, we obtain finally

$$\frac{\partial^2\varphi}{\partial t^2} = c^2 \frac{\partial^2\varphi}{\partial x^2}, \qquad c = \sqrt{\mu/S}\,. \tag{2.20}$$

This is again a wave equation identical to (2.2) describing longitudinal waves, but with another propagation velocity. Note that this velocity is independent of the radius of the rod.

At the rod's end $x = x_0$ the boundary conditions are the following: $\varphi(x_0, t) = 0$ at a fixed end, $M(x_0, t) = 0$ or $(\partial\varphi/\partial x)_{x_0} = 0$ according to (2.19), at a free end. In the case of two rods of equal radii with different material constants μ_1, ρ_1, and μ_2, ρ_2, respectively, glued together at $x = x_0$, we find equality of the twist angles $\varphi_1(x_0, t) = \varphi_2(x_0, t)$ and of the moments $\mu_1(\partial\varphi_1/\partial x)_{x_0} = \mu_2(\partial\mu_2/\partial x)_{x_0}$, being analogous to (2.8).

Thus, both the equations and thc boundary conditions for longitudinal and torsional waves are identical to within the notation. Therefore, all results of Sects. 2.2, 3 also hold with the only difference being that the twist angle φ plays the part of the longitudinal displacement and the shear modulus μ replaces of Young's modulus E. For example, (12.15) for the resonant frequencies remains valid but with a new value of c.

2.5 Bending Waves in Rods

2.5.1 The Equation for Bending Waves

Suppose we strike the rod in a transverse direction. This gives rise to local bending which propagates from the point of strike in the form of *a bending wave* shown in Fig. 2.3. Assume that the rod is symmetric about the figure's plane and that the displacements ζ occur in this plane. This displacement is assumed to be sufficiently small so that the neutral line can be considered one of constant length and, consequently, the rod extension can be neglected.

Consider a small segment Δx of the rod. We know that the shear force $F(x)$ acts at the section x. This is the force with which the rod acts from one side of

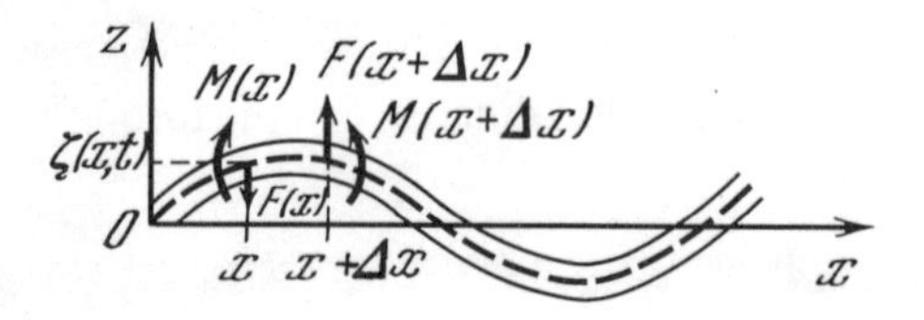

Fig. 2.3. Bending-wave propagation along a rod. Forces and moments acting on the rod

the section to the other one. Newton's second law for the element Δx can be written in the form

$$\rho S\,\Delta x\,\partial^2\zeta/\partial t^2 = F(x+\Delta x) - F(x) \simeq \Delta x\,\partial F/\partial x\,, \tag{2.21}$$

where ρ is the density and S is the sectional area of the rod.

Consider now the balance of moments in the rod. Bending produces the moment $M(x)$ in each rod section which, according to (1.15, 16), is equal to (if $|\partial\zeta/\partial x| \ll 1$)

$$M(x) = EI\,\partial^2\zeta/\partial x^2\,. \tag{2.22}$$

The total moment acting on the element Δx is the difference

$$M(x+\Delta x) - M(x) \simeq \Delta x\,\partial M/\partial x$$

plus the moment $F(x)\,\Delta x$ caused by the shear force. The total moment $(F + \partial M/\partial x)\,\Delta x$ must be balanced by the moment $\rho I\,\Delta x\,\partial^2\alpha/\partial t^2$ due to rotational acceleration of element Δx about an axis normal to Fig. 2.3. Here, $\rho I\,\Delta x$ is the moment of inertia of this segment and $\alpha \simeq \partial\zeta/\partial x$ the slope of the neutral line. Supposing the latter is small, the moment balance equation is

$$\rho I\,\frac{\partial^3\zeta}{\partial t^2\,\partial x} = \frac{\partial M}{\partial x} + F\,. \tag{2.23}$$

Note that in the static case ($\partial/\partial t = 0$) this formula is identical to (1.20).

Differentiating (2.23) with respect to x and substituting M from (2.22) and $\partial F/\partial x$ from (2.21), we obtain an equation for the displacement $\zeta(x,t)$:

$$\rho I\,\frac{\partial^4\zeta}{\partial t^2\,\partial x^2} = EI\,\frac{\partial^4\zeta}{\partial x^4} + \rho S\,\frac{\partial^2\zeta}{\partial t^2}\,. \tag{2.24}$$

In practice, a simplified equation is commonly used when the left-hand side of (2.24) is assumed to be zero (the rotational inertia is neglected). To justify this approximation, we introduced the so-called radius of inertia r_0 of the cross section determined by the relation $I = r_0^2 S$. To within an order of magnitude, r_0 equals the transverse dimensions of the rod. Furthermore, let T and λ be characteristic scales of time and length of the variation $\zeta(x,t)$. Then, the left-hand side of (2.24) will be of the order of: $\rho I\,\partial^4\zeta/\partial t^2\,\partial x^2 \sim r_0^2 S\rho\zeta/\lambda^2 T^2$ whereas for the second term on the right-hand side we have $\rho S\,\partial^2\zeta/\partial t^2 \sim \rho S\zeta/T^2$. The ratio of these quantities is of the order of r_0^2/λ^2. This quantity is usually negligibly small as compared with unity, since the transverse dimension of the rod r_0 are usually small compared with the wavelength λ. As a result, we obtain the following equation for the bending waves:

$$\frac{\partial^2\zeta}{\partial t^2} + r_0^2\,\frac{E}{\rho}\,\frac{\partial^4\zeta}{\partial x^4} = 0\,. \tag{2.25}$$

2.5.2 Boundary Conditions. Harmonic Waves

We have some interesting new features because the bending wave equation is of the fourth order. In particular, there must be more boundary conditions than in the case of a second-order equation for the longitudinal or torsional waves. Possible conditions at the rod's end $x = x_0$ are:

i) *rigidly clamped end* (Fig. 2.4a)—the displacement ζ and its derivative (slope) are both zero

$$\zeta(x_0, t) = 0\,, \quad (\partial\zeta/\partial x)_{x_0} = 0\,; \tag{2.26}$$

ii) *Free end* (Fig. 2.4b)—the moment M and shear force F are zero, or in few of (2.22) and (1.20)

$$(\partial^2\zeta/\partial x^2)_{x_0} = 0\,, \quad (\partial^3\zeta/\partial x^3)_{x_0} = 0\,; \tag{2.27}$$

iii) *supported end* (Fig. 2.4c)—both displacement and moment are zero

$$\zeta(x_0, t) = 0\,, \quad (\partial^2\zeta/\partial x^2)_{x_0} = 0\,. \tag{2.28}$$

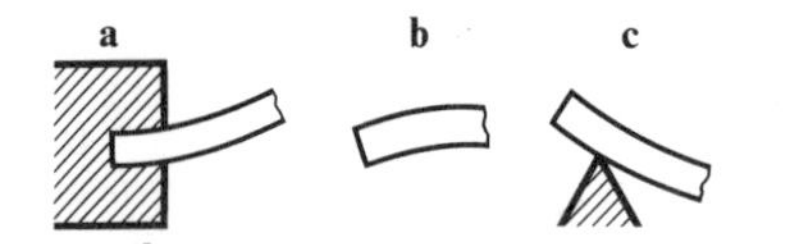

Fig. 2.4a – c. Possible conditions at the end of the rod under bending: (**a**) clamped end, (**b**) free end, (**c**) supported end

The general solution of (2.25) cannot be expressed by waves of arbitrary shape propagating at constant velocity without change of shape as for the longitudinal or torsional waves. Therefore, we consider harmonic waves at the very beginning, assuming

$$\zeta(x, t) = A\phi(x)\exp(-\mathrm{i}\omega t)\,. \tag{2.29}$$

Substituting this expression into (2.25) leads to the ordinary differential equation for $\phi(x)$:

$$d^4\phi/dx^4 - k^4\phi = 0\,, \quad k = (\rho/Er_0^2)^{1/4}\omega^{1/2}\,, \tag{2.30}$$

whose general solution can be written in the form

$$\phi(x) = A_+\exp(\mathrm{i}kx) + A_-\exp(-\mathrm{i}kx) + B_+\exp(kx) + B_-\exp(-kx) \tag{2.31}$$

and, correspondingly, for $\zeta(x, t)$:

$$\begin{aligned}\zeta(x, t) = {} & A_+\exp[\mathrm{i}(kx - \omega t)] + A_-\exp[-\mathrm{i}(kx + \omega t)] \\ & + B_+\exp(kx - \mathrm{i}\omega t) + B_-\exp(-kx - \mathrm{i}\omega t).\end{aligned} \tag{2.32}$$

Here, the first two terms correspond to undamped (in space) progressive waves propagating in the positive (A_+) and the negative (A_-) directions of x with the *phase velocity*

$$c_{\mathrm{ph}} = \pm\omega/k = \pm(Er_0^2/\rho)^{1/4}\omega^{1/2}\,. \tag{2.33}$$

Since c_{ph} depends on frequency, i.e., the dispersion takes place (Sect. 2.6), a wave of finite duration will change its shape while propagating.

Two other terms in (2.32) with coefficients B_+ and B_- correspond to non-propagating harmonic oscillations with exponentially decreasing or increasing amplitudes. We will see below that such solutions can occur only near the rod ends or some other inhomogeneities.

2.5.3 Reflection of Waves. Bending Vibrations

We consider the reflection of a harmonic wave $A\exp[\mathrm{i}(kx - \omega t)]$, $-\infty < x \le 0$ from the rod clamped at $x = 0$, for example. As in the case of longitudinal waves, the frequency ω does not change during the process of reflection. Since at $x \to -\infty$ the solution must be finite, we set $B_- = 0$ in (2.32). We assume also $A_+ \equiv A$, $A_- \equiv VA$, $B_+ \equiv WA$. Substitution of (2.32) into the boundary conditions (2.26) for $x_0 = 0$ yields $1 + V + W = 0$; $\mathrm{i}(1 - V) + W = 0$, and thus the coefficients V and W:

$$V = -\frac{1-\mathrm{i}}{1+\mathrm{i}}, \qquad W = -\frac{2\mathrm{i}}{1+\mathrm{i}}. \tag{2.34}$$

Hence, the solution of the problem under consideration (besides the incident wave) is a superposition of the reflected wave $AV\exp[-\mathrm{i}(kx + \omega t)]$ and of an oscillation $AW\exp(kx - \mathrm{i}\omega t)$, which decreases exponentially as $x \to -\infty$.

Let us now determine the bending normal modes of vibrations or a rod of finite length. For this purpose it is advisable to write the general solution of (2.30) in the form

$$\phi(x) = A_c\cos kx + A_s\sin kx + B_c\cosh kx + B_s\sinh kx \tag{2.31a}$$

which follows from (2.31) if the exponential functions are expressed in terms of trigonometrical and hyperbolic ones. The complex constants A_c, A_s, B_c, B_s can easily be expressed in terms of the wave amplitudes A_+, A_-, B_+, B_-.

As an example, consider a rod lying on two supports a distance l apart. From the boundary conditions (2.28) at $x_0 = 0$ we have $A_c + B_c = B_c - A_c = 0$, i.e., $A_c = B_c = 0$. Now again from (2.28) but for $x_0 = l$ we obtain a system of homogeneous equations for A_s and B_s:

$$A_s\sin kl + B_s\sinh kl = 0\,, \qquad -A_s\sin kl + B_s\sinh kl = 0\,.$$

A nonzero solution exists only if $\sin kl = 0$, or $k_n = n\pi/l$, $n = 1, 2, \ldots$. In this case $B_s = 0$, $A_s = A$ is arbitrary, and we have for the displacement $\zeta_n = A \sin k_n x \exp(-i\omega t)$. As in the case of longitudinal oscillations the rod length equals an integer number of halfwaves. For the eigenfrequencies of oscillations we obtain from (2.30)

$$\omega_n = \sqrt{Er_0^2/\rho}(n\pi/l)^2 . \tag{2.35}$$

In the general case of arbitrary boundary conditions at the rod ends, a system of four linear homogeneous equations for the coefficients A_c, B_c, A_s, B_s (considered as components of a four-dimensional vector) is obtained:

$$[C]\begin{pmatrix} A_c \\ B_c \\ A_s \\ B_s \end{pmatrix} = 0$$

where the matrix $[C] = \{c_{jm}\}$ with $j, m = 1, 2, 3, 4$ depends on the wave number k. For the existence of a nontrivial (nonzero) solution of the system, we require $g(k) = \det[C] = 0$ from which the eigenvalues k_n can be found. For example, in the case under consideration we have

$$[C] = \begin{bmatrix} 1 & 1 & 0 & 0 \\ -1 & 1 & 0 & 0 \\ \cos kl & \cosh kl & \sin kl & \sinh kl \\ -\cos kl & \cosh kl & -\sin kl & \sinh kl \end{bmatrix}, \qquad g(k) = 4 \sin kl \sinh kl .$$

2.6 Wave Dispersion and Group Velocity

2.6.1 Propagation of Nonharmonic Waves

We have already seen that the phase velocity of harmonic bending waves depends on frequency, i.e., *dispersion* takes place. The wave-frequency dependence on the wave number, $\omega = \omega(k)$, is called a *dispersion relationship.* We consider now some characteristic features of wave propagation with dispersion. In particular, we will see below that in the case under consideration a nonharmonic wave changes its shape during propagation.

Suppose we have some wave disturbance which at the instance $t = 0$ is specified by the function $f(x, 0)$. We write this function in the form of a Fourier integral

$$f(x,0) = \int_{-\infty}^{\infty} \tilde{f}(k) \exp(ikx)\, dk ,$$

where $\tilde{f}(k)$ describes the spatial Fourier spectrum of the initial disturbance. For an arbitrary t the disturbance $f(x, t)$ can be written in the form

$$f(x,t) = \int_{-\infty}^{\infty} \tilde{f}(k) \exp\{i[kx - \omega(k)t]\} \, dk$$

$$= \int_{-\infty}^{\infty} \tilde{f}(k) \exp[ik(x - c_{ph}t)] \, dk \,, \tag{2.36}$$

where $c_{ph}(k) = \omega(k)/k$ is the phase velocity. Indeed, (2.36) becomes $f(x, 0)$ at $t = 0$. It also satisfies a wave equation, [say, (2.25) in the case of bending waves with $c_{ph}(k)$ from (2.33)], since each harmonic wave $\exp\{i[kx - \omega(k)t]\}$ satisfies a wave equation.

Now consider (2.36) more thoroughly. If the propagation velocity of each harmonic wave, $c_{ph} = \omega/k$, were independent of k [$\omega(k)$ being a linear function of k], as for longitudinal waves in a rod, for example, the disturbance would travel in the x direction with velocity c_{ph} without changing its shape. In fact, for $c_{ph} = \text{const}$, it follows from (2.36) that

$$f(x,t) = \int_{-\infty}^{\infty} \tilde{f}(k) \exp[ik(x - c_{ph}t)] \, dk = f(x - c_{ph}t, 0). \tag{2.37}$$

Such waves are termed *nondispersive*.

When c_{ph} depends on k, each harmonic component in (2.36) propagates at its own velocity. They are superimposed at different moments with different phases leading to a change of shape compared to the original wave. As a rule, an original wave's profile broadens and the disturbance amplitude decays. At sufficiently large t, the disturbance resembles a sinusoidal wave train.

2.6.2 Propagation of Narrow-Band Disturbances

In wave theory so-called narrow-band disturbances $f(x, t)$ play an important role, their spatial spectrum $\tilde{f}(k)$ differs from zero only in a small vicinity near some k_0. We will show that such disturbances represent so-called modulated harmonic waves whose mathematical expression is

$$f(x,t) = F(x,t) \exp[i(k_0 x - \omega_0 t)] \,, \qquad \omega_0 \equiv \omega(k_0) \,, \tag{2.38}$$

where the function $F(x, t)$ (the envelope) changes slowly in spacc and time compared with $\exp[i(k_0 x - \omega_0 t)]$.

At the initial instance $t = 0$ we have [$F_0(x) \equiv F(x, 0)$]

$$f(x,0) = \int_{-\infty}^{\infty} \tilde{f}(k) \exp(ikx) \, dk = F_0(x) \exp(ik_0 x) \,,$$

where the function

$$\tilde{f}(k) \simeq 0 \quad \text{with} \quad |k - k_0| > \Delta k, \qquad \Delta k \ll k_0 .$$

If we set $k = k_0 + \varkappa$, then

$$F_0(x) = \int_{-\infty}^{\infty} \tilde{f}(k_0 + \varkappa) \exp(\mathrm{i}\varkappa x)\, d\varkappa . \tag{2.39}$$

To obtain the disturbance $f(x, t)$ at an arbitrary moment t we use (2.36), assuming $k = k_0 + \varkappa$ and expanding the function $\omega(k)$ in a Taylor series in the vicinity of k_0:

$$f(x, t) = \int_{-\infty}^{\infty} \tilde{f}(k) \exp\{\mathrm{i}[k_0 x + \varkappa x - \omega_0 t - (d\omega/dk)_{k_0} \varkappa t - \cdots]\}\, d\varkappa .$$

Since only small $\varkappa$ is of importance, for not too large t we can retain only the linear term in the expansion for $\omega(k)$:

$$f(x, t) \simeq \exp[\mathrm{i}(k_0 x - \omega_0 t)] \int_{-\infty}^{\infty} \tilde{f}(k_0 + \varkappa) \exp\{\mathrm{i}\varkappa[x - (d\omega/dk)_{k_0} t]\}\, d\varkappa .$$

The derivative $d\omega/dk = c_g$ is called the *group velocity* of the wave. Indeed we note that a narrow-band disturbance resembles the modulated harmonic wave:

$$f(x, t) = F_0(x - c_g t) \exp[\mathrm{i}(k_0 x - \omega_0 t)] \tag{2.40}$$

with the envelope $F_0(x - c_g t)$ propagating at the group velocity c_g without changing its form. The density of wave energy is concentrated in the area where $F_0(x - c_g t)$ is not small, i.e., the energy also moves with the group velocity.

Note that all this is true only for t not too large when the next omitted term in the frequency expansion yields a small correction to the phase:

$$|d^2\omega/dk^2|\, \Delta k^2 t/2 \ll \pi .$$

The concept of a group velocity is valid for waves of arbitrary nature. For longitudinal waves in the rod we have $c_g = c$ (dispersion is absent). For bending waves $\omega = (E r_0^2/\rho)^{1/2} k^2$ and

$$c_g = d\omega/dk = 2(E r_0^2/\rho)^{1/2} k = 2\omega/k = 2c_{ph} , \tag{2.41}$$

the group velocity is twice as large as the phase velocity.

2.7 Exercises

2.7.1. Directly determine (not applying the Fourier transform) the reflection coefficient of an arbitrary longitudinal wave at the end of a fixed rod.

Solution: Let the semi-infinite rod correspond to $-\infty < x \leq 0$. It is fixed at $x = 0$ where displacement is assumed to be zero. The general solution of the wave equation can be written as

$$u(x, t) = f(x - ct) + g(x + ct),$$

where $f(x - ct)$ is the incident wave. The reflected wave $g(x + ct)$ can be determined from the boundary condition $u(0, t) = 0$ which yields $f(-ct) + g(ct) = 0$. Therefore, the functions $g(\xi)$ and $f(\xi)$ of one variable ξ must be related by $g(\xi) = -f(-\xi)$. Thus, $u(x, t) = f(x - ct) - f(-x - ct)$, i.e., the reflection coefficient is $V = -1$ as for harmonic waves.

2.7.2. Determine the reflection and transmission coefficients of a longitudinal harmonic wave at an interface of two rods with the parameters ρ_1, E_1, S_1 and ρ_2, E_2, S_2, assuming that the rods are rigidly connected by a thin weightless plate (Fig. 2.5).

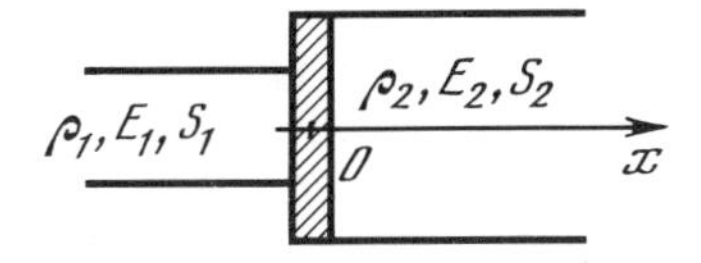

Fig. 2.5. Two rods connected by an absolutely rigid, thin, weightless plate

Solution: Look for the solution of the form

$$u_1 = A[\exp(ik_1 x) + V\exp(-ik_1 x)]\exp(-i\omega t), \quad x < 0,$$

$$u_2 = AW\exp[i(k_2 x - \omega t)], \quad x > 0,$$

where $k_j = \omega/c_j$, $c_j^2 = E_j/\rho_j$ with $j = 1, 2$. At the point of contact of the rods, $x = 0$, displacements are the same $u_1(0, t) = u_2(0, t)$ and forces on the plate exerted by each rod are equal:

$$F_1(0, t) = S_1 E_1(\partial u_1/\partial x)_0 = F_2(0, t) = S_2 E_2(\partial u_2/\partial x)_0 .$$

Substitution of u_1 and u_2 into the boundary conditions yields

$$1 + V = W, \quad 1 - V = W\sqrt{E_2\rho_2/E_1\rho_1}\, S_2/S_1 .$$

Hence, we have

$$V = \frac{S_1\sqrt{E_1\rho_1} - S_2\sqrt{E_2\rho_2}}{S_1\sqrt{E_1\rho_1} + S_2\sqrt{E_2\rho_2}}, \quad W = \frac{2S_1\sqrt{E_1\rho_1}}{S_1\sqrt{E_1\rho_1} + S_2\sqrt{E_2\rho_2}}.$$

If the rods are of the same material, then

$$V = (S_1 - S_2)/(S_1 + S_2), \quad W = 2S_1/(S_1 + S_2).$$

2.7.3. Determine the reflection coefficient of a longitudinal harmonic wave at the rod end where the impedance $Z = -F/Sv$, i.e., the ratio of tension (with an opposite sign) to the velocity is prescribed.

Solution: One has, according to Hooke's law, $F/S = E\,\partial u/\partial x$. Also taking into account that $v = \partial u/\partial t = -i\omega u$, and using the boundary condition we obtain $Z = (E/i\omega u)\,\partial u/\partial x$. Assuming that

$$u = A[\exp(ikx) + V\exp(-ikx)]\exp(-i\omega t),$$
$$k = \omega/c, \quad c = \sqrt{E/\rho},$$

yields

$$Z = \frac{ikE(1 - V)}{i\omega(1 + V)} = \sqrt{E\rho}\,\frac{1 - V}{1 + V},$$

whence

$$V = \frac{\sqrt{E\rho} - Z}{\sqrt{E\rho} + Z} = \frac{\rho c - Z}{\rho c + Z}.$$

2.7.4. Using the definition of the impedance in Exercise 2.7.3, determine the rod's impedance Z_0 at $x = 0$ if the impedance Z_l at $x = l$ is prescribed.

Solution: The harmonic solution of the wave equation in the rod is $u = (a\sin kx + b\cos kx)\exp(-i\omega t)$. Determine the relationship between the coefficients a and b, given Z_l:

$$Z_l = -\frac{F}{Sv}\bigg|_{x=l} = \frac{Ek(a\cos kl - b\sin kl)}{i\omega(a\sin kl + b\cos kl)} = -i\rho c\,\frac{(a/b) - \tan kl}{(a/b)\tan kl + 1}.$$

From the last equation we find the ratio a/b and after that the desired impedance:

$$Z_0 = \frac{Eka}{i\omega b} = -i\rho c\,\frac{i\rho c\tan kl - Z_l}{i\rho c + Z_l\tan kl} = \rho c\,\frac{Z_l - i\rho c\tan kl}{\rho c - iZ_l\tan kl}.$$

2.7.5. Determine the "input" impedance of a semi-infinite rod under the condition that only the wave leaving its end at $x = 0$ can exist (no sources in the rod).

Solution: The wave leaving the rod's end is $u = a\exp[i(kx - \omega t)]$. Therefore (Exercise 2.7.4),

$$Z_0 = \frac{E}{i\omega u}(\partial u/\partial x)_{x=0} = \frac{kE}{\omega} = \sqrt{E\rho} = \rho c.$$

2.7.6. Find the parameters of a rod of finite length (ρ, c, l) inserted between two

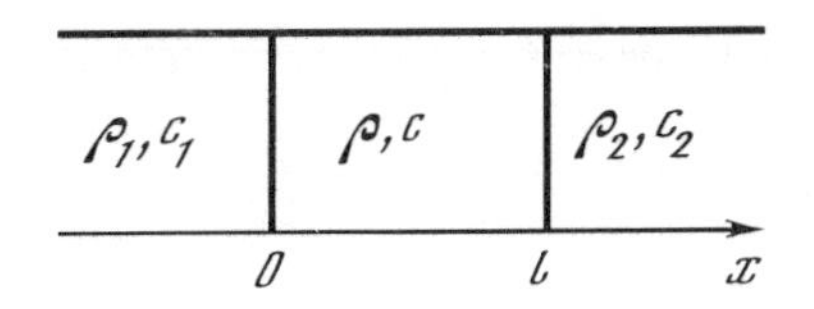

Fig. 2.6. Rod of finite length l inserted between two semi-infinite rods to prevent waves reflection

semi-infinite rods (Fig. 2.6) in order for the reflection coefficient of the harmonic longitudinal wave incident from the left to be zero.

Solution: According to the solution of the previous exercise, we have $Z_l = \rho_2 c_2$ and using the solution of Exercise 2.7.4 we find for the input impedance at $x = 0$,

$$Z_0 = Z|_{x=0} = \rho c \frac{\rho_2 c_2 - \mathrm{i}\rho c \tan kl}{\rho c - \mathrm{i}\rho_2 c_2 \tan kl}.$$

Now apply it to the expression for the reflection coefficient obtained in Problem 2.3, and it follows that $V = 0$ for $Z_0 = \rho_1 c_1$. It can be readily seen that this will be the case if $kl = \pi(n + 1/2)$ with $n = 0, 1, 2, \ldots$ and $\rho c = \sqrt{\rho_1 c_1 \rho_2 c_2}$. In fact, then $\tan kl \to \pm\infty$ and $Z_0 = (\rho c)^2/\rho_2 c_2 = \rho_1 c_1$.

2.7.7. Examine the propagation of longitudinal elastic disturbances in a rod of finite length l after this rod had hit (at $t = 0$) an absolutely rigid wall, with velocity v_0 (Fig. 2.7).

Solution: At $t > 0$ the wall acts on the rod with a force which we denote as $f(t)$ [$f(t) \equiv 0$ at $t < 0$]. The corresponding stress propagates along the rod with the velocity $c = \sqrt{E/\rho}$. Hence, at $0 < t < l/c$ we have for the force at the section x: $F(x, t) = f(x/c + t)$. At $t = l/c$ this disturbance reaches the rod's free end $x = -l$ and a reflected disturbance arises, so that $F(x, t) = f(x/c + t) + g(t - x/c)$ at

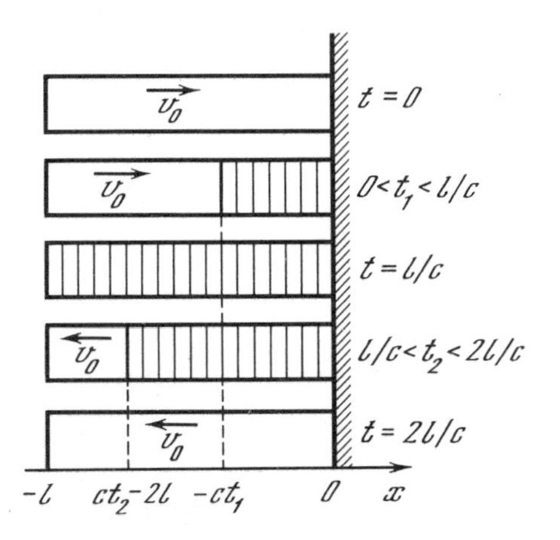

Fig. 2.7. Illustrating the impact of a rod at the wall for different phases of the process. Those parts of the rod which are under tension, are shaded

$l/c<t<2l/c$. Since $F(-l,t)\equiv 0$ at the free end, we have $f(t-l/c)+g(t+l/c)=0$ or $g(\xi) = -f(\xi - 2l/c)$. Therefore,

$$F(x,t) = f(t + x/c) - f(t - x/c - 2l/c) \quad \text{at} \quad 0 < t < 2l/c\,.$$

Now we find the velocity of the rod's particles. Using the relation $F(x,t) = SE\,\partial u/\partial x$ we obtain

$$v(x,t) = \partial u/\partial t = \frac{\partial}{\partial t}\int_0^x \frac{\partial u}{\partial x}\,dx + \alpha(t) = \frac{1}{ES}\frac{\partial}{\partial t}\int_0^x F(x,t)\,dx + \alpha(t)\,.$$

For $0 < t < 2l/c$ we have $\alpha(t) \equiv 0$, since $v(0,t) = 0$. Substituting the expression for $F(x,t)$ into $v(x,t)$ we obtain

$$\begin{aligned} v(x,t) &= \frac{1}{ES}\frac{\partial}{\partial t}\left[\int_0^x f(x/c + t)\,dx - \int_0^x f(t - x/c - 2l/c)\,dx\right] \\ &= \frac{c}{ES}\frac{\partial}{\partial t}\left[\int_t^{t+x/c} f(\xi)\,d\xi + \int_{t-2l/c}^{t-x/c-2l/c} f(\xi)\,d\xi\right] \\ &= \frac{c}{ES}\left[f(t + x/c) - f(t) + f(t - x/c - 2l/c)\right], \end{aligned}$$

since $f(t - 2l/c) = 0$ for $t < 2l/c$. We can easily find an explicit form of the function $f(\xi)$, using the condition that $v(-l,t) = v_0$ for $0 < t < l/c$ (the disturbance has not yet reached the left end of the rod):

$$v_0 = \frac{c}{ES}\left[f(t - l/c) - f(t) + f(t - l/c)\right] = -\frac{c}{ES}f(t)\,,$$

since $f(t - l/c) = 0$ for $t < l/c$. Hence, we find $f(t) = -ESv_0/c$—or, since $f(t) = 0$ at $t < 0$, $f(t) = -ESv_0\theta(t)/c$, where

$$\theta(t) = \begin{cases} 1, & t > 0 \\ 0, & t < 0 \end{cases}.$$

Finally we write for $0 < t < 2l/c$:

$$\begin{aligned} F(x,t) &= -ES(v_0/c)[\theta(t + x/c) - \theta(t - x/c - 2l/c)]\,, \\ v(x,t) &= -v_0[\theta(t + x/c) - \theta(t) + \theta(t - x/c - 2l/c)]\,. \end{aligned}$$

We infer from the last expressions that the part of the rod $-l < x < ct - 2l$ becomes stress-free for $t > l/c$ $[F(x,t) = 0]$, its particle's velocity being $-v_0$. At $t = 2l/c$ all of the rod becomes stress-free, moving as a whole with velocity $-v_0$ after its rebound. The disturbed portions of the rod at different times are shown in Fig. 2.7.

2.7.8. Prove the momentum-conservation law for the rod when rebounding.

Solution: During the contact time $2l/c$ the constant force $F_0 = F(0,t) = -ESv_0/c$ is exerted by the wall on the rod (Exercise 2.7.7). Expressing the total momentum of this force $F_0 2l/c$ in terms of velocity v_0 and the rod mass $m = \rho l S$, we obtain

$$2F_0 l/c = -2lESv_0/c^2 = -2\rho l S v_0 = -mv_0 - mv_0 = mv|_{t>2l/c} - mv|_{t<0}\,.$$

which is the momentum-conservation law.

2.7.9. A constant force F_0 of short duration $\tau \ll l/c$ is applied to the left end of the rod. Examine the resulting motion of the rod.

Solution: When the force ceases to act, the compression pulse will run along the rod and $F(x,t) = -F_0 \Pi(t - x/c)$ for $0 < t < l/c$ where

$$\Pi(\xi) = \begin{cases} 1, & 0 < \xi < \tau, \\ 0, & \xi < 0, \quad \xi > \tau \end{cases},$$

At $t = l/c$ the pulse reaches the right end of the rod and is reflected. Under the condition that $F(l,t) = 0$ (free end), we obtain for $0 < t < 2l/c$

$$F(x,t) = -F_0[\Pi(t - x/c) - \Pi(t + x/c - 2l/c)]\,.$$

For the velocity of rod particles we have (Exercise 2.7.7)

$$v(x,t) = \frac{1}{ES}\frac{\partial}{\partial t}\int_0^x F(x,t)\,dx + \alpha(t)$$

$$= \frac{F_0 c}{ES}\left[\frac{\partial}{\partial t}\int_t^{t-x/c} \Pi(\xi)\,d\xi + \alpha_1(t) + \frac{\partial}{\partial t}\int_{t-2l/c}^{t+x/c-2l/c} \Pi(\xi)\,d\xi\right]$$

$$= \frac{F_0 c}{ES}[\Pi(t - x/c) - \Pi(t) + \Pi(t + x/c - 2l/c) - \Pi(t - 2l/c) + \alpha_1(t)]\,.$$

Since $\Pi(t - 2l/c) = 0$ for $t < 2l/c$ and $v(l,t) = 0$ for $t < l/c$, we have $\alpha_1(t) = \Pi(t)$ for $t < l/c$. Taking this into account we have further $v(0,t) = F_0 c\Pi(t)/ES = 0$ for $\tau < t < l/c$ and, consequently, $v(0,t) = 0$ for $\tau < t < 2l/c$, since the reflected pulse has not yet reached the left end. Therefore, $\alpha_1(t) = 0$ for $l/c < t < 2l/c$ and finally we obtain

$$v(x,t) = \frac{F_0 c}{ES}[\Pi(t - x/c) + \Pi(t + x/c - 2l/c)], \quad 0 < t < 2l/c\,.$$

It follows from the last expression that in the interval of time $0 < t < 2l/c$, each element of the rod (except those in the vicinity of the right-hand end) will acquire twice the velocity $v_0 = F_0 c/ES$, the duration of the latter motion being τ.

Elements close to the right end during part of the time move at twice the velocity. As a result, each rod's particle is displaced by a distance of $u_0 = 2v_0\tau = 2F_0c\tau/ES$ during the time $2l/c$. Further, after reflection of the pulse from the already free left end, this process repeats.

2.7.10. Show that the motion of the rod considered in the previous problem satisfies the energy and momentum conservation laws.

Solution: The moment of external force $F_0\tau$ must be equal to that transported by the compression wave μv_0, $\mu = \rho S c\tau$ being the mass of the disturbed portion of the rod. Using the expression for v_0 in terms of F_0, we immediately prove the momentum conservation law:

$$\mu v_0 = \rho S c\tau F_0 c/ES = F_0\tau\,.$$

Introducing an average velocity of the rod travelling as a whole:

$$\bar{v} = u_0 c/2l = v_0\tau c/l = F_0 c^2\tau/ESl = F_0\tau/\rho Sl = F_0\tau/m\,,$$

where m is the rod mass, we obtain the classical law of momentum conservation: $F_0\tau = m\bar{v}$.

We now calculate the energy of the compression pulse. For the kinetic energy we have $\mathscr{E}_\text{k} = \mu v_0^2/2$, whence the potential energy, according to Exercise 1.4.8, will be

$$\mathscr{E}_\text{p} = \frac{1}{2} S c\tau \frac{1}{E}\left(\frac{F_0}{S}\right)^2 = \frac{1}{2} S c\tau \frac{E}{c^2} v_0^2 = \frac{1}{2}\mu v_0^2 = \mathscr{E}_\text{k}\,.$$

Thus the total energy of the pulse is

$$\mathscr{E} = \mathscr{E}_\text{k} + \mathscr{E}_\text{p} = \mu v_0^2 = \rho S c\tau v_0^2 = \rho S c\tau \frac{F_0 c}{ES} = F_0 v_0\tau\,.$$

On the other hand, under the action of a force F_0 during time τ the rod's left end is displaced by a distance of $u_0/2 = v_0\tau$, i.e., the work performed by the force is $A = F_0 v_0\tau = \mathscr{E}$. Hence, the energy conservation law holds, too. Note that the kinetic energy of the average motion

$$\mathscr{E}_\text{k} = \frac{m}{2}\bar{v}^2 = \frac{m}{2}\bar{v}\frac{F_0\tau}{m} = \frac{1}{2}F_0 v_0\tau\frac{\bar{v}}{v_0} = \frac{1}{2}\frac{\tau c}{l}\mathscr{E}$$

is much less than the total energy $\mathscr{E}$.

2.7.11. Determine the reflection and transmission coefficients of a bending wave at the rod's section $x = 0$ where the point mass m is placed.

Solution: We have for $x < 0$ and $x > 0$, respectively:

$$\zeta_1 = \exp[\mathrm{i}(kx - \omega t)] + V \exp[-\mathrm{i}(kx + \omega t)] + V_1 \exp(kx - \mathrm{i}\omega t),$$
$$\zeta_2 = W \exp[\mathrm{i}(kx - \omega t)] + W_1 \exp(-kx - \mathrm{i}\omega t).$$

At $x = 0$ the displacement, slope, and moment must be continuous:

$$\zeta_1(0, t) = \zeta_2(0, t), \quad (\partial\zeta_1/\partial x)_{x=0} = (\partial\zeta_2/\partial x)_{x=0}, \quad (\partial^2\zeta_1/\partial x^2)_{x=0} = (\partial^2\zeta_2/\partial x^2)_{x=0}.$$

The difference between shear forces at $x = 0$ causes the mass motion with acceleration $\partial^2\zeta_1/\partial t^2$: $EI(\partial^3\zeta_2/\partial x^3 - \partial^3\zeta_1/\partial x^3)_{x=0} = m(\partial^2\zeta_1/\partial t^2)_{x=0}$. Substituting the expressions for ζ_1 and ζ_2 into these conditions we obtain

$$1 + V + V_1 = W + W_1, \quad \mathrm{i}(1 - V) + V_1 = \mathrm{i}W - W_1,$$
$$-(1 + V) + V_1 = -W + W_1, \quad \mathrm{i}(1 - V) + V_1 - \mathrm{i}W - W_1 = -\frac{m}{EI}\omega^2(1 + V + V_1).$$

Hence we find

$$V = \frac{m\omega^2}{4\mathrm{i}EI - m\omega^2(1 + \mathrm{i})}, \quad V_1 = W_1 = \mathrm{i}V,$$
$$W = 1 + V = \frac{4\mathrm{i}EI - \mathrm{i}m\omega^2}{4\mathrm{i}EI - m\omega^2(1 + \mathrm{i})}.$$

2.7.12. Find the reflection and transmission coefficients of a bending wave for a device which forbids displacements but does not prevent rotations.

Solution: This problem differs from the previous one only by the boundary conditions which are now $\zeta_1(0, t) = \zeta_2(0, t) = 0$, $(\partial\zeta_1/\partial x)_{x=0} = (\partial\zeta_2/\partial x)_{x=0}$, $(\partial^2\zeta_1/\partial x^2)_{x=0} = (\partial^2\zeta_2/\partial x^2)_{x=0}$. There are no conditions for the shear force. The following system of equations results from these conditions:

$$1 + V + V_1 = W + W_1 = 0, \quad \mathrm{i}(1 - V) + V_1 = \mathrm{i}W - W_1,$$
$$-(1 + V) + V_1 = -W + W_1;$$

its solution is

$$V = -(1 - \mathrm{i})/2, \quad W = (1 + \mathrm{i})/2, \quad V_1 = W_1 = -W = -(1 + \mathrm{i})/2.$$

2.7.13. Determine the eigenfrequencies of bending vibrations of a rod of length l clamped at both ends.

Solution: Similar to the case of a rod supported at both ends considered in Sect. 2.5.3, we obtain an equation for the determination of the eigenfrequencies: $\cos kl = 1/\cosh kl$. For $kl \gg 1$ we have $k_n l \simeq \pi(n + 1/2)$ and

$$\omega_n \simeq \sqrt{\frac{Er_0^2}{\rho}\frac{\pi^2}{l^2}}(n + 1/2)^2 .$$

2.7.14. Determine the frequency band within which the modulus of the reflection coefficient of the longitudinal wave in Exercise 2.7.6 is less then $\varepsilon \ll 1$.

2.7.15. Find the parameters of a rod (ρ, c, l in Exercise 2.7.6) when the reflection coefficient remains the same as without the rod.

2.7.16. How must the ends of the rod be fixed in order to the eigenfrequency of bending vibrations to be minimal.

3. General Theory of Stress and Strain

Let us now proceed to the general theory of the behaviour of an elastic body under the action of external forces. Our discussion in this chapter is not restricted to isotropic bodies, i.e., bodies whose characteristics are the same in all directions, but is also applicable to crystals of arbitrary symmetry.

3.1 Description of the State of a Deformed Solid

3.1.1 Stress Tensor

Internal stresses arise in a body when it is deformed by the action of external forces. To specify these stresses we cut a deformed body into two parts by an arbitrary surface S. To retain the equilibrium of each part when removing the other, the forces applied to interface S must be equal to those acting prior to cutting. These forces characterize the stresses in the material. To describe them we choose a point of surface S and introduce the orthogonal coordinates x_1, x_2, x_3 with the axis x_1 normal to S. Take a small element of surface S with the area $\Delta S_1 = \Delta x_2 \Delta x_3$ containing the chosen point. It is reasonable to assume that the force $\Delta \boldsymbol{F}_1$ acting on this segment is proportional[1] to the area ΔS_1. Denote by ΔF_{1k} the component of this force along the axis x_k ($k = 1, 2, 3,$) and introduce a quantity independent of the area ΔS_1:

$$\sigma_{1k} = \Delta F_{1k}/\Delta S_1 \tag{3.1}$$

Analogously we can choose some $\Delta S_2 = \Delta x_1 \Delta x_3$ and $\Delta S_3 = \Delta x_1 \Delta x_2$ perpendicular to the axes x_2 and x_3, respectively, and introduce

$$\sigma_{2k} = \Delta F_{2k}/\Delta S_2 , \qquad \sigma_{3k} = \Delta F_{3k}/\Delta S_3 . \tag{3.2}$$

Nine quantities σ_{ik} ($i, k = 1, 2, 3$) specified by (3.1, 2) completely determine the stress state of the solid at the point considered. Using these quantities one can determine the force acting across an arbitrarily oriented small area at this point.

[1] Generally speaking, the distribution of forces over ΔS_1 can be nonuniform, then it will be equivalent both to the force $\Delta \boldsymbol{F}_1$ and the moment $\Delta \boldsymbol{M}_1$. We assume that

$$\lim_{\Delta S_1 \to 0} \Delta \boldsymbol{M}_1/\Delta S_1 = 0$$

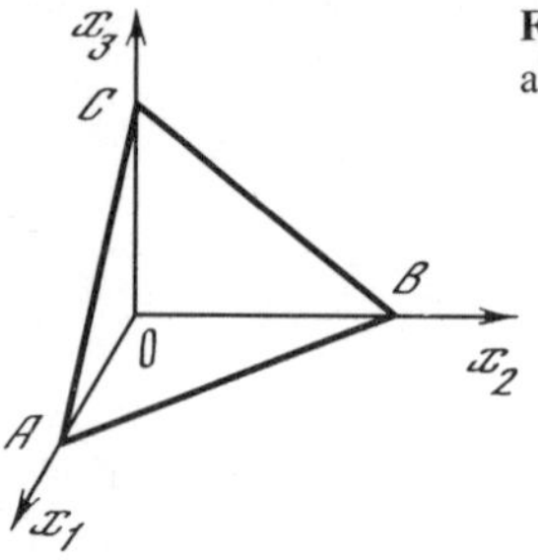

Fig. 3.1. To determine the force acting across an arbitrarily oriented area

To prove the last statement, consider an infinitesimal volume in the form of a tetrahedron separated from the body and shown in Fig. 3.1. It is bounded by i) the face ABC with an area ΔS and normal $\boldsymbol{n}$. The force $\Delta \boldsymbol{F}$ acting across this face is to be determined. Since it is proportional to ΔS we can write $\Delta \boldsymbol{F} = \boldsymbol{f} \Delta S$ with components along the axes $\Delta F_i = f_i \Delta S$; ii) the faces OBC, OAC and OAB whose outer normals are opposite to the unit vectors of the coordinate axes x_i ($i = 1, 2, 3$), respectively. Their areas are equal to $\Delta S_i = n_i \Delta S$, n_i being the components of the unit vector $\boldsymbol{n}$ along the x_i-axis. Taking into account the definition of σ_{ik}, we can write the k-component of forces acting across each of these faces as:

$$\begin{aligned} -\sigma_{1k} \Delta S_1 &= -\sigma_{1k} n_1 \Delta S, \\ -\sigma_{2k} \Delta S_2 &= -\sigma_{2k} n_2 \Delta S, \\ -\sigma_{3k} \Delta S_3 &= -\sigma_{3k} n_3 \Delta S. \end{aligned}$$

The minus sign is introduced here using the convention that a positive force corresponds to a tension or pull and is directed opposite to the coordinate axes in Fig. 3.1.

Since the volume considered is in equilibrium, the sum of all these surface forces must be zero: $f_i \Delta S - \sigma_{1i} n_1 \Delta S - \sigma_{2i} n_2 \Delta S - \sigma_{3i} n_3 \Delta S = 0$. Cancelling ΔS we obtain

$$f_i = \sigma_{ki} n_k \,. \tag{3.3}$$

As always, the summation over a repeated index is implied. The body forces (gravity force, for example) can be neglected here, for they are proportional to $\Delta x_1 \Delta x_2 \Delta x_3$ and as $\Delta x_i \to 0$ become infinitesimal of higher order as compared with the surface forces which are proportional to $\Delta x_i \Delta x_k$.

Thus, the quantities σ_{ik} determine completely the stress state of a solid by means of (3.3). Moreover, it follows from (3.3) (Appendix, Theorem A.1) that σ_{ik} is a second-order tensor, since $\boldsymbol{n}$ and $\boldsymbol{f}$ are vectors.

The tensor σ_{ik} is called the *stress tensor*. We now prove that it is *symmetric*, i.e., $\sigma_{ik} = \sigma_{ki}$. For this purpose, consider a parallelepiped of infinitesimal volume $\Delta V = \Delta x_1 \Delta x_2 \Delta x_3$ with edges parallel to the coordinate axes. A section of this parallelepiped given by the plane $x_3 = \text{const}$ is shown in Fig. 3.2. Arrows indicate

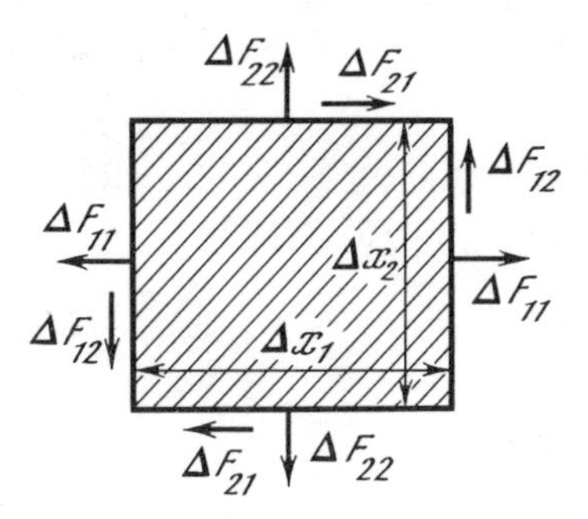

Fig. 3.2. Forces acting on the face of an infinitesimal parallelepiped

the forces acting on the parallelepiped faces. Note that ΔF_{ik} is the force acting along the x_k-axis and on the face which is perpendicular to the x_i-axis. Since the parallelepiped is assumed to be in equilibrium, the total force and moment acting on it must be zero. The total force is zero since the forces applied to the opposite faces are equal in magnitude and act in opposite directions. The sum of moments, say, about the x_3-axis will be zero if $\Delta F_{12}\Delta x_1 = \Delta F_{21}\Delta x_2$ since the moments of forces ΔF_{12} and ΔF_{21} are in opposite directions. Expressing ΔF_{12} and ΔF_{21} in terms of σ_{12} and σ_{21} using (3.1, 2) and cancelling the volume $\Delta x_1 \Delta x_2 \Delta x_3$, we obtain $\sigma_{21} = \sigma_{12}$. Similarly, the equalities $\sigma_{31} = \sigma_{13}$ and $\sigma_{32} = \sigma_{23}$ are proved. Consequently, the symmetry of the tensor σ_{ik} is proved. Using this, (3.3) can also be written as

$$f_i = \sigma_{ik} n_k . \tag{3.4}$$

The components of the stress tensor σ_{ik} are generally the functions of coordinates and are constant only in the case of homogeneous strain.

3.1.2 The Strain Tensor

Let the position of each point of the body prior to deformation be specified by the vector $\boldsymbol{r} = \{x_1, x_2, x_3\}$, and after deformation by $\boldsymbol{r}' = \{x_1', x_2', x_3'\}$. The quantity $\boldsymbol{u} = \boldsymbol{r}' - \boldsymbol{r}$, $u_i = x_i' - x_i$ is called the *displacement*. The latter depends on $\boldsymbol{r}$, since only in this case do stress arise in an elastic body. The case where $\boldsymbol{u}$ does not depend on $\boldsymbol{r}$ corresponds to the motion of a body in its undeformed state without internal stress.

If $\boldsymbol{u}(\boldsymbol{r})$ is known, it is possible to calculate the change in distance between two arbitrary points. We consider two infinitely close points x_i and $x_i + dx_i$, the squared distance between them before deformation being $dl^2 = \sum_{i=1}^{3} dx_i^2 \equiv dx_i dx_i \equiv dx_i^2$. After deformation the coordinates of these points will be, respectively, $x_i' = x_i + u_i(x_k)$[2] and

$$x_i + dx_i + u_i(x_k + dx_k) = x_i + dx_i + u_i(x_k) + (\partial u_i/\partial x_k)dx_k .$$

[2] Here and below, $u_i(x_k)$ stands for $u_i(x_1, x_2, x_3)$.

Consequently, the squared distance between the points chosen after deformation will be

$$(dl')^2 = \left(dx_i + \frac{\partial u_i}{\partial x_k}dx_k\right)^2 = dx_i^2 + 2\frac{\partial u_i}{\partial x_k}dx_i\,dx_k + \frac{\partial u_i}{\partial x_k}\frac{\partial u_i}{\partial x_m}dx_k\,dx_m\,.$$

Using the obvious identities

$$2\frac{\partial u_i}{\partial x_k}dx_i\,dx_k = \frac{\partial u_i}{\partial x_k}dx_i\,dx_k + \frac{\partial u_k}{\partial x_i}dx_i\,dx_k\,,$$

$$\frac{\partial u_i}{\partial x_k}\frac{\partial u_i}{\partial x_m}dx_k\,dx_m = \frac{\partial u_m}{\partial x_i}\frac{\partial u_m}{\partial x_k}dx_i\,dx_k\,,$$

we can write $(dl')^2$ in the form

$$(dl')^2 = dl^2 + 2e_{ik}\,dx_k\,dx_i\,, \qquad e_{ik} = \frac{1}{2}\left(\frac{\partial u_i}{\partial x_k} + \frac{\partial u_k}{\partial x_i} + \frac{\partial u_m}{\partial x_i}\frac{\partial u_m}{\partial x_k}\right). \tag{3.5}$$

Since $(dl')^2 - dl^2$ is a scalar, and dx_i and dx_k are vectors, then by virtue of Theorem A.2, the quantity e_{ik} is a second-order tensor called the *strain tensor*. Its components are generally functions of coordinates and are constant in the particular case of *homogeneous strain*. The symmetry of the tensor e_{ik} ($e_{ik} = e_{ki}$) is evident. In the case of small strain, we may confine ourselves to terms linear in u_i in the expression for e_{ik}, thus

$$e_{ik} = \frac{1}{2}\left(\frac{\partial u_i}{\partial x_k} + \frac{\partial u_k}{\partial x_i}\right). \tag{3.6}$$

3.1.3 The Physical Meaning of the Strain Tensor's Components

First, we consider the diagonal components, say, $e_{11} = \partial u_1/\partial x_1$. Let x_1 and $x_1 + \Delta x_1$ be two close points on the x_1-axis prior to deformation. In the process of deformation these points can be removed from the x_1-axis, their coordinates along this axis being $x_1 + u_1(x_1, x_2, x_3)$ and $x_1 + \Delta x_1 + u_1(x_1 + \Delta x_1, x_2, x_3)$, respectively. The distance between the points prior to deformation is Δx_1 and that along the x_1-axis after deformation will be

$$\Delta x_1 + u_1(x_1 + \Delta x_1, x_2, x_3) - u_1(x_1, x_2, x_3) \simeq \Delta x_1 + (\partial u_1/\partial x_1)\,\Delta x_1\,.$$

For the increment of the distance we have $(\partial u_1/\partial x_1)\,\Delta x_1$. The ratio of this quantity to the initial distance Δx_1 gives a relative elongation along the x_1-axis. The same is true for the other axes. Hence, the diagonal components of the strain tensor $e_{11} = \partial u_1/\partial x_1$, $e_{22} = \partial u_2/\partial x_2$, $e_{33} = \partial u_3/\partial x_3$ are *relative elongations* (*contractions*) along the coordinate axes.

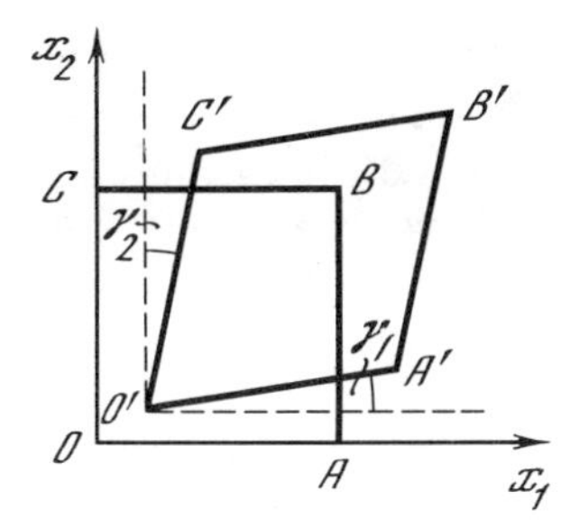

Fig. 3.3. Shear deformation of an infinitesimal parallelepiped

Let us now estimate the change of volume $V = \Delta x_1 \Delta x_2 \Delta x_3$ of an elementary parallelepiped with edge lengths Δx_1, Δx_2, Δx_3 prior to deformation. As we have seen, after deformation the edge lengths become $(1 + e_{11})\Delta x_1$, $(1 + e_{22})\Delta x_2$, $(1 + e_{33})\Delta x_3$, respectively. Retaining only the first-order terms with respect to u_i, we have for the parallelepiped volume after deformation $V' \simeq (1 + e_{11} + e_{22} + e_{33})\Delta x_1 \Delta x_2 \Delta x_3$. The relative increase in volume is

$$(V' - V)/V = e_{11} + e_{22} + e_{33} \equiv e_{kk}\,, \tag{3.7}$$

i.e., the *trace of the strain tensor*. The latter, according to Theorem A.3 is invariant. Consequently, the relative change of volume is independent of the choice of the coordinate system.

Now we proceed to clarify the physical meaning of nondiagonal components of a strain tensor and consider e_{12}, for example. We again take an elementary parallelepiped with edge lengths Δx_1, Δx_2, Δx_3 parallel to the coordinate axes. In Fig. 3.3, OCBA is the parallelepiped face lying in the coordinate plane x_1x_2 prior to deformation. In the process of deformation, the points O, A, B, C displace, in general leaving the plane x_1x_2. We denote by O′, A′, B′, C′ the projections of the new positions of these points upon the x_1x_2 plane. Taking into account that strains and, consequently, the angles γ_1, γ_2 (Fig. 3.3) are small, we have

$$\gamma_1 \simeq \tan\gamma_1 \simeq \frac{u_2(A) - u_2(O)}{\Delta x_1} \simeq \frac{\partial u_2}{\partial x_1}, \qquad \gamma_2 \simeq \frac{u_1(C) - u_1(O)}{\Delta x_2} \simeq \frac{\partial u_1}{\partial x_2},$$

so that

$$2e_{12} = (\partial u_2/\partial x_1) + (\partial u_1/\partial x_2) = \gamma_1 + \gamma_2 = e_{12} + e_{21}\,. \tag{3.8}$$

Thus, the symmetric sum of *nondiagonal components* of the strain tensor $e_{12} + e_{21}$ determines the *angle change* between the corresponding faces of the parallelepiped. As we know (Sect. 2.3); the angle distortion means shear. Consequently, the *nondiagonal elements of the strain tensor* e_{ik} describe the *shear strains.*

The strain tensor can be transformed into a diagonal form at any fixed point by an appropriate rotation of the coordinate system. Therefore, shear strain will be absent in the new system; only extensions and contractions take place. In particular, we have seen (Fig. 1.7) that uniform shear strain of a cube can be considered a combination of a contraction along one of the diagonals and an extension along the other.

3.2 Equations of Motion for a Continuous Medium

3.2.1 Derivation of the Equation of Motion

Consider a volume V in an elastic medium. By Newton's second law the inertial force $\int_V \rho \ddot{u}\, dV$ (ρ is the material density, a dot above a quantity means differentiation with respect to time) must be balanced by i) the forces $\boldsymbol{F}$ acting across the surface S bounding the volume. This force is due to the state of stress of the body; ii) the body forces $\int_V \boldsymbol{f}_b\, dV$. The gravity force with volume density ρg is an example of the latter. A mathematical expression for this balance is the equation motion

$$\int_V \rho \ddot{\boldsymbol{u}}\, dV = \int_V \boldsymbol{f}_b\, dV + \boldsymbol{F}. \tag{3.9}$$

Using (3.4) we obtain for the force $\boldsymbol{F}$ in terms of the stress tensor σ_{ik}:

$$F_i = \int_S f_i\, dS = \int_S \sigma_{ik} n_k\, dS, \tag{3.10}$$

where $\boldsymbol{n} = \{n_k\}$ is an outward normal to the surface S. Using the Gauss theorem (Theorem A.4), (3.10) can be rewritten as $F_i = \int_V (\partial \sigma_{ik}/\partial x_k)\, dV$. As a result, we get for the component of (3.9) along the i-axis:

$$\int_V [\rho \ddot{u}_i - (f_b)_i - \partial \sigma_{ik}/\partial x_k]\, dV = 0.$$

By virtue of the arbitrariness of V, the integrand in the last equation must be zero. Hence, we obtain the *equation of motion* for a deformable body:

$$\rho \ddot{u}_i = (f_b)_i + \partial \sigma_{ik}/\partial x_k. \tag{3.11}$$

In the static case ($\boldsymbol{u}$ being independent of time), Eq. (3.11) becomes

$$\partial \sigma_{ik}/\partial x_k + (f_b)_i = 0, \tag{3.12}$$

which is the *equation of equilibrium* for a deformed body.

Note that the state of stress of an elastic body in equilibrium can only be due to the action of external forces, particularly those applied to its surface. In the latter case, these forces must be included in the boundary conditions for

(3.12). Namely, with p_i being the surface density of external forces, we have, according to (3.3),

$$p_i = \sigma_{ik} n_k|_S . \tag{3.13}$$

3.2.2 Strain-Stress Relation. Elasticity Tensor

To solve (3.11) we need the stress tensor expressed in terms of the strain tensor. It is reasonable to assume that a one-to-one relationship between the stress and strain tensors of the type $\sigma_{ik} = G_{ik}(e_{jl})$ must occur for an elastic body in the general case. Expanding the functions G_{ik} in a Taylor series for sufficiently small strains, we obtain linear relations between σ_{ik} and e_{jl}, or the *generalized Hooke's law*.

$$\sigma_{ik} = C_{ikjl} e_{jl} , \tag{3.14}$$

where the fourth-order *elasticity tensor* $C_{ikjl} = (\partial G_{ik}/\partial e_{jl})_0$ is characteristic of the material considered. A general fourth-order tensor in three dimensions has $3^4 = 81$ components. Because of symmetry in σ_{ik} and e_{jl}, however, the tensor C_{ikjl} has considerably fewer independent components.

Indeed, we can write a sequence of identities:

$$\sigma_{ik} = \underline{C_{ikjl} e_{jl}} = C_{iklj} e_{lj} = \underline{C_{iklj} e_{jl}} = \sigma_{ki} = \underline{C_{kijl} e_{jl}} ,$$

or if we combine the underlined terms in pairs:

$$(C_{ikjl} - C_{iklj}) e_{jl} = 0 , \quad (C_{ikjl} - C_{kijl}) e_{jl} = 0 .$$

Since the tensor e_{jl} is arbitrary, we can choose the elasticity tensor as being a symmetric one, i.e.,

$$C_{ikjl} = C_{iklj} , \quad C_{ikjl} = C_{kijl} . \tag{3.15}$$

Consequently, an *elasticity tensor is invariant to the permutation of subscripts in the first and second pairs.* Hence, for fixed i and k there are only 6 independent components of C_{ikjl}. Analogously, only 6 of a possible nine combinations of subscripts i and k give an independent C_{ikjl}. Consequently, a number of independent components of an elasticity tensor cannot exceed $6 \times 6 = 36$. Moreover, one can show that this number must be even less if one applies energy considerations.

3.3 The Energy of a Deformed Body

3.3.1 The Energy Density

To deform an elastic body, some work must be done. We can calculate this work assuming that the body is deformed quasistatically, i.e., so slowly that at all times it is in mechanical as well as in thermodynamical equilibrium. Consider a volume V inside the body bounded by the surface S with the outer normal $\boldsymbol{n}$. The forces $\sigma_{ik} n_k$ act on this volume across the surface S, according to (3.4). Let the increment of the displacement $u_i(\boldsymbol{r})$ be $\delta u_i(\boldsymbol{r})$. Then the work of these forces will be $\delta A = \int_S \sigma_{ik} n_k \, \delta u_i \, dS$. Applying Gauss' divergence theorem to the vector $\sigma_{ik} \, \delta u_i$, we obtain

$$\delta A = \int_V \frac{\partial}{\partial x_k} (\sigma_{ik} \, \delta u_i) \, dV = \int_V \frac{\partial \sigma_{ik}}{\partial x_k} \delta u_i \, dV + \int_V \sigma_{ik} \, \delta \frac{\partial u_i}{\partial x_k} \, dV .$$

For a quasistatic process, the first integral on the right-hand side is zero according to (3.12) and the assumption that there are no body forces ($\boldsymbol{f}_b = 0$). In the second integral we have, by virtue of the symmetry of the tensor σ_{ik},

$$\sigma_{ik} \, \delta(\partial u_i / \partial x_k) = \sigma_{ik} \, \delta \left[\frac{1}{2} \left(\frac{\partial u_i}{\partial x_k} + \frac{\partial u_k}{\partial x_i} \right) \right] = \sigma_{ik} \, \delta e_{ik} .$$

On the other hand, the work δA increases the internal energy of the volume under consideration. If we also take into account a possible entropy change, then we have by the known thermodynamical identity,

$$\int_V \delta \varepsilon \, dV = \int_V T \, \delta s \, dV + \int_V \sigma_{ik} \, \delta e_{ik} \, dV ,$$

where ε and s are the internal energy and entropy of unit volume, respectively, and T is the temperature. Since the last relationship must be true for any V, it can be rewritten in differential form:

$$d\varepsilon = T \, ds + \sigma_{ik} \, de_{ik} . \tag{3.16}$$

If we use the free energy $f = \varepsilon - Ts$, $df = d\varepsilon - s \, dT - T \, ds$ instead of the internal energy ε, then (3.16) becomes

$$df = -s \, dT + \sigma_{ik} \, de_{ik} . \tag{3.16a}$$

This expression is convenient to use for isothermal processes. Thus, with ε and f known as functions of s and e_{ik}, or T and e_{ik}, the components of the stress tensor σ_{ik} can be found from the relations

$$\sigma_{ik} = (\partial \varepsilon / \partial e_{ik})_s = (\partial f / \partial e_{ik})_T . \tag{3.17}$$

If the deformation is purely adiabatic or purely isothermal, then $dw = \sigma_{ik} \, de_{ik}$ is the differential of the energy stored in the solid. The function $w(e_{ik})$ is called the *strain-energy function* or *elastic energy function*. Instead of (3.17) we now have

$$\sigma_{ik} = \partial w / \partial e_{ik} . \tag{3.17a}$$

According to the generalized Hooke's law (3.14), σ_{ik} is a linear function of e_{ik}. Consequently, $w(e_{ik})$ must be a *homogeneous quadratic* function of the components of the strain tensor:

$$W = \tfrac{1}{2} C_{ikjl} e_{ik} e_{jl} \,. \tag{3.18}$$

It is easily seen that (3.14) follows from (3.17a, 18), taking the symmetry relations (3.15) into account.

The components of the elasticity tensor C_{ikjl} will be different for different thermodynamic processes, since $w = \varepsilon$ for adiabatic and $w = f$ for isothermal processes. In accordance with this, one must distinguish between *adiabatic* and *isothermal elastic constants.*

3.3.2 The Number of Independent Components of the Elasticity Tensor

One can use (3.18) to determine a number of independent components of the tensor C_{ikjl}. First of all, note that this tensor can be determined in such a manner that it will be invariant to permutations of the subscript pairs *ik* and *jl*:

$$C_{ikjl} = C_{jlik} \,. \tag{3.19}$$

Indeed, taking into account that repeated subscripts can be permutated arbitrarily, we have

$$w = \frac{1}{2} C_{ikjl} e_{ik} e_{jl} = \frac{1}{2}\left(\frac{C_{ikjl} e_{ik} e_{jl} + C_{jlik} e_{ik} e_{jl}}{2}\right) = \frac{1}{2} C'_{ikjl} e_{ik} e_{jl} \,,$$

where the tensor $C'_{ikjl} = (C_{ikjl} + C_{jlik})/2$ satisfies the required symmetry property.

We now determine the number of relations (3.19). By virtue of symmetry with respect to the permutation of subscripts within each pair, the number of independent pairs *ik* as well as the pairs *jl* is six, not nine. Then the number of equations (3.19) equals the number of combinations of six pairs taken two at a time, i.e., $6 \times 5/2 = 15$. Consequently, the *maximum number of elastic constants of a crystal* with the most complex symmetry *cannot exceed* $36 - 15 = 21$. The higher the symmetry, the fewer different elastic constants exist.

Consider, for example, a crystal with cubic symmetry. In this crystal the directions x_1, x_2 and x_3 are equivalent. Therefore, we have, in particular,

$$C_{1111} = C_{2222} = C_{3333} \,. \tag{3.20}$$

Furthermore, we have mirror symmetry, i.e., invariance under the transformation $x_i \to -x_i$ $(u_i \to -u_i)$. But according to (3.6), we have $e_{12} \to -e_{12}$, etc., for this transformation. Hence, the quantity w can include no terms of the type $C_{1112}e_{11}e_{12}$. Generalizing this statement, it can be said that $C_{ikjl} = 0$ if any subscript of the four is encountered an uneven number of times (1 or 3 times). As a result, along with (3.20) the following components differ from zero: $C_{1122} = C_{2211} = C_{1133} = C_{3311} = C_{2233} = C_{3322}$, $C_{1212} = C_{2121} = C_{1221} = C_{2112} = \cdots = C_{3223}$. Taking into account these relations, the energy density (3.18) for a cubic crystal can be written as

$$w = \tfrac{1}{2}[C_{1111}(e_{11}^2 + e_{22}^2 + e_{33}^2) + 2C_{1122}(e_{11}e_{22} + e_{11}e_{33} + e_{22}e_{33}) + 4C_{1212}(e_{12}^2 + e_{13}^2 + e_{23}^2)]. \tag{3.21}$$

3.4 The Elastic Behaviour of Isotropic Bodies

3.4.1 The Generalized Hooke's Law for an Isotropic Body

The expression for the energy density in an isotropic elastic body must be invariant with respect to the rotation of the coordinate system. This means that w can be expressed as a function of invariants of the tensor e_{ik} (Appendix). Since w is a homogeneous quadratic function of e_{ik}, its form is determined in a unique manner:

$$w = \frac{\lambda}{2}(e_{11} + e_{22} + e_{33})^2 + \mu(e_{11}^2 + e_{22}^2 + e_{33}^2 + 2e_{12}^2 + 2e_{13}^2 + 2e_{23}^2),$$

where the scalars λ and μ are called the *Lamé's constants*. Now using (3.17a) we find

$$\sigma_{ik} = \partial w/\partial e_{ik} = \lambda e_{jj}\delta_{ik} + 2\mu e_{ik} \tag{3.22}$$

which is the *generalized Hooke's law for an isotropic medium*. Comparing (3.22) with the general formula (3.14) yields

$$C_{1111} = \lambda + 2\mu, \quad C_{1212} = 2\mu, \quad C_{1122} = \lambda. \tag{3.23}$$

The other elasticity tensor components different from zero can be obtained by cyclic permutation of subscripts taking account of symmetry properties. Note the equality $C_{1111} = C_{1122} + C_{1212}$.

3.4.2 The Relationship Between Lamé's Constants and E and ν

It is apparent that the constants λ and μ can be expressed in terms of Young's modulus E and the Poisson ratio ν introduced in Chap. 1. To find the corresponding relations, consider the two simplest types of strain:

i) *Extension of a rod with lateral displacements forbidden:* the effective Young's modulus $E_{\mathrm{eff}} = E(1-\nu)/(1+\nu)(1-2\nu)$ for this case was derived in Problem 1.3. Let us express it in terms of Lamé's constants. Here the strain is homogeneous and $u_2 = u_3 = 0$. According to the definition of the stress and strain tensors, we have

$$\sigma_{11} = F_{x_1}/S_{x_1}, \qquad \Delta l/l = [u_1(x_1 + \Delta x_1) - u_1(x_1)]/\Delta x_1 = \partial u_1/\partial x_1 = e_{11}.$$

In calculating $\Delta l/l$, it must be taken into account that due to strain homogeneity, the relative extension of the whole rod is the same as that of its infinitesimal element. As a result, we obtain from (3.22) and an expression for E_{eff},

$$F_{x_1}/S_{x_1} = \sigma_{11} = (\lambda + 2\mu)e_{11} = E_{\mathrm{eff}}\,\Delta l/l = E(1-\nu)e_{11}/(1+\nu)(1-2\nu).$$

Hence, the relation

$$\lambda + 2\mu = E(1-\nu)/(1+\nu)(1-2\nu) \tag{3.24}$$

holds.

ii) *Isotropic compression:* the coefficient of compressibility K is given by (1.5). The relations $\sigma_{11} = \sigma_{22} = \sigma_{33} = -p$, $\Delta V/V = e_{11} + e_{22} + e_{33} = 3e_{11}$ hold in this case. From (3.22) we have $\sigma_{11} = (2\mu + 3\lambda)e_{11}$ and, consequently, $p = -(2\mu + 3\lambda)\,\Delta V/3V$. Comparing the last relation with (1.5), we find

$$2\mu + 3\lambda = E(1 - 2\nu). \tag{3.25}$$

As a result, we obtain from (3.24, 25) the desired relations:

$$\lambda = \frac{E\nu}{(1+\nu)(1-2\nu)}, \qquad \mu = \frac{E}{2(1+\nu)},$$

$$E = \frac{\mu}{\lambda + \mu}(2\mu + 3\lambda), \qquad \nu = \frac{\lambda}{2(\lambda + \mu)}. \tag{3.26}$$

We also see that the Lamé constant μ is the same as the shear modulus determined by (1.9).

3.4.3 The Equations of Motion for an Isotropic Medium

We now return to the equations of motion (3.11). From (3.22) we find

$$\frac{\partial \sigma_{ik}}{\partial x_k} = 2\mu \frac{\partial e_{ik}}{\partial x_k} + \lambda\,\delta_{ik}\frac{\partial e_{jj}}{\partial x_k} = \mu \frac{\partial}{\partial x_k}\left(\frac{\partial u_i}{\partial x_k} + \frac{\partial u_k}{\partial x_i}\right) + \lambda \frac{\partial e_{jj}}{\partial x_i},$$

but $\partial e_{jj}/\partial x_i \equiv \partial e_{kk}/\partial x_i = \partial^2 u_k/\partial x_k\,\partial x_i$, where $\partial u_k/\partial x_k = e_{kk} = \theta$ is the *relative volume change*. Thus,

$$\partial \sigma_{ik}/\partial x_k = \mu\,\partial^2 u_i/\partial x_k^2 + (\lambda + \mu)\,\partial\theta/\partial x_i.$$

Also taking into account that $\partial^2 u_i/\partial x_k^2 = \Delta u_i$ where Δ is the *Laplacian operator*, one can write (3.11) in the form:

$$\rho \frac{\partial^2 u_i}{\partial t^2} = \mu \Delta u_i + (\lambda + \mu) \frac{\partial \theta}{\partial x_i} + (f_b)_i . \tag{3.27}$$

It can also be written in vector form, since $\theta = \operatorname{div} \boldsymbol{u}$ and $\partial/\partial x_i$ is the ith component of the gradient

$$\rho \frac{\partial^2 \boldsymbol{u}}{\partial t^2} = \mu \Delta \boldsymbol{u} + (\lambda + \mu) \operatorname{grad} \operatorname{div} \boldsymbol{u} + \boldsymbol{f}_b . \tag{3.28}$$

The equilibrium equations are obtained automatically if the left-hand side in (3.27 or 28) equals zero:

$$\mu \Delta \boldsymbol{u} + (\lambda + \mu) \operatorname{grad} \operatorname{div} \boldsymbol{u} + \boldsymbol{f}_b = 0. \tag{3.29}$$

Very often another form of the equation of motion is used, which is obtained from (3.27, 28) by use of the well-known vector identity $\Delta \boldsymbol{u} = \operatorname{grad} \operatorname{div} \boldsymbol{u} - \operatorname{curl} \operatorname{curl} \boldsymbol{u}$. Then, (3.29) becomes

$$(\lambda + 2\mu) \operatorname{grad} \operatorname{div} \boldsymbol{u} - \mu \operatorname{curl} \operatorname{curl} \boldsymbol{u} + \boldsymbol{f}_b = 0 . \tag{3.30}$$

3.5 Exercises

3.5.1. Find the types of deformation corresponding to the following stress tensors:

a) $\sigma_{ik} = \begin{pmatrix} \sigma & 0 & 0 \\ 0 & \sigma & 0 \\ 0 & 0 & \sigma \end{pmatrix}$ b) $\sigma_{ik} = \begin{pmatrix} \sigma & \sigma & \sigma \\ \sigma & \sigma & \sigma \\ \sigma & \sigma & \sigma \end{pmatrix}$

c) $\begin{pmatrix} 0 & \sigma & 0 \\ \sigma & 0 & 0 \\ 0 & 0 & 0 \end{pmatrix} \rightarrow \begin{pmatrix} \sigma & 0 & 0 \\ 0 & -\sigma & 0 \\ 0 & 0 & 0 \end{pmatrix}$

Solution a: Stresses are the same along every axis. Hence, isotropic compression (expansion) takes place.

Solution b: Refer the tensor to the principal axes. The characteristic equation, (see A.11), $(\sigma - \tau)^3 + 2\sigma^3 - 3\sigma^2(\sigma - \tau) = 0$ has the roots $\tau_1 = 3\sigma$, $\tau_2 = \tau_3 = 0$. Hence, in this coordinate system the tensor has the form $\sigma_{11} = 3\sigma$, $\sigma_{ik} = 0$, $i, k \neq 1$, which corresponds to the elongation (contraction) of a bar.

Solution c: pure shear

3.5.2. Determine the maximum normal and tangential stresses at a point where the stress state is specified by the tensor σ_{ik}.

Solution: Consider the problem in the principal axes where the stress tensor has the diagonal form

$$\begin{pmatrix} \tau_1 & 0 & 0 \\ 0 & \tau_2 & 0 \\ 0 & 0 & \tau_3 \end{pmatrix}.$$

The force acting upon an area whose normal is $\boldsymbol{n} = \{n_1, n_2, n_3\}$ is

$$\boldsymbol{F} = \{\tau_1 n_1, \tau_2 n_2, \tau_3 n_3\}.$$

Its squared normal and tangential components are $F_n = F_i n_i = \tau_1 n_1^2 + \tau_2 n_2^2 + \tau_3 n_3^2$ and $F_t^2 = F^2 - F_n^2 = \tau_1^2 n_1^2 + \tau_2^2 n_2^2 + \tau_3^2 n_3^2 - (\tau_1 n_1^2 + \tau_2 n_2^2 + \tau_3 n_3^2)^2$, respectively. To determine the maximum stresses we must find the conditional extrema of the functions $F_n(n_1, n_2, n_3)$ and $F_t^2\ (n_1, n_2, n_3)$ under the condition $n_i^2 = n_1^2 + n_2^2 + n_3^2 = 1$. Constructing the Lagrange function $\phi_n = F_n + \gamma(n_i^2 - 1)$ for F_n, for example, we find the extremum points as solutions of the system of equations:

$$\partial\phi_n/\partial n_1 = 2(\tau_1 + \gamma)n_1 = 0, \quad (\tau_2 + \gamma)n_2 = 0, \quad (\tau_3 + \gamma)n_3 = 0, \quad n_i^2 = 1.$$

Assuming that all τ_i are different, the solutions are

$$\begin{aligned}
&\text{a) } n_1 = \pm 1,\ n_2 = n_3 = 0, \quad F_n = \tau_1, \quad d\phi_n = 2(\tau_2 - \tau_1)\, dn_2^2 + 2(\tau_3 - \tau_1)\, dn_3^2, \\
&\text{b) } n_1 = n_3 = 0,\ n_2 = \pm 1, \quad F_n = \tau_2, \quad d\phi_n = 2(\tau_1 - \tau_2)\, dn_1^2 + 2(\tau_3 - \tau_2)\, dn_3^2, \\
&\text{c) } n_1 = n_2 = 0,\ n_3 = \pm 1, \quad F_n = \tau_3, \quad d\phi_n = 2(\tau_1 - \tau_3)\, dn_1^2 + 2(\tau_2 - \tau_3)\, dn_2^2.
\end{aligned}$$

Assuming, for definiteness, that $\tau_1 > \tau_2 > \tau_3$ and $|\tau_1| > |\tau_3|$, we obtain $\max|F_n| = |\tau_1|$.

An analogous system of equations for the tangential stresses is

$$\begin{aligned}
n_1[\tau_1^2 - 2\tau_1(\tau_1 n_1^2 + \tau_2 n_2^2 + \tau_3 n_3^2) + \gamma] &= 0, \\
n_2[\tau_2^2 - 2\tau_2(\tau_1 n_1^2 + \tau_2 n_2^2 + \tau_3 n_3^2) + \gamma] &= 0, \\
n_3[\tau_3^2 - 2\tau_3(\tau_1 n_1^2 + \tau_2 n_2^2 + \tau_3 n_3^2) + \gamma] &= 0.
\end{aligned}$$

Its solutions corresponding to the minimum of the tangential stress ($F_t^2 = 0$), is the direction of the principal axes. The maximum F_t^2 is obtained when

$$\begin{aligned}
n_3 &= 0, \quad n_1 = n_2 = \pm\sqrt{2}/2, \quad F_t = (\tau_1 - \tau_2)/2, \\
n_2 &= 0, \quad n_1 = n_3 = \pm\sqrt{2}/2, \quad F_t = (\tau_1 - \tau_3)/2, \\
n_1 &= 0, \quad n_2 = n_3 = \pm\sqrt{2}/2, \quad F_t = (\tau_2 - \tau_3)/2.
\end{aligned}$$

Consequently, the maximum tangential stress acts across the area which bisects the angle between the directions of the maximum and minimum principal stresses.

3.5.3 Express the tensor e_{ik} in terms of σ_{ik} in the generalized Hooke's law for an isotropic body.

Solution: Using (3.22), find for the trace of the stress tensor

$$\sigma_{kk} = 3\lambda e_{mm} + 2\mu e_{mm} = (3\lambda + 2\mu)e_{mm},$$

hence, $e_{mm} = \sigma_{mm}/(3\lambda + 2\mu)$. Now (3.22) can also be written as

$$\sigma_{ik} = \lambda\sigma_{mm}\delta_{ik}/(3\lambda + 2\mu) + 2\mu e_{ik}$$

and finally we obtain

$$e_{ik} = [\sigma_{ik} - \lambda\sigma_{mm}\delta_{ik}/(3\lambda + 2\mu)]/2\mu .$$

3.5.4. Using the generalized Hooke's law, calculate the unilateral extension of a rod by the action of force F; l is the rod's length and S is its cross-sectional area. Find also the relationship between the constants E, ν and λ and μ.

Solution: In the case under consideration, only one stress tensor component σ_{11} differs from zero. Consequently, $\sigma_{mm} = \sigma_{11}$ and using the formula obtained in the previous exercise, we have

$$e_{11} = \frac{\sigma_{11}}{2\mu}\left(1 - \frac{\lambda}{3\lambda + 2\mu}\right) = \frac{\sigma_{11}}{\mu}\frac{\lambda + \mu}{3\lambda + 2\mu}, \qquad e_{22} = \frac{\sigma_{11}}{2\mu}\frac{\lambda}{3\lambda + 2\mu} = e_{33} .$$

But $\sigma_{11} = F/S$, $e_{11} = \Delta l/l$, $e_{22} = \Delta h/h$, h being the rod's dimension in one of the transverse directions. Then, according to (1.1, 3), $e_{11} = \sigma_{11}/E$, $e_{22} = -\nu\sigma_{11}/E$. As a result, we obtain the relationships (3.26) between E, ν and λ and μ.

3.5.5. Determine the deformation of a homogeneous sphere of radius R immersed into a liquid under pressure p. Find the relative decrease of the volume of the sphere.

Solution: Write the equilibrium equation (3.30) for the sphere with $f_b = 0$:

$$(\lambda + 2\mu)\operatorname{grad}\operatorname{div}\boldsymbol{u} - \mu\operatorname{curl}\operatorname{curl}\boldsymbol{u} = 0 .$$

By virtue of the spherical symmetry, $\boldsymbol{u} = U(r)\boldsymbol{r}/r$. It is obvious that $\operatorname{curl}\boldsymbol{u} = 0$ in this case and $\operatorname{grad}\operatorname{div}\boldsymbol{u} = 0$, $\operatorname{div}\boldsymbol{u} = 3a$, a being a constant. But

$$\operatorname{div}\left[U(r)\frac{\boldsymbol{r}}{r}\right] = \boldsymbol{r}\operatorname{grad}\frac{U}{r} + 3\frac{U}{r} = 3\frac{U}{r} + \boldsymbol{r}\frac{rU' - U}{r^2}\frac{\boldsymbol{r}}{r} = U' + \frac{2U}{r} = \frac{(r^2U)'}{r^2},$$

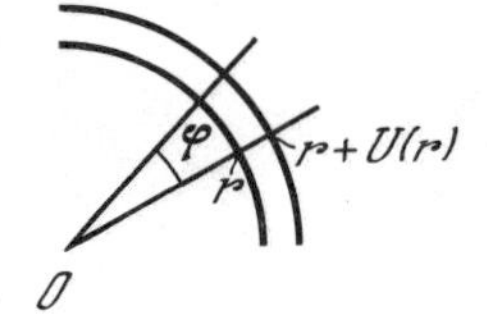

Fig. 3.4. Spherically simmetrical deformation

therefore, $U(r) = ar + b/r^2$. Since U must be finite at $r = 0$, we have $b = 0$, hence $U = ar$. We now determine the stress tensor component (Fig. 3.4):

$$u_{rr} = \frac{\partial u}{\partial r} = a\,, \qquad u_{\varphi\varphi} = \frac{(r + U)\,d\varphi - r\,d\varphi}{r\,d\varphi} = \frac{U}{r} = a = u_{\theta\theta} = u_{rr}\,,$$

$$\sigma_{rr} = \lambda(u_{rr} + u_{\varphi\varphi} + u_{\theta\theta}) + 2\mu u_{rr} = (3\lambda + 2\mu)u_{rr} = (3\lambda + 2\mu)a\,.$$

From the boundary condition $\sigma_{rr}|_R = -p$ we have finally

$$a = -\frac{p}{3\lambda + 2\mu}\,, \qquad U = -\frac{pr}{3\lambda + 2\mu}\,.$$

The relative volume change is

$$\frac{\Delta V}{V} = \frac{[R + U(R)]^3 - R^3}{R^3} = \left(1 + \frac{U(R)}{R}\right)^3 - 1$$

$$\simeq 3\frac{U(R)}{R} = -\frac{3p}{3\lambda + 2\mu} = -\frac{p}{K}\,,$$

a result which coincides with (1.5).

3.5.6. Determine the deformation of an elastic homogeneous space with a spherical cavity of radius R, the pressure inside being p.

Solution: As in Exercise 3.5.5, $\boldsymbol{u} = U(r)\boldsymbol{r}/r$, where $U = ar + b/r^2$. Since $\boldsymbol{u}$ must be finite at $\boldsymbol{r} \to \infty$, we have to set $a = 0$, i.e., $U(r) = b/r^2$. Further, $\sigma_{rr} = \lambda(-2b/r^3 + b/r^3 + b/r^3) - 4\mu b/r^3$. But $\sigma_{rr}|_R = -4\mu b/R^3 = -p$, whence $b = pR^3/4\mu$ and, consequently,

$$\boldsymbol{u} = pR^3\boldsymbol{r}/4\mu r^4\,.$$

3.5.7. Determine the deformation of an infinite hollow cylinder whose inner and outer radii are R_1 and R_2, respectively. The pressure is assumed to be p_1 inside and p_2 outside.

Solution: Like in Exercise 3.5.5, we have $\operatorname{div}\boldsymbol{u} = \alpha$,

$$\boldsymbol{u} = U(r)\frac{\boldsymbol{r}}{r}\,, \qquad \boldsymbol{r} = \{x_1, x_2\}\,,$$

$$\operatorname{div}\boldsymbol{u} = \boldsymbol{r}\nabla(U/r) + 2U/r = r^{-1}\,d(rU)/dr = a\,.$$

Hence,

$$U = ar/2 + b/r\,, \qquad u_{rr} = \partial U/\partial r = a/2 - b/r^2\,,$$

$$u_{\varphi\varphi} = U/r = a/2 + b/r^2\,, \qquad \sigma_{rr} = (\lambda + \mu)a - 2\mu b/r^2\,.$$

From the boundary conditions we find

$$(\lambda + \mu)a - 2\mu b/R_1^2 = -p_1\,, \qquad (\lambda + \mu)a - 2\mu b/R_2^2 = -p_2\,,$$

which gives

$$U = \frac{R_1^2 p_1 - R_2^2 p_2}{2(\lambda + \mu)(R_2^2 - R_1^2)}\, r + \frac{p_1 - p_2}{2\mu}\, \frac{R_1^2 R_2^2}{R_2^2 - R_1^2}\, \frac{1}{r}\,.$$

3.5.8. Determine the deformation of a cylinder of radius R rotating about its axis with frequency ω.

Solution: In cylindrical coordinates we have the body (centrifugal) force $\boldsymbol{f}_b = \rho\omega^2 \boldsymbol{r}$. The equilibrium equation assumes the form

$$(\lambda + 2\mu)\,\mathrm{grad}\,\mathrm{div}\,\boldsymbol{u} - \mu\,\mathrm{curl}\,\mathrm{curl}\,\boldsymbol{u} = -\rho\omega^2 \boldsymbol{r}\,.$$

As above, $\boldsymbol{u} = U(r)\boldsymbol{r}/r$, $\mathrm{curl}\,\boldsymbol{u} = 0$,

$$\mathrm{div}\,\boldsymbol{u} = (rU)'/r\,, \qquad \mathrm{grad}\,\mathrm{div}\,\boldsymbol{u} = [(rU)'/r]'\boldsymbol{r}/r = -\rho\omega^2 \boldsymbol{r}/(\lambda + 2\mu)\,,$$

so that

$$d[r^{-1}\, d(rU)/dr]/dr = -\rho\omega^2 r/(\lambda + 2\mu)\,.$$

Hence,

$$U = -\rho\omega^2 r^3/8(\lambda + 2\mu) + ar/2\,.$$

Furthermore,

$$\sigma_{rr} = -\lambda\rho\omega^2 r^2/2(\lambda + 2\mu) - 3\mu\rho\omega^2 r^2/4(\lambda + 2\mu) + (\lambda + \mu)a\,,$$

setting $\sigma_{rr}|_R = 0$ we find

$$a = \rho\omega^2 R^2 (2\lambda + 3\mu)[4(\lambda + 2\mu)(\lambda + \mu)]^{-1}$$

and finally,

$$U = \frac{1}{8}\frac{\rho\omega^2 r}{\lambda + 2\mu}\left(\frac{2\lambda + 3\mu}{\lambda + \mu} R^2 - r^2\right).$$

4. Elastic Waves in Solids

We will now continue to investigate elastic waves in solids. Using the general equation of motion obtained in the previous chapter, we will show that two kinds of waves can exist in an elastic medium—*longitudinal* (*compressional*) and *transverse* (*shear*) waves. The simplest case—plane harmonic waves—will be considered in detail as well as their reflection at a plane boundary. The surface Rayleigh and Love waves will be discussed, too.

4.1 Free Waves in a Homogeneous Isotropic Medium

4.1.1 Longitudinal and Transverse Waves

Let us consider what kinds of wave motion can occur in an elastic medium. We confine ourselves to free waves ($\boldsymbol{f}_\mathrm{b} = 0$) and isotropic media, then the equation of motion (3.28) becomes

$$\rho \frac{\partial^2 \boldsymbol{u}}{\partial t^2} = \mu \Delta \boldsymbol{u} + (\lambda + \mu) \operatorname{grad} \operatorname{div} \boldsymbol{u} . \tag{4.1}$$

This equation is different from the canonical wave equation and implicitly contains equations for two types of waves—*longitudinal* and *transverse*. In fact, let us write the displacement in the form

$$\boldsymbol{u} = \boldsymbol{u}_\mathrm{l} + \boldsymbol{u}_\mathrm{t} , \qquad \boldsymbol{u}_\mathrm{l} = \operatorname{grad} \varphi , \qquad \boldsymbol{u}_\mathrm{t} = \operatorname{curl} \boldsymbol{\psi} , \tag{4.2}$$

where φ and $\boldsymbol{\psi}$, called *scalar* and *vector potentials*, are functions of time and space variables. Obviously, φ is determined up to an arbitrary additive function of time, and $\boldsymbol{\psi}$ up to an arbitrary additive potential vector $\boldsymbol{A} = \operatorname{grad} f(x_1, x_2, x_3, t)$. Note that an arbitrary vector $\boldsymbol{u}$ can be represented as (4.2). Obviously, the vectors $\boldsymbol{u}_\mathrm{l}$ and $\boldsymbol{u}_\mathrm{t}$ satisfy

$$\operatorname{curl} \boldsymbol{u}_\mathrm{l} = 0 , \qquad \operatorname{div} \boldsymbol{u}_\mathrm{t} = 0 . \tag{4.2a}$$

Substituting (4.2) into (4.1), we obtain

$$\frac{\partial^2 \boldsymbol{u}_\mathrm{l}}{\partial t^2} - \frac{\lambda + 2\mu}{\rho} \Delta \boldsymbol{u}_\mathrm{l} + \frac{\partial^2 \boldsymbol{u}_\mathrm{t}}{\partial t^2} - \frac{\mu}{\rho} \Delta \boldsymbol{u}_\mathrm{t} = 0 . \tag{4.3}$$

Assuming the medium to be homogeneous (ρ, λ, μ are constant) and applying successively the operations div and curl to (4.3), we obtain, taking (4.2a) into account,

$$\operatorname{div}\left(\frac{\partial^2 \boldsymbol{u}_l}{\partial t^2} - \frac{\lambda + 2\mu}{\rho} \Delta \boldsymbol{u}_l\right) = 0, \quad \operatorname{curl}\left(\frac{\partial^2 \boldsymbol{u}_l}{\partial t^2} - \frac{\lambda + 2\mu}{\rho} \Delta \boldsymbol{u}_l\right) = 0,$$

$$\operatorname{div}\left(\frac{\partial^2 \boldsymbol{u}_t}{\partial t^2} - \frac{\mu}{\rho} \Delta \boldsymbol{u}_t\right) = 0, \quad \operatorname{curl}\left(\frac{\partial^2 \boldsymbol{u}_t}{\partial t^2} - \frac{\mu}{\rho} \Delta \boldsymbol{u}_t\right) = 0.$$

If div as well as curl of a certain vector are zero in the whole space, then without loss of generality this vector can be assumed to be zero; $\boldsymbol{u}_l$ and $\boldsymbol{u}_t$ are determined to within an arbitrary additive vector depending only on time. As a result, we obtain the two wave equations:

$$\frac{\partial^2 \boldsymbol{u}_l}{\partial t^2} - \frac{\lambda + 2\mu}{\rho} \Delta \boldsymbol{u}_l = 0, \quad \frac{\partial^2 \boldsymbol{u}_t}{\partial t^2} - \frac{\mu}{\rho} \Delta \boldsymbol{u}_t = 0.$$

Similar equations also hold for the potentials φ and $\boldsymbol{\psi}$. They can readily be obtained from equations for $\boldsymbol{u}_l$ and $\boldsymbol{u}_t$, respectively, and are usually written in the form

$$\frac{\partial^2 \varphi}{\partial t^2} = c_l^2 \, \Delta \varphi, \quad c_l^2 = \frac{\lambda + 2\mu}{\rho},$$
$$\frac{\partial^2 \boldsymbol{\psi}}{\partial t^2} = c_t^2 \, \Delta \boldsymbol{\psi}, \quad c_t^2 = \frac{\mu}{\rho}. \tag{4.4}$$

It can be easily proved that a solution of rather general form

$$\varphi = f(\boldsymbol{n} \cdot \boldsymbol{r} - c_l t) \tag{4.5}$$

satisfies the equation for φ where f is an arbitrary function; $\boldsymbol{n}$ is an arbitrary unit vector and $\boldsymbol{r} = \{x_1, x_2, x_3\}$. In fact, denoting $\xi = \boldsymbol{n} \cdot \boldsymbol{r} - c_l t$, we have $\dot{\varphi} = (d\varphi/d\xi)\, d\xi/dt = -c_l d\varphi/d\xi$, $\ddot{\varphi} = c_l^2\, d^2\varphi/d\xi^2$. On the other hand, taking $\boldsymbol{n} \cdot \boldsymbol{r} = n_i x_i$ into account we obtain

$$\frac{\partial \varphi}{\partial x_1} = \frac{d\varphi}{d\xi} \frac{\partial \xi}{\partial x_1} = n_1 \frac{d\varphi}{d\xi}, \quad \frac{\partial^2 \varphi}{\partial x_1^2} = n_1^2 \frac{d^2 \varphi}{d\xi^2}$$

and similarly for the derivatives with respect to x_2 and x_3. Now,

$$\Delta \varphi = \frac{\partial^2 \varphi}{\partial x_i^2} = n_i^2 \frac{d^2 \varphi}{d\xi^2} = \frac{d^2 \varphi}{d\xi^2}, \quad \text{since} \quad n_i^2 = \sum_{i=1}^{3} n_i^2 = 1$$

and, φ satisfies (4.4).

We show that (4.5) describes a wave which propagates in the direction of $\boldsymbol{n}$ with the velocity c_l without changing its form. Indeed, at fixed t the function

φ remains constant at the planes

$$\boldsymbol{n} \cdot \boldsymbol{r} - c_1 t = \text{const}, \tag{4.6}$$

which are called the *wave fronts*. These fronts are perpendicular to $\boldsymbol{n}$. The propagation velocity can easily be found by differentiating (4.6) with respect to time: $\boldsymbol{n} \cdot d\boldsymbol{r}/dt = c_1$. Such waves are called *plane waves.*

It can be shown also that (4.5) represents a *longitudinal wave* where particle displacements are parallel to $\boldsymbol{n}$—the propagation direction. In fact, we have

$$\boldsymbol{u} = \operatorname{grad} \varphi, \qquad u_i = \partial\varphi/\partial x_i = n_i\, d\varphi/d\xi, \qquad \boldsymbol{u} = \boldsymbol{n}\, d\varphi/d\xi.$$

Quite analogously, the solution of the second equation (4.4) is

$$\boldsymbol{\psi} = \boldsymbol{F}(\boldsymbol{n} \cdot \boldsymbol{r} - c_t t), \tag{4.7}$$

where $\boldsymbol{F}$ is an arbitrary vector function of the scalar argument $\xi = \boldsymbol{n} \cdot \boldsymbol{r} - c_t t$. This is a plane wave too, with wave fronts perpendicular to $\boldsymbol{n}$. The particle displacements in this wave, however, are perpendicular to $\boldsymbol{n}$ (*transverse wave*). In fact,

$$\boldsymbol{u} = \operatorname{curl} \boldsymbol{\psi} = \nabla \times \boldsymbol{\psi} = \boldsymbol{n} \times d\boldsymbol{\psi}/d\xi \perp \boldsymbol{n}.$$

This wave is also called a *shear wave*, since of all the strain tensor components, only nondiagonal ones describing shear strains differ from zero in this wave. This can be proved using a coordinate system with one of the axes along $\boldsymbol{n}$. The propagation velocity of the transverse wave is c_t, the same as the velocity of torsional waves in a rod.

From (4.4) we have $c_1/c_t = \sqrt{(\lambda + 2\mu)/\mu}$. Since $\mu > 0$ always and $\lambda \geq 0$ for real materials, we find $c_1 \geq \sqrt{2}c_t$.

Note also that

$$c_1^2 = \frac{\lambda + 2\mu}{\rho} = \frac{(1 - \nu)E}{(1 + \nu)(1 - 2\nu)\rho} = \frac{E_{\text{eff}}}{\rho},$$

where E_{eff} is the effective Young's modulus for the case of extension (contraction) of a rod with lateral displacements forbidden (Exercise 1.4.3). This is only natural because the lateral displacements in the longitudinal wave are also zero.

Harmonic plane waves represent an important case of (4.5, 7):

$$\begin{aligned} \varphi &= A \exp[\mathrm{i}\omega(\boldsymbol{n} \cdot \boldsymbol{r} - c_1 t)/c_1] = A \exp[\mathrm{i}(\boldsymbol{k} \cdot \boldsymbol{r} - \omega t)], \quad \boldsymbol{k} = \omega \boldsymbol{n}/c_1, \\ \boldsymbol{\psi} &= \boldsymbol{B} \exp[\mathrm{i}\omega(\boldsymbol{n} \cdot \boldsymbol{r} - c_t t)/c_t] = \boldsymbol{B} \exp[\mathrm{i}(\boldsymbol{\varkappa} \cdot \boldsymbol{r} - \omega t)], \quad \boldsymbol{\varkappa} = \omega \boldsymbol{n}/c_t, \end{aligned} \tag{4.8}$$

where A and $\boldsymbol{B}$ are arbitrary complex constants, $\boldsymbol{k}$ and $\boldsymbol{\varkappa}$ are wave vectors ($k = |\boldsymbol{k}|$ and $\varkappa = |\boldsymbol{\varkappa}|$), are the wave numbers, and ω is the wave frequency (the period is $T = 2\pi/\omega$). The wavelengths are, respectively, $\lambda_1 = 2\pi/k$ and $\lambda_t = 2\pi/\varkappa$. The following inequalities are readily obtained:

$$k/\varkappa = c_t/c_l \leq \sqrt{2}/2, \quad \lambda_l/\lambda_t \geq \sqrt{2}.$$

It is always possible to choose the coordinate axes so that $k_1 = k$, $k_2 = k_3 = 0$ or $\varkappa_1 = \varkappa$, $\varkappa_2 = \varkappa_3 = 0$. Then,

$$\varphi = A\exp[\mathrm{i}(kx_1 - \omega t)] \quad \text{or} \quad \boldsymbol{\psi} = \boldsymbol{B}\exp[\mathrm{i}(\varkappa x_1 - \omega t)]. \tag{4.9}$$

The waves propagating in opposite directions are

$$\varphi = A\exp[-\mathrm{i}(kx_1 + \omega t)], \quad \boldsymbol{\psi} = \boldsymbol{B}\exp[-\mathrm{i}(\varkappa x_1 + \omega t)]. \tag{4.10}$$

Note that the longitudinal and transverse waves *propagate without dispersion* (the propagation velocity is independent of the frequency). Therefore, plane waves of arbitrary shapes (4.5, 7) propagate without distortion.

4.1.2 Boundary Conditions for Elastic Waves

In an infinite homogeneous elastic medium, the longitudinal and transverse waves propagate independently, i.e., they do not interact with each other. If the parameters λ, μ, and ρ vary in space, then in the general case, a longitudinal wave while propagating generates transverse waves, and vice versa. The same takes place at a boundary of two different homogeneous media. In the latter case, the wave interaction is included implicitly in the boundary conditions. We will consider this case for different kinds of interfaces:

i) *contact between two elastic halfspaces without slip* (*welded contact*). The displacement as well as the force per the unit area of the interface S must be continuous in this case:

$$(u_j^{(1)} - u_j^{(2)})_S = 0, \quad (\sigma_{nj}^{(1)} - \sigma_{nj}^{(2)})_S = 0,$$

where $j = 1, 2, 3$ and n is the index of the axis normal to S. The superscripts 1 or 2 denote the halfspaces.

ii) *contact between two elastic halfspaces with slip*. The normal displacement and the normal component of the force are continuous, the tangential component of the force being zero in this case:

$$(u_n^{(1)} - u_n^{(2)})_S = 0, \quad (\sigma_{nn}^{(1)} - \sigma_{nn}^{(2)})_S = 0, \quad \sigma_{nj}^{(1)}|_S = \sigma_{nj}^{(2)}|_S = 0, \quad j \neq n.$$

There is no summation in the second equation in spite of the repeated index.

iii) *the welded contact with an infinitely rigid wall:*

$$u_j|_S = 0, \quad j = 1, 2, 3.$$

iv) *contact with an infinitely rigid wall with slip:*

$$u_n|_S = 0, \quad \sigma_{nj}|_S = 0, \quad j \neq n.$$

v) *the boundary of an elastic body with a vacuum (stress free surface):*

$$\sigma_{nj}|_S = 0, \quad j = 1, 2, 3.$$

4.2 Wave Reflection at a Stress-Free Boundary

4.2.1 Boundary Conditions

Consider, for example, the reflection of elastic waves at a stress-free boundary plane. Let the latter be $x_3 = 0$ and the x_3-axis be directed toward the vacuum. The boundary conditions are

$$\sigma_{31} = \sigma_{32} = \sigma_{33} = 0 \quad \text{at} \quad x_3 = 0. \tag{4.11}$$

Choose the directions of the axes x_1, x_2 so that the normal to the front of the incident wave lies in the x_1x_3 plane (the plane of incidence). Then the problem becomes two-dimensional, namely nothing depends on the coordinate x_2.

Express the stress tensor components in (4.11) in terms of the displacement vector $\boldsymbol{u}$ using Hooke's law (3.22) and (3.6). Then the boundary conditions become

$$\left(\frac{\partial u_1}{\partial x_3} + \frac{\partial u_3}{\partial x_1}\right)_{x_3=0} = \left.\frac{\partial u_2}{\partial x_3}\right|_{x_3=0} = \left[(\lambda + 2\mu)\frac{\partial u_3}{\partial x_3} + \lambda\frac{\partial u_1}{\partial x_1}\right]_{x_3=0} = 0. \tag{4.12}$$

Let us consider three possible cases of an incident wave:

i) *transverse wave where the particle motion is perpendicular to the plane of incidence:* $u_2 \neq 0$, $u_1 = u_3 = 0$. In seismology, while considering reflections of waves at the earth surface, this wave is called a *horizontally polarized wave* or *SH-wave;*

ii) *transverse wave where the particle motion is in the plane of incidence:* $u_1 \neq 0$, $u_3 \neq 0$, $u_2 = 0$;

iii) *longitudinal wave where the particle motion is in the plane of incidence:* $u_1 \neq 0$, $u_3 \neq 0$, $u_2 = 0$.

The waves (ii), (iii) are called vertically polarized.

Note that in (4.12) the derivatives of u_1 and u_3 are interrelated by the first and third conditions. These conditions do not contain u_2. On the contrary, the second condition contains only u_2. This means that when wave (i) is incident, waves (ii), (iii) do not arise. At the same time, the last two waves can interact at the interface and they should be considered together. In other words, the *horizontally* and *vertically polarized* waves can be considered *separately*. In what follows we also confine ourselves to the case of harmonic waves.

4.2.2 Reflection of a Horizontally Polarized Wave

In this case displacements in incident ("+") and reflected ("−") waves are written in the form

$$\begin{aligned} u_{2+} &= b_+ \exp[\mathrm{i}(\varkappa_1 x_1 + \varkappa_3 x_3 - \omega t)], \\ u_{2-} &= b_- \exp[\mathrm{i}(\varkappa_1' x_1 - \varkappa_3' x_3 - \omega' t)], \end{aligned} \tag{4.13}$$

where b_+ and b_- are the complex amplitudes of the waves. Since $\varkappa_1$, $\varkappa_3$ and $\varkappa_1'$, $\varkappa_3'$ are the components of the wave vectors in the incident and reflected waves, respectively, then

$$\varkappa_1^2 + \varkappa_3^2 = \varkappa^2 = \omega^2/c_t^2, \quad (\varkappa_1')^2 + (\varkappa_3')^2 = (\varkappa')^2 = (\omega')^2/c_t^2. \tag{4.14}$$

Substituting the sum of the displacements (4.13) into the boundary condition $(\partial u_2/\partial x_3)_{x_3=0} = 0$, we obtain

$$b_- = (\varkappa_3/\varkappa_3')b_+ \exp\{\mathrm{i}[(\varkappa_1 - \varkappa_1')x_1 - (\omega - \omega')t]\} \tag{4.15}$$

which can be satisfied for all x, and t only if $\omega' = \omega$ and $\varkappa_1' = \varkappa_1$. Thus, the *frequency of wave* and the *projection of the wave vector on the boundary* are *conserved* when the wave under consideration is reflected. According to (4.14), we also have $\varkappa_3' = \varkappa_3$ and (4.15) yields

$$b_- = b_+, \tag{4.16}$$

i.e., the amplitude of the reflected wave is equal to that of the incident one. The *reflection coefficient* $V = b_-/b_+$ is unity.

The equality $\varkappa_1' = \varkappa_1$ has a simple physical meaning. We write (4.13) in the form

$$\begin{aligned} u_{2+} &= b_+ \exp(\mathrm{i}\varkappa_3 x_3)\exp[\mathrm{i}(\xi x_1 - \omega t)], \\ u_{2-} &= b_- \exp(-\mathrm{i}\varkappa_3 x_3)\exp[\mathrm{i}(\xi x_1 - \omega t)], \end{aligned}$$

where $\xi = \varkappa_1 = \varkappa_1'$. Thus, the incident and reflected waves have the same dependence on x_1 and t given by the factor $\exp[\mathrm{i}(\xi x_1 - \omega t)]$. The velocity of propagation along the boundary, $v = \omega/\xi$, is the same for both waves. We denote by γ the angle of incidence or the angle between the normal to the incident wave front and the x_3-axis, and by γ' the *angle of reflection*—the same angle for

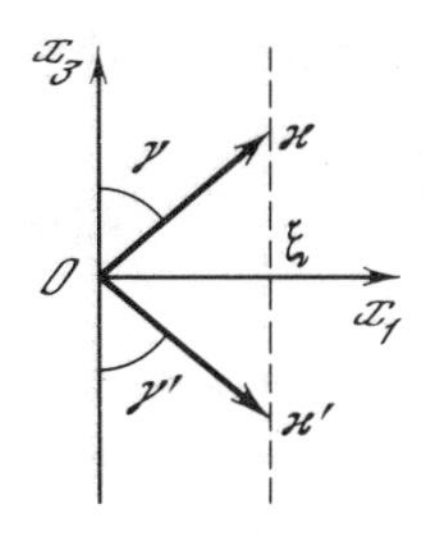

Fig. 4.1. Wave vectors for the case of a reflection of a horizontally polarized wave

the reflected wave. Then, evidently $\xi = \varkappa_1 = \varkappa \sin\gamma = \varkappa' \sin\gamma'$ and the equality $\varkappa_1' = \varkappa_1$ means also that $\gamma' = \gamma$, i.e., the *angle of incidence is equal to the angle of reflection* (*Snell's law*) (Fig. 4.1).

Equality of the frequencies and the projections of the wave vectors on the interface for the incident and reflected waves, is a general rule for the reflection and refraction phenomena (except at moving boundaries). It can be proved in each case as above. We shall use this rule throughout the following text.

4.2.3 The Reflection of Vertically Polarized Waves

In this case we have $u_2 = 0$ and it is expedient to express u_1 and u_3 in the boundary conditions (4.12) in terms of potentials according to (4.2):

$$u_1 = \partial\varphi/\partial x_1 - \partial\psi_2/\partial x_3\,, \qquad u_3 = \partial\varphi/\partial x_3 + \partial\psi_2/\partial x_1\,. \tag{4.17}$$

Of the three components of $\boldsymbol{\psi}$, only ψ_2 appears here, therefore its subscript will henceforth be omitted. Substitution of the latter expressions into the first and third boundary conditions (4.12) yields

$$\begin{aligned} &\left(2\frac{\partial^2\varphi}{\partial x_1\,\partial x_3} + \frac{\partial^2\psi}{\partial x_1^2} - \frac{\partial^2\psi}{\partial x_3^2}\right)_{x_3=0} = 0 \\ &\left[2\mu\left(\frac{\partial^2\varphi}{\partial x_3^2} + \frac{\partial^2\psi}{\partial x_1\,\partial x_3}\right) + \lambda\left(\frac{\partial^2\varphi}{\partial x_1^2} + \frac{\partial^2\varphi}{\partial x_3^2}\right)\right]_{x_3=0} = 0\,. \end{aligned} \tag{4.18}$$

Now we take into account that the incident and reflected waves depend on time t and the coordinate x_1 (along the boundary) in the same manner through the factor $\exp[\mathrm{i}(\xi x_1 - \omega t)]$. Therefore, we have $\partial/\partial x_1 = \mathrm{i}\xi$, $\partial/\partial t = -\mathrm{i}\omega$; hence the wave equation (4.4) and the boundary conditions (4.18) can be written as

$$\begin{aligned} &\frac{\partial^2\varphi}{\partial x_3^2} + (k^2 - \xi^2)\varphi = 0\,, \qquad \frac{\partial^2\psi}{\partial x_3^2} + (\varkappa^2 - \xi^2)\psi = 0\,, \\ &\left(\frac{\partial\varphi}{\partial x_3} + \mathrm{i}p\psi\right)_{x_3=0} = \left(\frac{\partial\psi}{\partial x_3} - \mathrm{i}p\varphi\right)_{x_3=0} = 0\,, \quad k = \frac{\omega}{c_l}\,, \quad \varkappa = \frac{\omega}{c_t}\,, \\ &p = (\xi^2 - \varkappa^2/2)/\xi\,. \end{aligned} \tag{4.19}$$

General solutions of (4.19) are [the factor $\exp[i(\xi x_1 - \omega t)]$ is omitted for brevity]

$$\begin{aligned} \varphi &= a_- \exp(-ik_3x_3) + a_+ \exp(ik_3x_3), \quad k_3 = \sqrt{k^2 - \xi^2}, \\ \psi &= b_- \exp(-i\varkappa_3x_3) + b_+ \exp(i\varkappa_3x_3), \quad \varkappa_3 = \sqrt{\varkappa^2 - \xi^2}. \end{aligned} \tag{4.20}$$

Here a_-, a_+ and b_-, b_+ are the constant amplitudes of the longitudinal and transverse waves propagating in the negative and positive x_3 directions, respectively.

Substitution of (4.20) into the boundary conditions (4.19) yields a relationship between the amplitudes:

$$\begin{aligned} &k_3(a_+ - a_-) + p(b_+ + b_-) = 0 \\ &\varkappa_3(b_+ - b_-) - p(a_+ + a_-) = 0. \end{aligned} \tag{4.21}$$

This system of equations holds for all cases of reflection, at a free boundary, of plane harmonic waves polarized in the plane of incidence. Suppose, for example, that only a longitudinal wave is incident at the boundary ($b_+ = 0$). Then, denoting by $V_{ll} = a_-/a_+$ the reflection coefficient for the longitudinal wave and by $V_{lt} = b_-/a_+$ the *coefficient of transformation* of the longitudinal wave into a transverse one, we find

$$V_{ll} = \frac{k_3\varkappa_3 - p^2}{k_3\varkappa_3 + p^2}, \quad V_{lt} = \frac{2k_3p}{k_3\varkappa_3 + p^2}. \tag{4.22}$$

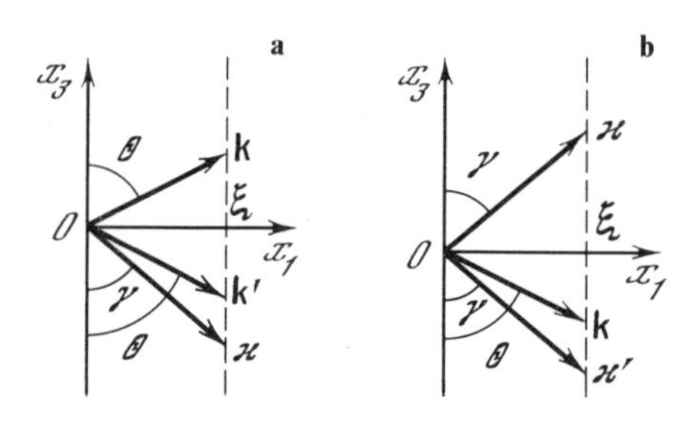

Fig. 4.2a, b. Wave vectors for the case of a reflection of a vertically polarized waves (**a**) longitudinal, (**b**) transverse

The coefficients V_{ll} and V_{lt} can also be expressed in terms of the angle of incidence. It should be remembered that ξ is the projection of the wave vector of the longitudinal or transverse wave onto the x_1-axis, i.e. (Fig. 4.2a),

$$\xi = k \sin\theta = \varkappa \sin\gamma, \tag{4.23}$$

where θ is the angle of incidence of the longitudinal wave, or the angle between the normal to the wave front and the x_3-axis, and γ is the angle of reflection of the transverse wave defined in an analogous way. According to the definition of k and $\varkappa$ in (4.19), we obtain

$$\sin\gamma = (c_t/c_l)\sin\theta = (k/\varkappa)\sin\theta\,. \tag{4.24}$$

Since we always have $c_t < c_l$, then $\gamma < \theta$ always. We can easily obtain $k_3 = k\cos\theta$, $\varkappa_3 = \varkappa\cos\gamma$, $p = -\varkappa_3\cot 2\gamma$ and

$$V_{ll} = \frac{\cos\theta\tan^2 2\gamma - (c_l/c_t)\cos\gamma}{\cos\theta\tan^2 2\gamma + (c_l/c_t)\cos\gamma}\,, \qquad V_{lt} = \frac{-2\cos\theta\tan 2\gamma}{\cos\theta\tan^2 2\gamma + (c_l/c_t)\cos\gamma}\,. \tag{4.25}$$

Similar formulae can be obtained for the case where the incident wave is transverse ($a_+ = 0$). There will be both longitudinal and transverse waves reflected from the boundary (Fig. 4.26b). For the reflection coefficient of the transverse wave $V_{tt} = b_-/b_+$ and for the coefficient of transformation of the transverse wave into the longitudinal one $V_{tl} = a_-/b_+$, we get

$$V_{tt} = V_{ll}\,, \qquad V_{tl} = -\frac{\varkappa_3}{k_3}V_{lt} = -\frac{c_l\cos\gamma}{c_t\cos\theta}V_{lt}\,. \tag{4.26}$$

4.2.4 Particular Cases of Reflection

For normal incidence we have $\theta = \gamma = 0$, $V_{ll} = V_{tt} = -1$, $V_{lt} = V_{tl} = 0$, i.e., a complete reflection with a phase shift by π takes place; the wave transformation is absent. Note that the case of normal incidence of a transverse wave is similar to that of the reflection of a horizontally polarized wave considered in Sect. 4.2.2, but the reflection coefficient in the latter is $+1$, not -1 as in the present subsection. This is because the reflection coefficient in Sect. 4.2.2 was defined as the ratio of displacements in reflected and incident waves, whereas in this subsection as that of the potentials ψ in the two waves. Since u_2 is obtained from ψ by the operation $\partial/\partial x_3$, this yields an opposite sign for the reflection coefficient.

An interesting pecularity in reflection is observed when

$$\cos\theta\tan^2 2\gamma = (c_l/c_t)\cos\gamma\,. \tag{4.27}$$

In this case $V_{ll} = V_{tt} = 0$, $V_{lt} = -\cot 2\gamma = -(c_t\cos\theta/c_l\cos\gamma)\tan 2\gamma$, i.e., the incident longitudinal wave transforms completely into a transverse one, and vice versa. Excluding γ from (4.27) with the help of (4.24), we obtain an equation for the angle of incidence θ at which this transformation takes place. Analysis of this equation shows that inequalitities $37° < \theta < 90°$, $25° < \gamma < 45°$ hold for all possible combinations of elastic constants.

We consider the opposite case when no transformation of waves occurs, and the reflected wave is of the same kind as the incident one. It is the case where $V_{lt} = 0$ and, according to (4.22), we have $p = (\xi^2 - \varkappa^2/2)/\xi = 0$. Since $\xi = \varkappa\sin\gamma$ then $\sin^2\gamma = 1/2$, $\gamma = 45°$. Consequently, the transformation is absent when the transverse wave is incident at the angle $\gamma = 45^\circ$. Note that the result would be the same if the longitudinal wave were incident and a transverse wave were reflected at the 45° angle. This is, however, impossible[3] since, according to (4.24) and the condition $c_l/c_t > \sqrt{2}$, we would have $\sin\theta = (c_l/c_t)\sin\gamma > 1$.

Expressions (4.25) for the reflection and transformation coefficients are valid for real θ and γ, not only for harmonic but for arbitrary plane waves as well. In fact, we represent the latter as a superposition of plane harmonic waves using a Fourier integral. While reflecting, each of the harmonic components is multiplied by the real reflection coefficient, independent of frequency. As a result, the reflected field will be the same Fourier integral multiplied by this coefficient. Note that θ and γ as well as V_{ll} and V_{lt} are always real when a longitudinal wave is incident on the boundary.

4.2.5 Inhomogeneous Waves

The situation is different if the transverse wave is incident at an angle γ and $\sin\gamma > c_t/c_l$. We have, according to (4.24, 20), in this case,

$$\begin{aligned} &\sin\theta > 1\,, \quad \xi > k\,, \quad k_3 = \mathrm{i}|k_3|\,, \\ &|k_3| = \sqrt{\xi^2 - k^2} = \varkappa\sqrt{\sin^2\gamma - (c_t/c_l)^2}\,. \end{aligned} \tag{4.28}$$

The square root is supposed to be positive here. The potential of the longitudinal wave (4.20) taking account of the factor $\exp[\mathrm{i}(\xi x_1 - \omega t)]$ and the condition $a_+ = 0$ has the form

$$\varphi = V_{tl} b_+ \exp[|k_3| x_3 + \mathrm{i}(\xi x_1 - \omega t)]\,. \tag{4.29}$$

This is a so-called *inhomogeneous wave* which propagates along the boundary with the velocity $v = \omega/\xi$; its amplitude decreases exponentially as the distance from the boundary increases ($x_3 \to -\infty$).

The reflection coefficient for the transverse wave V_{tt} is, according to (4.22, 26),

$$V_{tt} = \frac{\mathrm{i}|k_3|\varkappa_3 - p^2}{\mathrm{i}|k_3|\varkappa_3 + p^2} = \exp(-\mathrm{i}\alpha)\,, \qquad \alpha = 2\arctan\frac{|k_3|\varkappa_3}{p^2}\,. \tag{4.30}$$

Hence, we have a phase lag α of the reflected wave as with respect to the incident one. The reflection is total ($|V_{tt}| = 1$) and this is only natural, since a longitudinal inhomogeneous wave (4.29) does not carry energy in the x_3-direction. Note also that a nonharmonic plane wave changes its form while reflecting in this case, due to the phase lag α of each harmonic component.

[3] Introducing inhomogeneous longitudinal waves (Sect. 4.2.5) does not help either, for in this case one of the waves (incident or reflected) will increase exponentially at infinity

The phenomenon under consideration is somewhat analogous to the phenomena of total internal reflection in acoustics or optics, observed when a wave is reflected at the boundary with a medium leading to a higher propagation velocity. The only difference is that in the case under consideration we have two types of waves (longitudinal and transverse) instead of two halfspaces in contact. It is essential for the total reflection that $c_l > c_t$. Therefore, it is clear why an analogous effect does not occur when the incident wave is longitudinal. In the latter case, a reflected transverse wave always goes away from the boundary and carries some portion of energy. Indeed, if θ is the angle of incidence of the longitudinal wave we have $\xi = k \sin\theta$. Then, according to (4.20), $\varkappa_3 = \sqrt{\varkappa^2 - k^2 \sin^2\theta} = k\sqrt{(c_l/c_t)^2 - \sin^2\theta}$ is always real.

It can easily be seen [see, for example, (4.28)] that the total internal reflection of a transverse wave occurs only if $\gamma > \gamma_{cr}$ where $\sin\gamma_{cr} = c_t/c_l < \sqrt{2}/2$, i.e., $\gamma_{cr} < 45°$. The transformation coefficient is

$$V_{tl} = -\frac{2\varkappa_3 p}{k_3\varkappa_3 + p^2}.$$

At the angle of incidence $\gamma = 45°$ we have $V_{tl} = 0$; the amplitude of the inhomogeneous longitudinal wave is zero.

4.3 Surface Waves

4.3.1 The Rayleigh Wave

Under certain conditions, [see (4.27)], only two waves can satisfy the boundary conditions at a free boundary. Now we shall generalize these conditions, also taking into consideration inhomogeneous waves. Requiring $V_{ll} = V_{tt} = 0$, we obtain from (4.22)

$$k_3\varkappa_3 - p^2 = 0. \tag{4.31}$$

Substitution of p, k_3, $\varkappa_3$ from (4.19, 20) yields

$$4\xi^2\sqrt{k^2 - \xi^2}\sqrt{\varkappa^2 - \xi^2} = (2\xi^2 - \varkappa^2)^2 \tag{4.32}$$

whose roots ξ_n determine the required waves. Let us now discuss this equation.

Simple transformations lead to the equation of third power in $s = \varkappa^2/\xi^2$:

$$f(s) = s^3 - 8s^2 + 16(3/2 - q^2)s - 16(1 - q^2) = 0, \tag{4.33}$$

where $q = c_t/c_l < 1$ is the ratio of the velocities of the transverse and longitudinal waves. The right-hand side of (4.32) is real. If this equation has a root $\xi^2 < \varkappa^2$ $(s > 1)$, then it must be $\xi^2 < k^2$ for this root in order for the left-hand side to be real, too. This is the case of ordinary propagating waves making specific angles with the wave boundary. It can be shown (Exercise 4.4.6) that for the Poisson ratio $\nu < 0.26$, there is a pair of such roots as considered in (4.27) (Sect. 4.2.4).

On the other hand, (4.33) always has the root $s = s_0 < 1$. In fact, we have $f(0) = -16(1 - q^2) < 0$, $f(1) = 1 > 0$, i.e., the function $f(s)$ changes its sign in the interval (0, 1). A corresponding quantity ξ_0 satisfies the inequality $\xi_0 > \varkappa > k$. Hence, $\varkappa_3$ and k_3 are imaginary, i.e., both transverse and longitudinal waves are inhomogeneous. As a result, the solution of the wave equations (4.4) satisfying the boundary conditions (4.18) can be written as a superposition of two waves, the scalar and vector potentials of which are

$$\varphi = a\exp[|k_3|x_3 + \mathrm{i}(\xi_0 x_1 - \omega t)]$$

and

$$\psi = aV_{\mathrm{lt}}\exp[|\varkappa_3|x_3 + \mathrm{i}(\xi_0 x_1 - \omega t)]\,. \tag{4.34}$$

The combination of these waves is called a *surface Rayleigh wave*. The disturbance of this wave is concentrated in a narrow near-surface layer of thickness $\sim 1/|k_3|$. The displacement components u_1 and u_3 of the Rayleigh wave are easily found from the formula $\boldsymbol{u} = \nabla\varphi + \mathrm{curl}\,\boldsymbol{\psi}$, where $\boldsymbol{\psi} = \{0, \psi\, 0\}$. Rayleigh waves propagate along the boundary at the velocity $v = \omega/\xi_0 = (\omega/\varkappa)\varkappa/\xi_0 = \sqrt{s_0}c_t$. It is independent of frequency since the coefficients of (4.33) determining s_0 are independent of frequency. Consequently, a Rayleigh wave propagates *without dispersion* (an arbitrary pulse preserves its form). As was shown above, the inequality $0 < q < \sqrt{2}/2$ holds for all elastic bodies. Taking this into account, an analysis of (4.33) yields that $0.8741 \leq v/c_t \leq 0.9554$. Thus, the velocity of Rayleigh waves is slightly less than that of shear waves.

The Rayleigh wave plays an important role in seismology since it can be traced over large distances from the earthquake's epicenter. It is also used extensively in ultrasonic techniques.

4.3.2 The Surface Love Wave

Now we consider a horizontally polarized surface wave with particle motion parallel to the boundary and perpendicular to the propagation direction. A wave of this kind cannot occur in a homogeneous halfspace with a free boundary, but it can exist in the case of a layered medium, say, in a layer covering an elastic halfspace (Fig. 4.3). The theory of these waves was given by Love in 1911 after whom they were named.

We denote by H the thickness of a layer, by λ and μ its elastic constants and by ρ its density. Let the corresponding parameters in the underlying halfspace be λ_1, μ_1 and ρ_1. A harmonic wave of frequency ω is assumed to propagate in the x_1-direction. Displacements in the x_2-direction are denoted by u without an index. It must satisfy the wave equation in both media obtained from (4.4) for

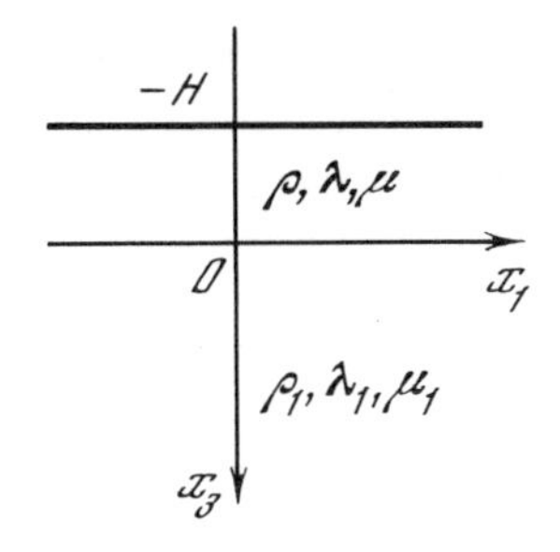

Fig. 4.3. Formulating the theory of Love's waves

ψ by the operation curl

$$\begin{aligned} &\Delta u + \varkappa^2 u = 0\,, \quad \varkappa^2 = \rho\omega^2/\mu\,, \quad -H \leq x_3 \leq 0 \\ &\Delta u + \varkappa_1 u = 0\,, \quad \varkappa_1^2 = \rho_1\omega^2/\mu_1\,, \quad x_3 \geq 0 \end{aligned} \tag{4.35}$$

under the following boundary conditions:

i) the stresses are zero at a free boundary $x_3 = -H$ [the condition (4.12) for $u_2 \equiv u$];
ii) the displacement u and the component $\sigma_{23} = 2\mu e_{23} = \mu\,\partial u/\partial x_3$ ($\sigma_{13} = \sigma_{33} = 0$) of the stress tensor are continuous at the interface $x_3 = 0$.

In terms of u these conditions take the form

$$\begin{aligned} &u|_{x_3=+0} = u|_{x_3=-0}\,, \\ &\mu_1 \left.\frac{\partial u}{\partial x_3}\right|_{x_3=+0} = \mu \left.\frac{\partial u}{\partial x_3}\right|_{x_3=-0}\,, \quad \left.\frac{\partial u}{\partial x_3}\right|_{x_3=-H} = 0\,. \end{aligned} \tag{4.36}$$

Expressions for u both in the layer and in the elastic half-space contain the same factor $\exp[\mathrm{i}(\xi x_1 - \omega t)]$ which, for brevity, we will be omitted. Then (4.35) with $\partial/\partial t = -\mathrm{i}\omega$ and $\partial/\partial x_1 = \mathrm{i}\xi$ can be written as

$$\begin{aligned} &\frac{\partial^2 u}{\partial x_3^2} + \alpha^2 u = 0\,, \quad \alpha^2 = \varkappa^2 - \xi^2\,, \quad -H \leq x_3 \leq 0\,, \\ &\frac{\partial^2 u}{\partial x_3^2} - \beta^2 u = 0\,, \quad \beta^2 = \xi^2 - \varkappa_1^2\,, \quad x_3 \geq 0\,. \end{aligned} \tag{4.37}$$

In a surface wave the displacement u must decrease exponentially with increasing x_3 into the homogeneous halfspace. This occurs when $\beta^2 > 0$ or $\xi^2 > \varkappa_1^2$ and

$$u = B\exp(-\beta x_3)\,, \quad x_3 \geq 0\,. \tag{4.38}$$

For a homogeneous layer, the general solution of (4.37) satisfying condition (4.36) at $x_3 = -H$ is

$$u = A\cos[\alpha(x_3 + H)], \qquad -H \le x_3 \le 0. \tag{4.39}$$

Substituting (4.38, 39) into the boundary conditions (4.36) at $x_3 = 0$, we obtain

$$A\cos\alpha H = B, \qquad -\mu A\alpha\sin\alpha H = -\mu_1\beta B. \tag{4.40}$$

Substituting the ratio A/B from the first relation into the second one yields

$$\tan\alpha H = \mu_1\beta/\mu\alpha. \tag{4.41}$$

Introducing a new variable $\eta = \alpha H$ so that

$$\xi^2 = \varkappa^2 - (\eta/H)^2, \qquad \beta H = \sqrt{(\varkappa H)^2 - (\varkappa_1 H)^2 - \eta^2},$$

we can write (4.41) in the form

$$\tan\eta = \frac{\mu_1}{\mu}\frac{\sqrt{(\varkappa H)^2 - (\varkappa_1 H)^2 - \eta^2}}{\eta}. \tag{4.42}$$

The roots η_n of this equation and the corresponding values ξ_n determine the main properties of Love's waves.

4.3.3 Some Features of Love's Waves

Let us consider the graphical solution of (4.42). When η is real, i.e., $\varkappa^2 > \xi^2$, existence of at least one root η_1 is obvious [see Fig. 4.4a where two branches of the function $\tan\eta$ and the right-hand side of (4.42) are plotted]. If $\varkappa^2 < \xi^2$,

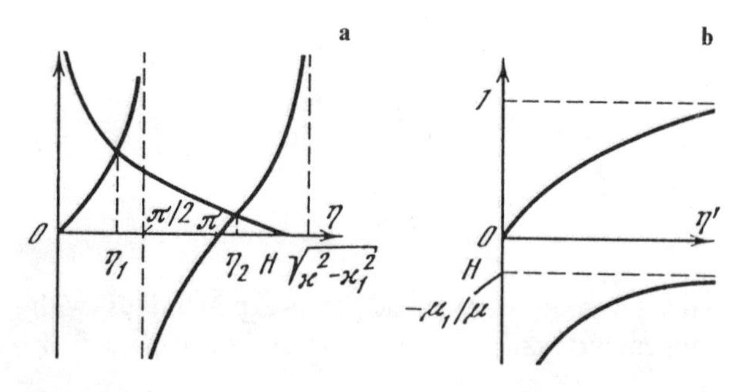

Fig. 4.4a, b. The graphical solution of the dispersion relation for Love's waves: **(a)** $\xi^2 < \varkappa^2$, at least one root η_n exists; **(b)** $\xi^2 > \varkappa^2$ there are no real roots

then $\alpha^2 < 0$, and $\eta = i\eta'$ is purely imaginary (inhomogeneous wave in a layer) so that (4.42) becomes

$$\tanh \eta' = -\frac{\mu_1}{\mu} \frac{\sqrt{(\varkappa H)^2 - (\varkappa_1 H)^2 + (\eta')^2}}{\eta'}. \tag{4.42a}$$

This equation has no real roots (Fig. 4.4b) where the upper curve is $\tanh \eta'$ and the lower one is a plot of the right-hand side of (4.42a).

Thus, for the existence of surface Love's waves it is necessary that

$$\varkappa_1 < \xi < \varkappa. \tag{4.43}$$

N modes of Love's waves exist in this case where $N = 1 + [H\sqrt{\varkappa^2 - \varkappa_1^2}/\pi]$; $[a]$ stands for the integral part of a. The phase velocity of the Love waves $v_n = \omega/\xi_n$ depends on frequency, i.e., a Love wave has dispersion. It follows from (4.43) that this velocity is greater than that of shear waves in a layer, but less than the velocity of shear waves in a halfspace:

$$c_{t1} > v_n > c_t.$$

For the particle displacement in the nth mode, we have from (4.38, 39), also taking into account (4.40) and the factor $\exp[i(\xi_n x_1 - \omega t)]$,

$$u_n = \begin{cases} A_n \cos[\alpha_n(x_3 + H)] \exp[i(\xi_n x_1 - \omega t)], & -H \le x_3 \le 0, \\ A_n \cos \alpha_n H \exp[i(\xi_n x_1 - \omega t) - \beta_n x_3], & x_3 \ge 0, \end{cases} \tag{4.44}$$

where A_n is the constant amplitude of the mode and $\alpha_n = \eta_n/H$.

Figure 4.5 shows schematically u_n for the first three modes in terms of the coordinate x_3 when $H\sqrt{\varkappa^2 - \varkappa_1^2}/\pi = 2.5$. The displacement vanishes at the nodal planes: $\alpha_n(x_3 + H) = \pi/2, 3\pi/2, \ldots$.

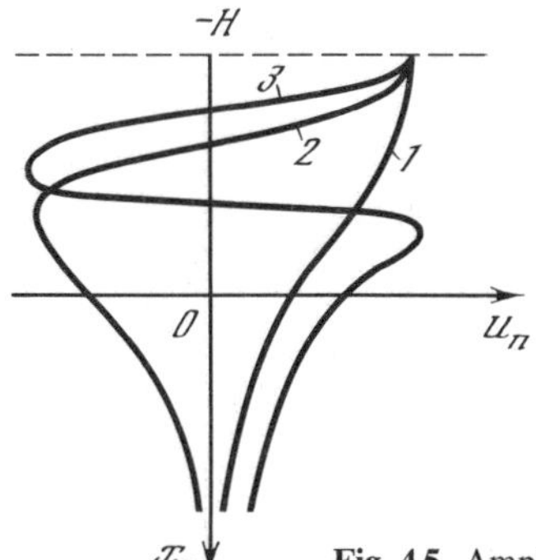

Fig. 4.5. Amplitude distribution of the first three Love's modes

4.4 Exercises

4.4.1. The following displacements are prescribed at the plane $x_3 = 0$: $u_1 = A\exp[\mathrm{i}(\xi x_1 - \omega t)]$, $u_2 = B\exp[\mathrm{i}(\xi x_1 - \omega t)]$, $u_3 = C\exp[\mathrm{i}(\xi x_1 - \omega t)]$. Find the wave field in the elastic half-space $x_3 > 0$.

Solution: The displacement $u_2|_{x_3=0}$ gives rise to the transverse horizontally polarized wave $u_2 = B\exp[\mathrm{i}(\xi x_1 + \sqrt{\varkappa^2 - \xi^2}\,x_3 - \omega t)]$ with the prescribed value of u_2 at $x_3 = 0$. The displacements $u_1|_{x_3=0}$ and $u_3|_{x_3=0}$ determine vertically polarized waves with the potentials

$$\varphi = A_1 \exp[\mathrm{i}(\xi x_1 + k_3 x_3 - \omega t)]\,, \qquad k_3 = \sqrt{k^2 - \xi^2}\,,$$

$$\psi_2 = \psi = A_2 \exp[\mathrm{i}(\xi x_1 + \varkappa_3 x_3 - \omega t)]\,, \qquad \varkappa_3 = \sqrt{\varkappa^2 - \xi^2}\,.$$

Taking (4.17) into account, we obtain conditions for the displacement continuity at $x_3 = 0$:

$$\mathrm{i}\xi A_1 - \mathrm{i}\varkappa_3 A_2 = A\,, \qquad \mathrm{i}k_3 A_1 + \mathrm{i}\xi A_2 = C\,.$$

Hence the wave amplitudes are

$$A_1 = -\mathrm{i}\frac{\xi A + \varkappa_3 C}{\xi^2 + \varkappa_3 k_3}\,, \qquad A_2 = -\mathrm{i}\frac{\xi C - k_3 A}{\xi^2 + \varkappa_3 k_3}\,.$$

4.4.2. Find the reflection coefficient V of the horizontally polarized wave at the boundary where displacements are prohibited [Item (iii) in Sect. 4.1.2].

Solution: For a horizontal displacement $u = u_2(x_1, x_3, t)$, we have the wave equation $\partial^2 u/\partial t^2 = c_\mathrm{t}^2\,\Delta u$. Its solution is the sum of incident and reflected harmonic waves of the form

$$u = A_+ \exp[\mathrm{i}(\xi x_1 + \varkappa_3 x_3 - \omega t)] + A_+ V \exp[\mathrm{i}(\xi x_1 - \varkappa_3 x_3 - \omega t)]\,,$$

where $\varkappa_3 = \sqrt{\omega^2/c_\mathrm{t}^2 - \xi^2}$. The boundary condition is $u|_{x_3=0} = 0$, whence $1 + V = 0$ whence $V = -1$.

4.4.3. Consider the reflection of a harmonic vertically polarized transverse wave at the contact with an absolutely rigid wall with slip [Item (iv) in Sect. 4.1.2].

Solution: The potentials φ and $\boldsymbol{\psi} = \{0, \psi, 0\}$ satisfy the wave equations (4.4). For the case where the incident wave is transverse, their solution can be written as (B is the amplitude of the incident wave)

$$\varphi = BV_{\mathrm{tl}} \exp[\mathrm{i}(\xi x_1 - k_3 x_3 - \omega t)]\,,$$

$$\psi = B\exp[\mathrm{i}(\xi x_1 + \varkappa_3 x_3 - \omega t)] + BV_{\mathrm{tt}} \exp[\mathrm{i}(\xi x_1 - \varkappa_3 x_3 - \omega t)]\,,$$

where $k_3 = \sqrt{\omega^2/c_l^2 - \xi^2}$, $\varkappa = \sqrt{\omega^2/c_t^2 - \xi^2}$. The boundary conditions at $x_3 = 0$ are (Sect. 4.1.2)

$$u_3|_{x_3=0} = \left(\frac{\partial \varphi}{\partial x_3} + \frac{\partial \psi}{\partial x_1}\right)_{x_3=0} = 0\,,$$

$$\sigma_{13}|_{x_3=0} = 2\mu e_{13}|_{x_3=0} = 2\mu\left(\frac{\partial^2 \varphi}{\partial x_1 \partial x_3} + \frac{1}{2}\frac{\partial^2 \psi}{\partial x_1^2} - \frac{1}{2}\frac{\partial^2 \psi}{\partial x_3^2}\right)_{x_3=0} = 0\,.$$

Substitution of φ and ψ into these conditions yields

$$\mathrm{i}\xi(1 + V_{tt}) - ik_3 V_{tl} = 0\,, \qquad \xi k_3 V_{tl} - (\xi^2 - \varkappa^2/2)(1 + V_{tt}) = 0\,,$$

or $V_{tt} = -1$ and $V_{tl} = 0$. Hence, we have total reflection with phase inversion.

4.4.4. Discuss qualitatively the reflection of a longitudinal wave at a plane interface separating two elastic media. Determine the types of waves produced and their wave vectors.

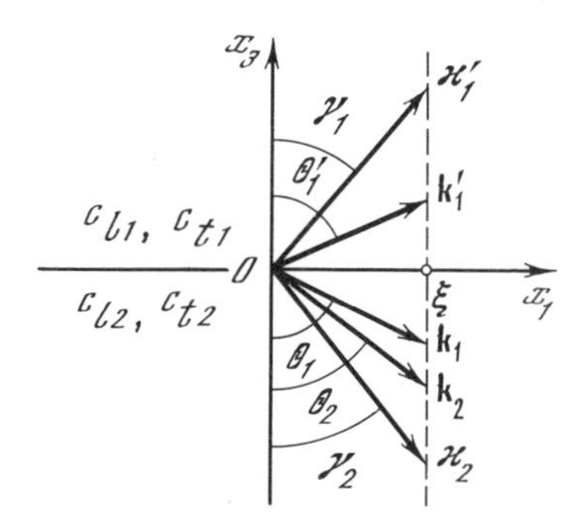

Fig. 4.6. Wave vectors of a system of reflected and refracted waves generated by longitudinal wave incident upon the interface of two elastic halfspace

Solution: The incident longitudinal wave in the halfspace $x_3 > 0$ has a wave vector $\boldsymbol{k}_1$ (Fig. 4.6) which lies in the x_1x_3 plane. Wave vectors of reflected longitudinal and transverse waves are $\boldsymbol{k}_1'$ and $\boldsymbol{\varkappa}_1'$, respectively, and those of refracted ones are $\boldsymbol{k}_2$ and $\boldsymbol{\varkappa}_2$. The fulfillment of the boundary conditions at all times and at all points of the interface is possible only if the frequency and x_1-components of the wave vectors are the same for all the waves. Hence, all wave vectors must end at the same straight line perpendicular to the x_1-axis, crossing it at a distance ξ from the origin. As a result, for reflection θ_1', γ_1 and refraction θ_2, γ_2 angles, we have the relations

$$\theta_1' = \theta\,, \qquad k_1 \sin\theta_1 = \varkappa_1 \sin\gamma_1 = k_2 \sin\theta_2 = \varkappa_2 \sin\gamma_2\,,$$

where θ_1 is the incidence angle.

4.4.5. Two similar elastic media are in slipping contact at an interface $x_3 = 0$. Find the reflection and refraction coefficients of a longitudinal harmonic wave at this interface.

Solution: A complete set of waves in this case is shown in Fig. 4.6. The corresponding fields can be written (the factor $\exp[\mathrm{i}(\xi x_1 - \omega t)]$ is the same in all expressions and is omitted):

$$\varphi_1 = A_1[\exp(-\mathrm{i}k_3x_3) + V_{ll}\exp(\mathrm{i}k_3x_3)], \quad \psi_1 = A_1V_{lt}\exp(\mathrm{i}\varkappa_3x_3), \quad x_3 > 0$$

$$\varphi_2 = A_1W_{ll}\exp(-\mathrm{i}k_3x_3), \quad \psi_2 = A_1W_{lt}\exp(-\mathrm{i}\varkappa_3x_3), \quad x_3 < 0$$

where $k_3 = \sqrt{k^2 - \xi^2}$, $k = \omega/c_l$, $\varkappa_3 = \sqrt{\varkappa^2 - \xi^2}$, $\varkappa = \omega/c_t$; A_1 is the amplitude of the incident wave. The boundary conditions at $x_3 = 0$ are [Item (ii) in Sect. 4.1.2]

$$u_3|_{x_3=+0} = u_3|_{x_3=-0},$$

$$\sigma_{33}|_{x_3=+0} = \sigma_{33}|_{x_3=-0},$$

$$\sigma_{31}|_{x_3=+0} = \sigma_{31}|_{x_3=-0} = 0.$$

Rewriting this in terms of potentials and substituting the expressions for φ_j and ψ_j, $j = 1, 2$ yields

$$k_3(1 - V_{ll}) - \xi V_{lt} = k_3W_{ll} - \xi W_{lt}, \quad p(1 + V_{ll}) - \varkappa_3V_{lt} = pW_{ll} + \varkappa_3W_{lt},$$

$$k_3(1 - V_{ll}) = pV_{lt}, \quad k_3W_{ll} = pW_{lt}, \quad p = \xi^{-1}(\xi^2 - \varkappa^2/2).$$

Solving the system of algebraic equations, we find

$$V_{ll} = \frac{k_3\varkappa_3}{p^2 + k_3\varkappa_3}, \quad W_{ll} = \frac{p^2}{p^2 + k_3\varkappa_3}, \quad V_{lt} = W_{lt} = \frac{k_3p}{p^2 + k_3\varkappa_3}.$$

4.4.6. It was shown in Sec. 4.2.4 that in the case of a longitudinal wave incident at a free boundary of elastic halfspace, no longitudinal reflected wave is produced under certain conditions, [see (4.27, 33)]. Show that this is possible only if the Poisson ratio $\nu \leq \nu_m$ and find ν_m.

Solution: We have to show that for $\nu > \nu_m$, (4.33) has no roots $s > 1$ corresponding to homogeneous waves. A plot of the function $f(s)$ is given in Fig. 4.7. It is evident that for the occurrence of real roots $s_{1,2} > 1$ of a polynomial $f(s) = s^3 - 8s^2 + 16(3/2 - q^2)s - 16(1 - q^2)$, two extrema of the function $f(s)$ are needed at $s > 1$. Since $f'(s) = 3s^2 - 16s + 16(3/2 - q^2)$, we have for the two extremum points $s_\pm = 4(2 \pm \sqrt{3q^2 - 1/2})/3$. The latter will be real under the condition $q^2 = c_t^2/c_l^2 \geq 1/6$. Expressing now the ratio $c_t^2/c_l^2 = \mu/(\lambda + 2\mu)$ in terms of E and ν by (3.26), we obtain $(1 - 2\nu)/(1 - \nu) \geq 1/3$ or $\nu < 0.4$.

However, as is seen from Fig. 4.7, the condition for the existence of extrema is not sufficient for the existence of the roots $s_{1,2} > 1$. For this purpose, $f(s_+) \leq 0$ is also required. Introducing the parameters $\alpha = \sqrt{3q^2 - 1/2}$, $q^2 = (1 + 2\alpha^2)/6$, $s_+ = 4(2 + \alpha)/3$ and calculating $f(s_+)$, we write the last condition in the form $19 - 30\alpha^2 - 16\alpha^3 \leq 0$ or $\alpha \geq \alpha_0$, where $\alpha_0 \simeq 0.68$. This gives $q^2 \geq 0.32$ and finally $v \leq v_m \simeq 0.26$.

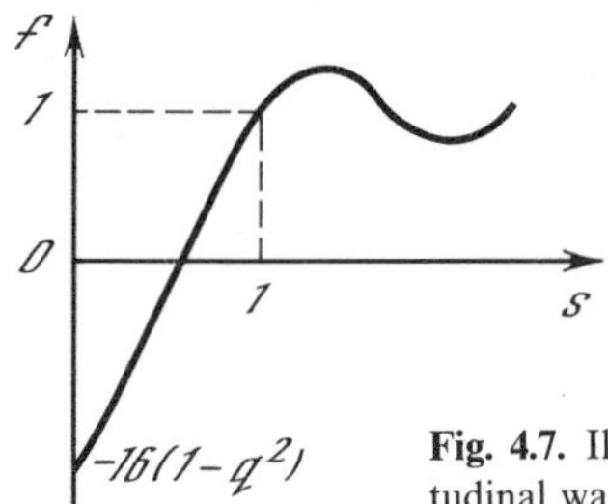

Fig. 4.7. Illustrating the case where the total transformation of a longitudinal wave into a transverse one and vice versa is possible

4.4.7. Consider horizontally polarized waves in a layer where one of its boundaries is stress-free and the other is absolutely rigid.

Solution: An absolutely rigid boundary can be considered as the boundary of the halfspace with an infinitely large velocity of wave propagation. Consequently, we obtain the required case by setting $\mu_1 \to \infty$ in the expressions for the Love waves (4.42, 44). This yields $\tan \eta \to \infty$, whence $\eta_n = (2n + 1)\pi/2$, $n = 1, 2, \ldots$, $\xi_n = \sqrt{\varkappa^2 - (2n - 1)^2\pi^2/4H^2}$. For the displacement we have now from (4.44),

$$u_n = A_n \cos[(2n - 1)\pi(x_3 + H)/2H] \exp[\mathrm{i}(\xi_n x_1 - \omega t)] .$$

The nodal planes are specified by

$$x_3/H = -1 + (2n - 1)^{-1}, \quad -1 + 3(2n - 1)^{-1}, \ldots, 0 .$$

There are n of these planes. One of these planes coincides with the boundary $x_3 = 0$.

4.4.8. Determine the parameters of harmonic, horizontally polarized waves propagating in a layer of elastic material of thickness H when both boundaries of the layer are stress-free.

Solution: For the horizontal displacement $u_2 = u(x_1, x_3, t)$, we have the wave equation $\partial^2 u/\partial t^2 = c_t^2 \Delta u$ with the conditions at the free boundaries $(\partial u/\partial x_3)_{x_3 = 0, H} = 0$. Look for a solution in the form: $u = \phi(x_3) \exp[\mathrm{i}\xi x_1 - \omega t)]$. For the function $\phi(x_3)$ we obtain an ordinary differential equation $\phi'' + (\varkappa^2 - \xi^2)\phi = 0$, $\varkappa = \omega/c_t$ and the boundary conditions $\phi'(0) = \phi'(H) = 0$. A solution of the type $\phi = A\cos\alpha x_3$, $\alpha = \sqrt{\varkappa^2 - \xi^2}$ satisfies both the equation and the condition at $x_3 = 0$. Substitution of this solution into the condition at $x_3 = H$ leads to an equation for possible values of α: $\sin\alpha H = 0$. Its solutions are $\alpha_n = n\pi/H$, $n = 0, 1, 2, \ldots$; hence, $\xi_n = \sqrt{\varkappa^2 - n^2\pi^2/H^2}$. There is a finite number $n_{\max} < \varkappa H/\pi$ of waves propagating along a layer. For $n > n_{\max}$, ξ_n is imaginary and we have just oscillations with an amplitude exponentially decreasing along x_1. The displacement $u_n(x_1, x_3, t)$ in the waves is written as $u_n = A_n \cos(n\pi x_3/H)\exp[i(\xi_n x_1 - \omega t)]$. In particular, for $n = 0$ we obtain a wave with an amplitude constant across a layer and propagating along x_1 at a velocity c_t. For $n = 1, 2, \ldots$ the wave amplitudes vary across the layer; there are n nodal planes at which $u_n = 0$.

4.4.9. Find the group velocities of horizontally polarized waves in a layer with free boundaries.

Solution: In Exercise 4.4.8, the horizontal wave number of the nth mode $\xi_n = \sqrt{\varkappa^2 - n^2\pi^2/H^2}$ has been found. For $n < n_{\max}$ we have for the phase velocity.

$$v_n = \omega/\xi_n = c_t[1 - (n\pi/\varkappa H)^2]^{-1/2} > c_t$$

and for the group velocity

$$(c_g)_n = (d\xi_n/d\omega)^{-1} = c_t(d\xi_n/d\varkappa)^{-1} = c_t[1 - (n\pi/\varkappa H)^2]^{1/2} < c_t.$$

Note that the simple relation $(c_g)_n v_n = c_t^2$ arises.

4.4.10. An elastic layer of thickness H with the parameters (ρ, c_l, c_t) is placed between two elastic half spaces with the parameters (ρ_1, c_{l1}, c_{t1}) and (ρ_2, c_{l2}, c_{t2}). Determine the reflection and transformation coefficients for an incidence of the longitudinal (or transverse) plane wave from one of the half spaces. Consider a different type of the boundary conditions.

5. Waves in Plates

A solid layer with stress-free, plane and parallel boundaries is called a *plate*. Horizontally polarized waves in plates where the particle displacements are parallel to the boundaries have been considered above (Exercise 4.4.8, 9). More important, however, is the case of vertically polarized waves in plates. In this case, displacements lie in the "vertical" plane, i.e., the plane which includes the normal to the boundaries as well as the propagation direction. Below we will consider different kinds of vertically polarized waves in plates. Waves in thin plates will be considered in more detail.

5.1 Classification of Waves

5.1.1 Dispersion Relations

We choose the x_3-axis normal to the plate boundaries and the x_1-axis along the propagation direction of the wave. Suppose that the plate boundaries are placed at $x_3 = \pm h$, so that its thickness is $2h$. It was shown in Sect. 4.2.1 that displacements of vertically polarized waves can be described by two scalar functions which are independent of x_2: a scalar potential $\varphi(x_1, x_3, t)$ and the component $\psi(x_1, x_3, t)$ of a vector potential $\boldsymbol{\psi} = \{0, \psi, 0\}$. For the components of the displacement vector we have (4.17)

$$u_1 = \partial\varphi/\partial x_1 - \partial\psi/\partial x_3\,, \qquad u_2 = 0\,, \qquad u_3 = \partial\varphi/\partial x_3 + \partial\psi/\partial x_1\,. \tag{5.1}$$

The functions φ and ψ satisfy the wave equations (4.4). The frequency as well as the horizontal component of the wave vector are conserved when waves are reflected at the boundaries. Therefore, it is reasonable to choose the x_1 and t dependence of φ and ψ in the form $\exp[\mathrm{i}(\xi x_1 - \omega t)]$. As a result, we obtain the two equations (4.19) for φ and ψ with a general solution (4.20) which can also be written (after redefinition of arbitrary constants) in the form

$$\begin{aligned}
&\varphi = (C_1 \cos k_3 x_3 + C_2 \sin k_3 x_3)\exp[\mathrm{i}(\xi x_1 - \omega t)]\,,\\
&\psi = (D_1 \cos \varkappa_3 x_3 + D_2 \sin \varkappa_3 x_3)\exp[\mathrm{i}(\xi x_1 - \omega t)]\,,\\
&k_3 = \sqrt{k^2 - \xi^2}\,, \quad k = \omega/c_1\,, \quad \varkappa_3 = \sqrt{\varkappa^2 - \xi^2}\,, \quad \varkappa = \omega/c_t\,.
\end{aligned} \tag{5.2}$$

Now, substituting these expressions into the boundary conditions (4.19) for $x_3 = \pm h$, we obtain a homogeneous system of equations for the complex constants C_1, C_2, D_1, D_2:

$$\begin{pmatrix} \sigma_1 \sin\sigma_1 & -iph\sin\sigma_t & \sigma_1\cos\sigma_1 & iph\cos\sigma_t \\ -\sigma_1\sin\sigma_1 & iph\sin\sigma_t & \sigma_1\cos\sigma_1 & iph\cos\sigma_t \\ -iph\cos\sigma_1 & \sigma_t\cos\sigma_t & iph\sin\sigma_1 & \sigma_t\sin\sigma_t \\ -iph\cos\sigma_1 & \sigma_t\cos\sigma_t & -iph\sin\sigma_1 & -\sigma_t\sin\sigma_t \end{pmatrix} \begin{pmatrix} C_1 \\ D_2 \\ C_2 \\ D_1 \end{pmatrix} = 0,$$

where $\sigma_1 = k_3 h = \sqrt{k^2 - \xi^2}\,h$, $\sigma_t = \varkappa_3 h = \sqrt{\varkappa^2 - \xi^2}\,h$, $p = \xi^{-1}(\xi^2 - \varkappa^2/2)$.

Subtract and add the first two equations term by term, then do the same with the second pair. As a result, we obtain a new equivalent system:

$$\begin{pmatrix} \sigma_1\sin\sigma_1 & -iph\sin\sigma_t & 0 & 0 \\ -iph\cos\sigma_1 & \sigma_t\cos\sigma_t & 0 & 0 \\ 0 & 0 & \sigma_1\cos\sigma_1 & iph\cos\sigma_t \\ 0 & 0 & iph\sin\sigma_1 & \sigma_t\sin\sigma_t \end{pmatrix} \begin{pmatrix} C_1 \\ D_2 \\ C_2 \\ D_1 \end{pmatrix} = 0. \tag{5.3}$$

A nonzero solution of this system exists only if its determinant is zero:

$$(\sigma_1\sigma_t\sin\sigma_1\cos\sigma_t + p^2h^2\cos\sigma_1\sin\sigma_t) \times (\sigma_1\sigma_t\cos\sigma_1\sin\sigma_t + p^2h^2\sin\sigma_1\cos\sigma_t) = 0. \tag{5.4}$$

For a fixed frequency ω and, consequently, fixed $k = \omega/c_1$ and $\varkappa = \omega/c_t$, (5.4) may be regarded as an equation for permissible values of x_1 projections of the wave vector $\xi = \xi_n(\omega)$. Conversely, for a fixed ξ this equation determines possible frequencies $\omega = \omega_n(\xi)$. Such waves are called *eigenwaves* or *modes*, and the corresponding values ξ_n and ω_n are *the eigenvalues*. The expressions $\omega = \omega_n(\xi)$ are the *dispersion relations* for the modes; the quantity $c_{ph} = \omega/\xi$ is the *phase velocity* of the wave and $c_g = d\omega/d\xi$ is its *group velocity*.

5.1.2 Symmetric and Asymmetric Modes

Equation (5.4) is satisfied when one of the factor is zero. Let us first consider

$$\sigma_1\sigma_t\tan\sigma_1 + p^2h^2\tan\sigma_t = 0. \tag{5.5}$$

It follows from (5.3) that $C_2 = D_1 = 0$, $D_2 = \mathrm{i}C_1 ph\cos\sigma_1(\sigma_t\cos\sigma_t)^{-1}$ in this case. Therefore, by (5.2),

$$\varphi = C_1\cos k_3x_3, \qquad \psi = D_2\sin\varkappa_3x_3. \tag{5.6}$$

Here and below the factor $\exp[\mathrm{i}(\xi x_1 - \omega t)]$ is omitted. Now, with (5.1) we find

$$
\begin{aligned}
u_1 &= \mathrm{i}\xi C_1 \cos k_3 x_3 - \varkappa_3 D_2 \cos \varkappa_3 x_3\,, \qquad u_1(-x_3) = u_1(x_3)\,,\\
u_3 &= -k_3 C_1 \sin k_3 x_3 + \mathrm{i}\xi D_2 \sin \varkappa_3 x_3\,, \qquad u_3(-x_3) = -u_3(x_3)\,.
\end{aligned}
\tag{5.7}
$$

The displacements are *symmetric* with respect to the plane $x_3 = 0$ for this type of mode. Equation (5.5) is the dispersion relation for the symmetric modes.

We now consider the other case

$$\sigma_l \sigma_t \tan \sigma_t + p^2 h^2 \tan \sigma_l = 0\,. \tag{5.8}$$

It follows from (5.3) that

$$C_1 = D_2 = 0\,, \qquad D_1 = \mathrm{i} C_2 \sigma_l \cos \sigma_l (ph \cos \sigma_t)^{-1}$$

and

$$\varphi = C_2 \sin k_3 x_3\,, \qquad \psi = D_1 \cos \varkappa_3 x_3\,. \tag{5.9}$$

The displacement components are

$$
\begin{aligned}
u_1 &= \mathrm{i}\xi C_2 \sin k_3 x_3 + \varkappa_3 D_1 \sin \varkappa_3 x_3\,, \qquad u_1(-x_3) = -u_1(x_3)\,,\\
u_3 &= k_3 C_2 \cos k_3 x_3 + \mathrm{i}\xi D_1 \cos \varkappa_3 x_3\,, \qquad u_3(-x_3) = u_3(x_3)\,.
\end{aligned}
\tag{5.10}
$$

The displacement vector is *asymmetric* with respect to the plane $x_3 = 0$. These are the *asymmetric modes.*

There is a countable number of eigenvalues $\omega_n(\xi)$, $n = 1, 2, \ldots$ for any fixed ξ because of the periodicity of the trigonometrical functions in the dispersion relation. On the other hand, when the frequency is prescribed, the dispersion equations have only a finite number of real roots $\xi_n(\omega)$. In fact, if $\xi^2 > \omega^2/c_t^2$, the quantities σ_t and σ_l become purely imaginary; the trigonometrical functions become hyperbolic and not more than one real root ξ_n exists for each of (5.5, 6). Note also that there is a countable infinite number of roots $\xi_n^2 < 0$ for any given ω which do not represent propagating modes but only oscillations with amplitudes varying exponentially φ, $\psi \sim \exp(\pm|\xi_n|x_1 - \mathrm{i}\omega t)$ and a constant phase—a sort of *inhomogeneous* wave.

5.1.3 Cut-Off Frequencies of the Modes

With the frequency ω decreasing, the number of propagating modes (where $\xi_n^2 > 0$) also decrease. There exists a cut-off frequency $(\omega_c)_n$ for the nth mode so that at $\omega < (\omega_c)_n$, this mode becomes non-propagating, i.e., $\xi_n(\omega)$ becomes purely imaginary. Since the transition of ξ_n from a real quantity to a purely imaginary one is continuous, we have $\xi_n(\omega_c) = 0$, and the phase velocity $c_{\mathrm{ph}} = \omega/\xi_n$ is infinite. Using this fact we can evaluate ω_c. Indeed, assuming $\xi \to 0$ $(p \to \infty)$ in (5.5) for symmetric modes, we obtain $\tan \varkappa h/\tan kh = 0$ which yields

$$\varkappa h = h\omega/c_t = n_1 \pi\,, \qquad (\omega_c)_{n_1} = n_1 \pi c_t/h\,, \qquad n_1 = 0, 1, \ldots$$

or

$$kh = h\omega/c_1 = \pi(2n_2 - 1)/2\,, \qquad (\omega_c)_{n_2} = (n_2 - 1/2)\pi c_1/h\,, \qquad n_2 = 1, 2, \ldots \tag{5.11}$$

From (5.3, 7) we have for $\omega = (\omega_c)_{n_1}$, $C_1 = 0$, $u_3 = 0$,

$$u_1 = -D_2\varkappa\cos\varkappa x_3\exp(-\mathrm{i}\omega t)\,.$$

It is a standing transverse wave with constant phase along a layer. Since $\varkappa = 2\pi/\lambda$ we have $2h/(\lambda/2) = 2n_1$. Hence, the plate's thickness $2h$ contains an even number of halfwaves. If $\omega = (\omega_c)_{n_2}$, we have $u_1 = 0$, and $u_3 = -C_1 k\sin kx_3\exp(-\mathrm{i}\omega t)$ is a standing longitudinal wave with an odd number of halfwaves over the plate thickness.

Analogous results can also be obtained for asymmetric modes:

$$\begin{aligned}&(\omega_c)_{n_3} = (n_3 - 1/2)\pi c_t/h\,,\ \ u_3 = 0\,,\ \ u_1 = D_1\sin\varkappa x_3\exp(-\mathrm{i}\omega t)\,,\ \ n_3 = 1, 2, \ldots\\&(\omega_c)_{n_4} = n_4\pi c_1/h\,,\ \ u_1 = 0\,,\ \ u_3 = C_2 k\cos kx_3\exp(-\mathrm{i}\omega t)\,,\ \ n_4 = 0, 1, \ldots\end{aligned} \tag{5.12}$$

The thickness of plate contains an odd number of shear halfwaves and an even number of longitudinal ones.

In the case of arbitrary frequency, modes have both longitudinal and transverse components. Sometimes it is convenient to consider such a mode as a superposition of a pair of longitudinal and a pair of transverse plane waves transforming into each other when reflecting at the plate boundaries. Such representation becomes natural if we rewrite the expressions for the potentials (5.6, 9) in the form

$$\begin{aligned}\varphi &= C\{\exp[\mathrm{i}(\xi x_1 + k_3 x_3 - \omega t)] \pm \exp[\mathrm{i}(\xi x_1 - k_3 x_3 - \omega t)]\}\,,\\ \psi &= D\{\exp[\mathrm{i}(\xi x_1 + \varkappa_3 x_3 - \omega t)] \mp \exp[\mathrm{i}(\xi x_1 - \varkappa_3 x_3 - \omega t)]\}\,,\end{aligned} \tag{5.13}$$

where the upper signs refer to the symmetric mode and the lower ones to the asymmetric mode. Wave vectors of all four waves $\boldsymbol{k} = \{\xi, k_3\}$, $\boldsymbol{k}' = \{\xi, -k_3\}$, $\boldsymbol{\varkappa} = \{\xi, \varkappa_3\}$, $\boldsymbol{\varkappa}' = \{\xi, -\varkappa_3\}$ are shown in Fig. 5.1; the angles θ and γ are related according to (4.24) by $c_1\sin\gamma = c_t\sin\theta$.

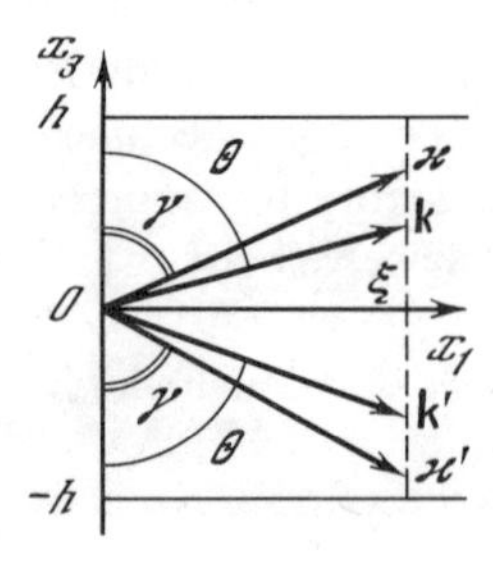

Fig. 5.1. Representation of propagating modes in a plate as a sum of four plane waves

5.1.4 Some Special Cases

The representation of a normal mode by a superposition of four plane waves (5.13) is useful, particularly for discussing some special cases. For instance, it was shown in Sect. 4.2.4 that if the transverse wave is incident upon a free boundary at an angle of 45°, no longitudinal wave arises. It means that at certain frequencies (see Exercise 5.4.2), a mode will be formed by a pair of plane transverse waves propagating at the angle $\gamma = 45°$.

Another interesting case is when a mode is formed by plane waves incident upon the boundaries at the angles θ_0 and γ_0 at which the reflection coefficient is zero and the transformation coefficient of a longitudinal wave into a transverse one and vice versa is unity. As has been shown in Sect. 4.2.4 (see also Exercise 4.4.6), there are two pairs of such angles for the Poisson ratio $\nu < 0.26$. Therefore, it is always possible to find the frequency at which the combination of two plane waves with wave vectors shown in Fig. 5.2a produces the nth mode (see Exercise 5.4.3). By virtue of symmetry, a mode at the same frequency can be formed by a system of plane waves with the wave vectors given in Fig. 5.2b. The amplitudes of plane waves $\boldsymbol{\varkappa}$ and $\boldsymbol{k}'$ are related by the transformation coefficient V_{lt}. The same is true for waves $\boldsymbol{k}$ and $\boldsymbol{\varkappa}'$. Neither of these pairs of waves makes up a symmetric or asymmetric mode. However, since frequencies and wave vector components along the boundary are the same for all the waves, any combination with complex weight coefficients forms a mode. In particular, these coefficients can be chosen to obtain symmetric or asymmetric modes.

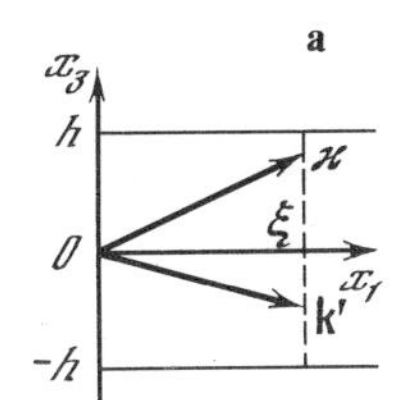

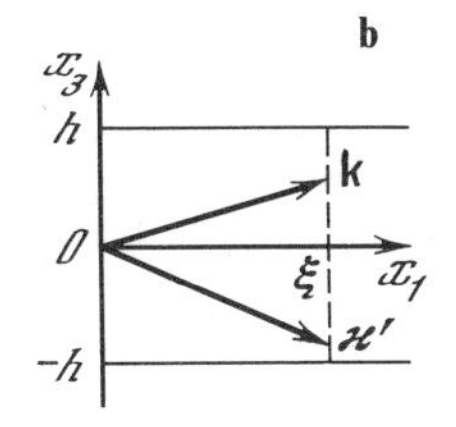

Fig. 5.2. Mode degeneration in the plate

5.2 Normal Modes of the Lowest Order

5.2.1 Quasi-Rayleigh Waves at the Plate's Boundaries

Among the normal waves in the plates, symmetric and antisymmetric modes of the lowest order with zero cut-off frequency [$n_1 = 0$ in (5.11) and $n_4 = 0$ in (5.12)] are in a special position. The distribution of the particles' displacements across the plate has a minimum number of sign changes for these modes. Longitudinal displacement in a symmetric mode of lowest order does not change the sign at all; for an antisymmetric mode, the vertical particles' displacement behaves analogously.[4] From this it follows that the maximum possible roots ξ^2 of the dispersion equations (5.5, 8) and, consequently, [see (5.2)] the minimum positive (and purely imaginary) values of the vertical wave numbers $\varkappa_3$ and k_3, correspond to these modes. In this case, the particles' displacements (5.7) have the least number of oscillations across the plate.

At high frequencies the large (by modulus) negative values of $\sigma_t^2(\xi^2 > \omega^2/c_t^2 > \omega^2/c_l^2)$ correspond to these modes. Indeed, assuming in the dispersion equations (5.5, 8)

$$\tan\sigma_l = \mathrm{i}\tanh|\sigma_l| \simeq \mathrm{i}[1 - 2\exp(-2|\sigma_l|)],$$
$$\tan\sigma_t = \mathrm{i}\tanh|\sigma_t| \simeq \mathrm{i}[1 - 2\exp(-2|\sigma_t|)],$$

we obtain for the symmetric and antisymmetric modes, respectively,

$$\begin{aligned} f_r(\xi) &\simeq 2[\sigma_l\sigma_t\exp(-2|\sigma_l|) + p^2h^2\exp(-2|\sigma_t|)], \\ f_r(\xi) &\simeq 2[p^2h^2\exp(-2|\sigma_l|) + \sigma_l\sigma_t\exp(-2|\sigma_t|)], \end{aligned} \tag{5.14}$$

where $f_r(\xi) = \sigma_l\sigma_t + p^2h^2$. The right-hand sides of (5.14) are small, therefore the eigenvalues ξ_s and ξ_a are close to the root ξ_r of $f_r(\xi) = 0$, corresponding to a Rayleigh wave, cf. (4.31). For $\xi = \xi_r f_r(\xi) \simeq f_r'(\xi_r)(\xi - \xi_r)$, so that for ξ_s and ξ_a we have

$$\begin{aligned} &\xi_s = \xi_r - \varepsilon, \quad \xi_a = \xi_r + \varepsilon, \\ &\varepsilon \simeq -2\{(f')^{-1}p^2h^2[\exp(-2|\sigma_t|) - \exp(-2|\sigma_l|)]\}_{\xi=\xi_r}, \end{aligned} \tag{5.15}$$

where $\varepsilon > 0$ by virtue of $f_r'(\xi_r) > 0$ and $|\sigma_t| < |\sigma_l|$.

Thus, at high frequencies the phase velocity ω/ξ_s of the lowest symmetric mode considered is somewhat greater than that of a Rayleigh wave. In turn it exceeds slightly the phase velocity of the lowest asymmetric mode.

The expressions for displacement, say, for u_1 are, according to (5.7, 10),

$$\begin{aligned} u_{1s} &= C_s\left[\frac{\cosh|k_3|x_3}{\cosh|\sigma_l|} + \left(1 - \frac{\varkappa^2}{2\xi_r^2}\right)\frac{\cosh|\varkappa_3|x_3}{\cosh|\sigma_t|}\right]\exp\{\mathrm{i}[(\xi_r - \varepsilon)x_1 - \omega t]\}, \\ u_{1a} &= C_a\left[\frac{\sinh|k_3|x_3}{\cosh|\sigma_l|} - \left(1 - \frac{\varkappa^2}{2\xi_r^2}\right)\frac{\sinh|\varkappa_3|x_3}{\cosh|\sigma_t|}\right]\exp\{\mathrm{i}[(\xi_r + \varepsilon)x_1 - \omega t]\}. \end{aligned} \tag{5.16}$$

Figure 5.3 illustrates the distributions $u_1(x_3)$ for these modes which are similar to the displacements of Rayleigh waves near each boundary. Some interesting phenomena arise due to the interaction of these quasi-Rayleigh waves. Suppose

[4] At cut-off frequencies, this fact can be easily established on the basis of (5.11, 12) and the corresponding expressions for the particles' displacements

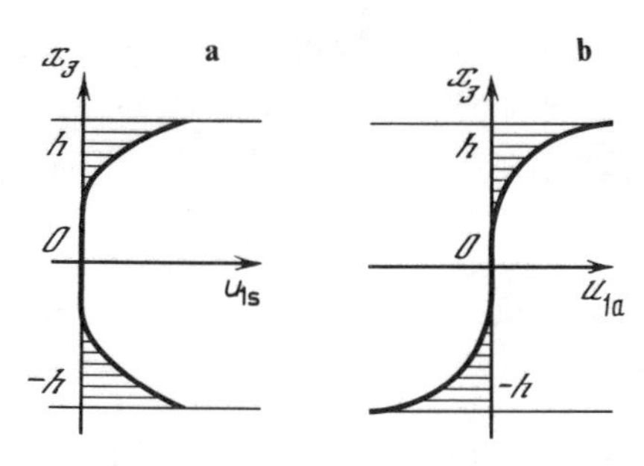

Fig. 5.3a, b. Lower normal modes of high frequency: (**a**) symmetrical (**b**) asymmetrical

a Rayleigh wave is excited at the upper boundary ($x_3 = h$) of the plate. This wave may be represented as a half sum of symmetric and asymmetric modes with equal phases. Indeed, the field of this sum in the vicinity of the source differs from zero only near the upper boundary, since near the lower one, the modes almost cancel each other. However, by virtue of the difference between the phase velocities, their phases differ by π at the distance $\pi/2\varepsilon$ from the excitation point. As a result, only near the lower boundary at this distance is the total excitation not zero. At greater distances the excitation will concentrate again near the upper boundary and so on. This phenomenon is similar to that of the energy transfer in a system of two weakly-connected oscillators of equal resonance frequencies.

5.2.2 The Young and Bending Waves

These waves can exist at any low frequencies (their cut-off frequencies are zero). We consider the symmetric mode first and analyse the dispersion equation (5.5) for the case $|\sigma_l| = |k_3|h \ll 1$, $|\sigma_t| = |\varkappa_3|h \ll 1$ assuming the plate to be thin. The dispersion equation becomes

$$\sigma_l^2/h^2 + p^2 = k^2 - \xi^2 + p^2 = 0\,.$$

Hence, $\sigma_l^2 < 0$, $\xi^2 > k^2$, i.e., we have a longitudinal inhomogeneous wave over the thickness of the plate. Taking into account the value of p, we obtain from the last equation

$$\xi^2 = \frac{\varkappa^4}{4(\varkappa^2 - k^2)} = \frac{\rho\omega^2(\lambda + 2\mu)}{4\mu(\lambda + \mu)} = \frac{\omega^2}{c_p^2}\,, \tag{5.17}$$

where the squared phase velocity is

$$c_p^2 = \frac{4\mu(\lambda + \mu)}{\rho(\lambda + 2\mu)} = \frac{E}{\rho(1 - \nu^2)} = \frac{E_{ef}}{\rho} \tag{5.18}$$

and E_{ef} is the effective Young's modulus for the bar when lateral displacements in one direction are forbidden (1.6).

Keeping in mind that $|k_3x_3| \ll 1$ and $|\varkappa_3x_3| \ll 1$ we obtain from (5.7, 8) for the particle displacement in wave under consideration

$$u_1 = iC_1\sqrt{\varkappa^2 - k^2}\,, \quad u_3 = -C_1(k^2 - \varkappa^2/2)x_3\,, \quad |u_3| \ll |u_1|\,. \tag{5.19}$$

Displacements are mainly longitudinal and constant across the plate. That is why this symmetric wave at low frequencies is often called "*Young's*" *longitudinal wave* in a plate. Because the phase velocity is independent of frequency (c_p is a material constant), this wave propagates without dispersion.

We now proceed to the asymmetric wave at low frequencies. In the expansion of the dispersion equation (5.8) for $|\sigma_l| \ll 1$ and $|\sigma_t| \ll 1$, the terms of higher order should be kept, otherwise ξ^2 drops out. Setting $\tan\sigma_l \simeq \sigma_l + \sigma_l^3/3$ and analogously for $\tan\sigma_t$, we rewrite (5.8) as

$$\sigma_t^2 + p^2h^2[1 + (\sigma_l^2 - \sigma_t^2)/3] = 0\,.$$

It follows from this that $\sigma_t^2 < 0$ and hence $\xi^2 > \varkappa^2 > k^2$, i.e., we have transverse as well as longitudinal inhomogeneous waves across the plate. Solving the last equation with respect to ξ^2 (retaining only the linear terms $\sqrt{\varkappa^2 - k^2h}$), we obtain the dispersion relation

$$\xi^2 = \pm(\sqrt{3}/2h)\varkappa^2(\varkappa^2 - k^2)^{-1/2}\,. \tag{5.20}$$

Negative ξ^2, corresponding to nonpropagating waves, is of no interest to us. The propagating mode with $\xi^2 > 0$ is called a *bending wave* in a thin plate. One can see from (5.10), since k_3x_3 and $\varkappa_3x_3$ are small, that the particle displacements occur mostly in a transverse direction and are independent of x_3, as in Young's wave.

A bending wave has a dispersion and its phase and group velocities are, respectively,

$$\begin{aligned} c_{ph} &= \omega/\xi = (4/3)^{1/4}(\varkappa h)^{1/2}(1 - c_t^2/c_l^2)^{1/4}c_t\,, \\ c_g &= d\omega/d\xi = 2(4/3)^{1/4}(\varkappa h)^{1/2}(1 - c_t^2/c_l^2)^{1/4}c_t = 2c_{ph}\,. \end{aligned} \tag{5.21}$$

5.3 Equations Describing the Bending of a Thin Plate

5.3.1 Thin-Plate Approximation

Bending waves are rather important. However, their separation from the whole set of modes in plates often leads to unnecessary difficulties. It is possible, however, to obtain an approximate equation for the plate bending if the condition that the plate is thin is taken into account at the very beginning.

Sophie Germain was the first to obtain such an equation in 1815. Later, Kirchhoff developed a theory of plate bending under the following assumptions:

i) the displacement of any point of the plate is small compared with its thickness. The slopes of the plate boundaries are to be small, too;
ii) the middle plane undergoes neither extension nor contraction;
iii) linear elements perpendicular to the middle plane prior to deformation remain straight and perpendicular to it after deformation;
iv) forces of internal stresses in the plate perpendicular to the middle plane are small as compared with the forces acting in this plane.

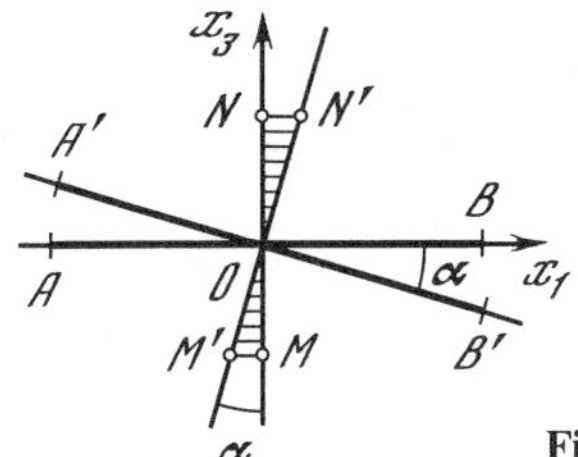

Fig. 5.4. Illustrating the theory of thin-plate bending.

On the basis of these assumptions, we obtain an approximate equation for the middle-plane displacement $\zeta(x_1, x_2, t) = u_3(x_1, x_2, 0, t)$. Taking into account that, according to (ii), $u_1|_{x_3=0} = u_2|_{x_3=0} = 0$, we expand $u_1(x_3)$ in a series and confine ourselves to terms linear with respect to the x_3: $u_1 = \alpha x_3$ where $\alpha \simeq (\partial u_1/\partial x_3)_{x_3=0}$. In Fig. 5.4, AB is the position of a small element of the middle plane prior to deformation; A'B' is the same after deformation and MN and M'N' are similar positions of a linear element originally perpendicular to the middle plane. Obviously, $\alpha \simeq \partial u_1/\partial x_3$ is the deviation angle of a linear element from its original vertical position. According to (iii), this element must remain perpendicular to the middle plane whose slope is $-\partial\zeta/\partial x_1$, then $\alpha = -\partial\zeta/\partial x_1$ and $u_1 \simeq -x_3\partial\zeta/\partial x_1$. Similarly, we obtain $u_2 \simeq -x_3\,\partial\zeta/\partial x_2$. Now the following components of the strain tensor can be easily found:

$$\begin{aligned} e_{11} &= \partial u_1/\partial x_1 = -x_3\partial^2\zeta/\partial x_1^2, \\ e_{22} &= -x_3\,\partial^2\zeta/\partial x_2^2, \quad e_{12} = -x_3\,\partial^2\zeta/\partial x_1\,\partial x_2 . \end{aligned} \tag{5.22}$$

Applying the generalized Hooke's law we write

$$\sigma_{33} = (\lambda + 2\mu)e_{33} + \lambda(e_{11} + e_{22}) \quad \text{or} \quad e_{33} = -\frac{\lambda}{\lambda + 2\mu}(e_{11} + e_{22}) + \frac{\sigma_{33}}{\lambda + 2\mu} .$$

Using this, we also find from (3.22)

$$\sigma_{11} = (\lambda + 2\mu)e_{11} + \lambda(e_{22} + e_{33}) = \frac{2\mu}{\lambda + 2\mu}[2(\lambda + \mu)e_{11} + \lambda e_{22}] + \frac{\lambda}{\lambda + 2\mu}\sigma_{33}.$$

But by (iv) $|\lambda\sigma_{33}/(\lambda + 2\mu)| < \sigma_{33}| \ll \sigma_{11}|$; hence, neglecting the term with σ_{33} and taking (5.22), into account we obtain

$$\sigma_{11} = -\frac{2\mu x_3}{\lambda + 2\mu}\left[2(\lambda + \mu)\frac{\partial^2 \xi}{\partial x_1^2} + \lambda\frac{\partial^2 \xi}{\partial x_2^2}\right] = -\frac{E x_3}{1 - \nu^2}\left(\frac{\partial^2 \xi}{\partial x_1^2} + \nu\frac{\partial^2 \xi}{\partial x_2^2}\right), \quad (5.23)$$

where the relation (3.26) of λ and μ to Young's modulus E and the Poisson coefficient ν has been used. In a similar way one finds

$$\sigma_{22} = -\frac{E x_3}{1 - \nu^2}\left(\frac{\partial^2 \zeta}{\partial x_2^2} + \nu\frac{\partial^2 \zeta}{\partial x_1^2}\right), \qquad \sigma_{12} = 2\mu e_{12} = -2\mu x_3\frac{\partial^2 \xi}{\partial x_1 \partial x_2}. \quad (5.24)$$

The remaining components of the stress tensor will be found from the conditions of equilibrium of the elastic element (3.12) when the volume force is absent: $\partial\sigma_{i1}/\partial x_1 + \partial\sigma_{i2}/\partial x_2 + \partial\sigma_{i3}/\partial x_3 = 0$. For $i = 1, 2$, we obtain with (5.23, 24),

$$\partial\sigma_{i3}/\partial x_3 = E x_3(1 - \nu^2)^{-1}\,\partial\Delta_{-}\zeta/\partial x_i,$$

where $\Delta_{-} = \partial^2/\partial x_1^2 + \partial^2/\partial x_2^2$ is the Laplace operator with respect to the horizontal variables. Integrating the latter equation over x_3 with $\sigma_{i3}|_{x_3 = \pm h} = 0$, we obtain

$$\sigma_{i3} = \frac{E(h^2 - x_3^2)}{2(1 - \nu^2)}\frac{\partial}{\partial x_i}\Delta_{-}\zeta, \qquad i = 1, 2. \quad (5.25)$$

5.3.2 Sophie Germain Equation

We are now in a position to determine transverse forces accompanying the plate deformation. From the equilibrium equation $\partial\sigma_{3k}/\partial x_k = 0$, we have

$$\frac{\partial\sigma_{33}}{\partial x_3} = -\frac{\partial\sigma_{31}}{\partial x_1} - \frac{\partial\sigma_{32}}{\partial x_2} = \frac{E(h^2 - x_3^2)}{2(1 - \nu^2)}\Delta_{-}^2\zeta.$$

Integrating this equation over x_3 in the limits $(-h, h)$, we obtain the difference between the normal stresses at the plate boundaries $q \equiv \sigma_{33}|_{x_3 = h} - \sigma_{33}|_{x_3 = -h}$:

$$q = \frac{2Eh^3}{3(1 - \nu^2)}\Delta_{-}^2\zeta = D\Delta_{-}^2\zeta. \quad (5.26)$$

The quantity

$$D = 2Eh^3/3(1 - \nu^2) \tag{5.27}$$

is called the *flexural rigidity* or *cylindric rigidity* of a plate and q is the external force normal to the plate per unit area. Equation (5.26) describes the equilibrium of the plate when the distributed external force q is applied. It is called the *Sophie Germain equation.*

We determine the bending moments accompanying the plate deformation. In view of (5.23) we obtain for the moment along the x_2-axis per unit area:

$$M_2 = \int_{-h}^{h} \sigma_{11} x_3\, dx_3 = -E(1 - \nu^2)^{-1}[(\partial^2\zeta/\partial x_1^2)$$

$$+ \nu(\partial^2\zeta/\partial x_2^2)] \int_{-h}^{h} x_3^2\, dx_3 = -D[(\partial^2\zeta/\partial x_1^2) + \nu(\partial^2\zeta/\partial x_2^2)]\,. \tag{5.28}$$

A similar expression with an interchange of indices $1 \leftrightarrows 2$ holds for the moment along the x_1-axis.

5.3.3 Bending Waves in a Thin Plate

In the case of bending vibrations of a plate or simply bending waves, the inertia of the plate per unit area taken with an opposite sign $-2\rho h\, \partial^2\zeta/\partial t^2$ is equal to the distributed force q in (5.26). Hence, the equation describing the propagation of bending waves in a plate has the form

$$\frac{\partial^2\zeta}{\partial t^2} + \frac{D}{2\rho h} \Delta_-^2 \zeta = 0\,. \tag{5.29}$$

In the case of a one-dimensional wave, choosing the axis x_1 along the propagation direction we obtain

$$\frac{\partial^2\zeta}{\partial t^2} + \frac{D}{2\rho h}\frac{\partial^4\zeta}{\partial x_1^4} = 0\,. \tag{5.30}$$

This equation differs from that for bending waves in a rod, (2.25), only by the coefficient in the second term. Hence, all the results obtained for the rod are valid in the case of one-dimensional waves in a plate. In particular, the dispersion relation for the bending waves in a plate is

$$\xi^2 = \pm(2\rho h/D)^{1/2}\omega\,. \tag{5.31}$$

Note that the latter coincides with (5.20) for asymmetric modes in a thin plate (Exercise 5.4.7). Taking into account what was said before obtaining (5.20), we conclude that the plate's thickness h must be small compared with the bending wavelength for the applicability of the theory of bending waves developed in this sub-section.

5.4 Exercises

5.4.1. Find the dispersion relation and the eigenfunctions of modes in a plate at $\omega \to \infty$.

Solution: Apply the dispersion equation for symmetric modes (5.5), for example. Excluding the superfluous modes considered in Sect. 5.2, note that σ_t must remain real and finite as $\omega \to \infty$, say a_n:

$$\sigma_t = h\sqrt{\varkappa^2 - \xi_n^2} = h\omega\sqrt{c_t^{-2} - c_{ph}^{-2}} = a_n .$$

Hence, $c_{ph}^{-2} = c_t^{-2} - a_n^2/h^2\omega^2 \to c_t^{-2}$. The phase velocity of modes is close to c_t and $\xi_n \simeq \varkappa$, $p \simeq \varkappa/2$, $\sigma_l \simeq h(k^2 - \varkappa^2)^{1/2} = \mathrm{i}|\sigma_l|$, $|\sigma_l| \to \infty$, $\tan\sigma_l = \mathrm{i}\tanh|\sigma_l| = \mathrm{i}$. Equation (5.5) becomes

$$\tan a_n \simeq 4a_n|\sigma_l|/(\varkappa h)^2 = 4a_n(1 - c_t^2/c_l^2)^{1/2}/\varkappa h \to 0 .$$

Its solution is $\sigma_t = a_n = n\pi$. A similar result is obtained for asymmetric modes with the only difference that $\sigma_t \simeq (n + 1/2)\pi$. Thus, for $\omega \to \infty$, the modes considered are the sum of two transverse plane waves propagating at small angles to the horizontal and inhomogeneous longitudinal waves in close vicinity of the boundaries of the plate.

5.4.2. Determine the frequencies at which modes in a plate are just pairs of transverse waves propagating at a 45° angle. Determine also the particle displacements in these modes.

Solution: For the incidence angle $\gamma = 45°$, we have $\xi = \varkappa\sin\gamma = \varkappa/\sqrt{2}$, $\sigma_t = \varkappa h/\sqrt{2} = \omega h/\sqrt{2}c_t$, $p = (\xi^2 - \varkappa^2/2)\xi^{-1} = 0$ and in (5.2) $C_1 = C_2 = 0$ according to the assumptions. Now the dispersion relations (5.5, 8) and the expressions for displacements (5.7, 10) become for a symmetric wave

$$\cos\sigma_t = 0 , \quad \varkappa h = (2n-1)\pi/\sqrt{2} , \quad \omega_n = (2n-1)\pi c_t/\sqrt{2}h ,$$

$$u_1 = -\varkappa_3 D_2 \cos\left(\frac{2n-1}{\sqrt{2}}\pi\frac{x_3}{h}\right), \quad u_3 = \mathrm{i}\xi D_2 \sin\left(\frac{2n-1}{\sqrt{2}}\pi\frac{x_3}{h}\right),$$

and for an asymmetric wave

$$\sin\sigma_t = 0 , \quad \varkappa h = \sqrt{2}n\pi , \quad \omega_n = \sqrt{2}n\pi c_t/h ,$$

$$u_1 = \varkappa_3 D_1 \sin\left(\sqrt{2}n\pi\frac{x_3}{h}\right), \quad u_3 = \mathrm{i}\xi D_1 \cos\left(\sqrt{2}n\pi\frac{x_3}{h}\right).$$

5.4.3. In Sect. 5.1.4 modes were described, which are formed by plane waves with such incidence angles θ_0 and γ_0 that the longitudinal wave transforms completely into a transverse one, and vice versa. Determine the frequencies ω_n corresponding to these modes.

Solution: Taking into account the relation $\sigma_l \sigma_t = p^2 h^2$ which holds for the incidence angles θ_0 and γ_0, see (4.31), note that the dispersion equations (5.5, 8) coincide in this case:

$$\tan \sigma_t + \tan \sigma_l = \frac{\sin(\sigma_l + \sigma_t)}{\cos \sigma_l \cos \sigma_t} = 0 .$$

Hence, $\sigma_l + \sigma_t = n\pi$ and with $\sigma_l = (\omega h/c_l) \cos \theta_0$ and $\sigma_t = (\omega h/c_t) \cos \gamma_0$, we obtain

$$\omega_n = \frac{n\pi}{h} \frac{c_l c_t}{c_l \cos \gamma_0 + c_t \cos \theta_0} .$$

5.4.4. A thin plate rests on parallel-line supports a distance l apart. Determine the shape of the plate under the action of its own weight.

Solution: With the plate thickness $2h$ and the material density ρ, a transverse force $q = -2\rho h g$ acts per unit area. Choose the x_1-axis perpendicular to the supports and write the Sophie Germain equation

$$d^4\zeta/dx_1^4 = -\alpha = -2\rho g h/D .$$

Integrating it we find

$$\zeta = -\alpha x_1^4/24 + a x_1^3 + b x_1^2 + c x_1 + d .$$

The integration constants can be determined from the boundary conditions, i.e. vanishing displacements and moments at the supports $x_1 = 0, l$. Taking (5.28) into account, we obtain

$$d = 0 , \quad b = 0 , \quad a l^3 + c l = \alpha l^4/24 , \quad 6 a l = \alpha l^2/2 ,$$

whence

$$a = \alpha l/12 , \quad c = -\alpha l^2/24 , \quad \zeta = -\alpha x_1(x_1^3 - 2 l x_1^2 + l^3)/24 .$$

5.4.5. Determine the shape of a thin circular plate with clamped edges under the action of its own weight.

Solution: We have $q = -2\rho h g$ as in the previous exercise. Due to the radial symmetry the function $\zeta(x_1, x_2)$ depends upon $r = \sqrt{x_1^2 + x_2^2}$ only. The Sophie

Germain equation in polar coordinates assumes the form

$$r^{-1}\frac{d}{dr}\left\{r\frac{d}{dr}\left[r^{-1}\frac{d}{dr}\left(r\frac{d\zeta}{dr}\right)\right]\right\} = -\alpha = -\frac{2\rho g h}{D}.$$

Integrating it twice over r yields

$$\Delta_{-}\zeta = r^{-1}\frac{d}{dr}\left(r\frac{d\zeta}{dr}\right) = -\frac{\alpha r^2}{4} + a\ln\frac{r}{R} + b.$$

The constant a must be zero, otherwise σ_{i3} is infinite according to (5.25). Integrating twice more over r, we find $\zeta = -\alpha r^4/64 + br^2/4 + c\ln(r/R) + d$. Here again $c = 0$ since $\zeta(0)$ must be finite. Also, for clamped edges we have $\zeta(R) = d\zeta(R)/dr = 0$ which yields

$$-\alpha R^3/16 + bR/2 = 0, \qquad -\alpha R^4/64 + bR^2/4 + d = 0$$

whence we find

$$b = \alpha R^2/8, \qquad d = -\alpha R^4/64,$$

$$\zeta = \alpha(r^4 - 2R^2r^2 + R^4)/64 = -\alpha(R^2 - r^2)^2/64.$$

5.4.6. Keep the statement of Exercise 5.4.5 but assume the plate to be weightless and that the force F_0 is applied at its center.

Solution: The reaction force $-F_0/2\pi R$ acts on the support of the plate at $r = R$ per unit length. Consequently, a tangent stress σ_{r3} must act at the edge of plate so that $\int_{-h}^{h} \sigma_{r3}\,dx_3 = -F_0/2\pi R$. Taking (5.25) into account we obtain

$$\frac{F_0}{2\pi R} = \frac{E}{2(1-\nu^2)}\left.\frac{d\Delta_{-}\zeta}{dr}\right|_{r=R}\int_{-h}^{h}(h^2 - x_3^2)\,dx_3 = D\left.\frac{d\Delta_{-}\zeta}{dr}\right|_{r=R}.$$

Everywhere, except the point $r = 0$, $\Delta_{-}^2\zeta = 0$ holds, (5.26), or in the case of cylindrical symmetry, $d(r\,d\Delta_{-}\zeta/dr)/dr = 0$. Hence, $d\Delta_{-}\zeta/dr = ar^{-1}$. In particular,

$$D\left.\frac{d\Delta_{-}\zeta}{dr}\right|_{r=R} = \frac{aD}{R} = \frac{F_0}{2\pi R} \quad \text{or} \quad a = \frac{F_0}{2\pi D}.$$

Integrating again over r, we obtain

$$\zeta = (F_0/8\pi D)r^2\ln(r/R) + br^2 + c\ln(r/R) + d.$$

Since $\zeta(0)$ is finite, we have $c = 0$. The conditions $\zeta(R) = d\zeta(R)/dr = 0$ yield $bR^2 + d = 0$, $2bR + F_0R/8\pi D = 0$ whence

$$b = -F_0/16\pi D, \qquad d = F_0R^2/16\pi D, \qquad \zeta = F_0[2r^2\ln(r/R) + R^2 - r^2]/16\pi D.$$

5.4.7. Show that the dispersion equation (5.31) for bending waves in a thin plate coincides with (5.20) for an asymmetric mode at a low frequency.

Solution: Taking into account the relations $\varkappa = \omega/c_t$, $k = \omega/c_l$, $c_t^2 = \mu/\rho$, $c_l^2 = (\lambda + 2\mu)/\rho$, write (5.20) as

$$\xi^2 = \pm\frac{\sqrt{3}}{2h}\frac{c_l}{c_t}\frac{\omega}{\sqrt{c_l^2 - c_t^2}} = \pm\frac{\sqrt{3}}{2h}\,\omega\sqrt{\frac{\rho(\lambda + 2\mu)}{\mu(\lambda + \mu)}}.$$

Then express λ and μ in terms of E and ν according to (3.26). As a result, we obtain

$$\xi^2 = \pm\frac{\sqrt{3}}{2h}\,\omega\sqrt{\frac{\rho(1 - \nu^2)}{E}} = \pm\omega\sqrt{\frac{2\rho h}{D}}$$

coinciding with (5.31).

5.4.8. Consider the reflection of a harmonic bending wave at a supported edge of a thin plate $x_1 = 0$.

Solution: Substitution of the expression for harmonic waves

$$\zeta = \exp[\mathrm{i}(\xi x_1 + \eta x_2 - \omega t)]$$

into (5.29) yields the relationship between ξ, η and ω: $(\xi^2 + \eta^2)^2 = k^4$, where $k^4 = 2\rho h\omega^2 D^{-1}$. Hence, we have the four possible values of ξ:

$$\xi_1 = k_1 = \sqrt{k^2 - \eta^2}, \quad \xi_2 = -k_1, \quad \xi_3 = \mathrm{i}\varkappa_1 = \mathrm{i}\sqrt{k^2 + \eta^2}, \quad \xi_4 = -\mathrm{i}\varkappa_1.$$

The general expression for the displacement at $x_1 < 0$ can now be written as the factor $\exp[\mathrm{i}(\eta x_2 - \omega t)]$ being omitted:

$$\zeta = A[\exp(\mathrm{i}k_1 x_1) + V\exp(-\mathrm{i}k_1 x_1) + W\exp(\varkappa_1 x_1)].$$

The first term represents the incident wave, and the second one the reflected wave. The term $\exp(-\varkappa_1 x_1)$ was excluded since it tends to infinity at $x_1 \to -\infty$. The boundary conditions at the plate edge are

$$\zeta|_{x_1=0} = 0, \quad M_2|_{x_1=0} = -D\left(\frac{\partial^2\zeta}{\partial x_1^2} + \nu\frac{\partial^2\zeta}{\partial x_2^2}\right)_{x_1=0} = 0.$$

Substitution of ζ yields $1 + V + W = 0$, $-k_1^2(1 + V) + \varkappa_1^2 W = 0$ and finally $V = -1$, $W = 0$.

5.4.9. The same as Exercise 5.4.8 but the plate edge is assumed to be free.

Solution: According to the boundary conditions, the tangential stress and moment are zero at $x_1 = 0$ in this case:

$$\sigma_{13}|_{x_1=0} \sim \frac{\partial}{\partial x_1} \Delta_- \zeta|_{x_1=0} = 0, \qquad M_2|_{x_1=0} \sim \left(\frac{\partial^2 \zeta}{\partial x_1^2} + \nu \frac{\partial^2 \zeta}{\partial x_2^2}\right)_{x_1=0} = 0.$$

In the same way as in the previous problem, we obtain equations for the coefficients V and W:

$$-\mathrm{i}k_1(1 - V) + \varkappa_1 W = 0, \qquad -(k_1^2 + \nu\eta^2)(1 + V) + (\varkappa_1^2 - \nu\eta^2)W = 0.$$

Their solutions are:

$$W = \frac{2\mathrm{i}\sqrt{k^2 - \eta^2}[k^2 - (1 - \nu)\eta^2]}{\mathrm{i}\sqrt{k^2 - \eta^2}[k^2 + (1 - \nu)\eta^2] + \sqrt{k^2 + \eta^2}[k^2 - (1 - \nu)\eta^2]},$$

$$V = \frac{\mathrm{i}\sqrt{k^2 - \eta^2}[k^2 + (1 - \nu)\eta^2] - \sqrt{k^2 + \eta^2}[k^2 - (1 - \nu)\eta^2]}{\mathrm{i}\sqrt{k^2 - \eta^2}[k^2 + (1 - \nu)\eta^2] + \sqrt{k^2 + \eta^2}[k^2 - (1 - \nu)\eta^2]}, \qquad |V| = 1.$$

5.4.10. Show that a wave can propagate along a free edge of a plate, the disturbances differing from zero only near the edge. Find the phase velocity of this wave.

Solution: Suppose that the incident wave in the previous problem is an inhomogeneous one: $\eta > k$, $k_1 = -\mathrm{i}\alpha$, $\alpha = \sqrt{\eta^2 - k^2}$. A reflected wave $V\exp(-\mathrm{i}k_1 x_1)$ would increase exponentially for $x_1 \to -\infty$. Hence, we set $V = 0$ or

$$\sqrt{\eta^2 - k^2}\,[k^2 + (1 - \nu)\eta^2] = \sqrt{k^2 + \eta^2}\,[k^2 - (1 - \nu)\eta^2].$$

Squaring both sides of this equality, we easily find

$$\eta^4 = k^4(1 - \nu^2)^{-1} = 3\rho\omega^2/Eh^2$$

whence for the phase velocity of the wave we have $c_{\text{ph}} = \omega/\eta = \eta h\sqrt{E/3\rho}$. Note that the formula for W in the previous problem gives

$$W = \sqrt{\frac{\eta^2 - k^2}{\eta^2 + k^2}} = \frac{1 - \sqrt{1 - \nu^2}}{\nu}.$$

Hence, the general expression for ζ (Exercise 5.4.8) yields in this case (if the factor $\exp[\mathrm{i}(\eta x_2 - \omega t)]$ is also taken into account):

$$\zeta = A[\exp(\sqrt{\eta^2 - k^2}x_1) + \sqrt{\frac{\eta^2 - k^2}{\eta^2 + k^2}}\exp(\sqrt{\eta^2 + k^2}x_1)\exp[\mathrm{i}(\eta x_2 - \omega t)].$$

Part II Fluid Mechanics

The dynamics of a fluid differ from that of elastic bodies by some important features:

i) an ideal (inviscid) fluid does not support shearing stresses;
ii) displacements of particles in a fluid flow may be large even when forces are small;
iii) viscosity (internal friction) fundamentally influences the fluid flow.

The first point simplifies the equation of motion of a fluid compared with that of an elastic body, while the two others make it more complicated. Complication greatly exceeds simplification and, as a result, fluid dynamics appears to be a rather sophisticated science. Some of its aspects, for example, the theory of turbulence, are not completely developed yet despite the fact that the basic principles of fluid dynamics were already established by Euler, Bernoulli and Lagrange.

6. Basic Laws of Ideal Fluid Dynamics

We begin with ideal fluids, i.e. with those without internal friction, and hence without the transformation of mechanical energy into heat. Moreover, we assume that heat transfer between volumes of the fluid can also be neglected. Hence, the entropy of a material volume of the fluid is assumed to be constant and stresses in the fluid can be specified in terms of only one scalar quantity, namely the pressure p.

6.1 Kinematics of Fluids

6.1.1 Eulerian and Lagrangian Representations of Fluid Motion

Fluid motion is determined completely if its velocity $\boldsymbol{v}$, pressure p, density ρ, temperature and other possible parameters are specified as functions of coordinates and time. This is an *Eulerian* way of representing fluid motion. Here we fix our attention a point in space and follow the time variation of $\boldsymbol{v}$, p, etc., at this point. Alternatively, if we fix time we can determine the spatial variation of the same parameters. However, there is no direct information about the motion of a given "fluid particle" in this representation. In a *Lagrangian* representation, on the other hand, all parameters including the coordinates of the fluid's particles x_i $(i = 1, 2, 3)$ are specified in terms of time t and some vector $\{\xi_1, \xi_2, \xi_3\}$ which identifies a given particle of the fluid: $x_i = x_i(\xi_k, t)$, $p = p(\xi_k, t)$, $\rho = \rho(\xi_k, t)$, etc. The variables (ξ_1, ξ_2, ξ_3) are usually assumed to be the coordinates of a fluid particle at some time t_0 (initial coordinates), i.e. $\xi_i = x_i(\xi_k, t_0)$. Therefore, in a Lagrangian representation, our attention is fixed on a chosen fluid particle and we trace this particle in its motion specifying the variation of $\boldsymbol{v}$, p, ρ, etc., in its vicinity.

The choice of representation depends on the problem's character and is mostly a question of convenience. Thus, for example, most devices measure the fluid parameters and its velocity at a fixed point, i.e. they supply an Eulerian information. On the other hand, Lagrangian information can be obtained if we label the fluid element with dye and observe the spread of this dye. The Lagrangian representation is also preferable for describing diffusion processes connected directly with the transfer of matter.

6.1.2 Transition from One Representation to Another

The relationship between two representations is, in general, rather intricate. For example, if the pressure $p_L(\xi_k, t)$ is known in a Lagrangian representation and we want to determine this quantity $p_E(x_k, t)$ at the point $\{x_i\}$ (Eulerian representation), we have to know the initial coordinates $\{\xi_k\}$ of the fluid particle at this point at time t. In other words, we need to solve the system of equations $x_i = x_i(\xi_k, t), i = 1, 2, 3$ for ξ_k leading to $p_E(x_k, t) = p_L[\xi_i(x_k, t), t]$. It is well known that the solution $\xi_i(x_k, t)$ exists if the Jacobian of the transformation

$$J = \frac{\partial(x_1, x_2, x_3)}{\partial(\xi_1, \xi_2, \xi_3)} = \det\left\{\frac{\partial x_i}{\partial \xi_k}\right\} \tag{6.1}$$

is not zero.

In some cases the time rate of J is of interest. It is shown in Exercise 6.7.1 that

$$\frac{1}{J}\frac{dJ}{dt} = \frac{\partial v_k}{\partial x_k} = \operatorname{div} \boldsymbol{v}. \tag{6.2}$$

6.1.3 Convected and Local Time Derivatives

In a Lagrangian representation, the partial time derivative of some function is its change of rate for a fixed fluid particle. In particular, $v_i = \partial x_i/\partial t$ and $\partial v_i/\partial t$ are the velocity and acceleration of a particle, respectively. On the other hand, in a Eulerian representation the partial derivatives with respect to x_i and t are the component of the gradient along the x_i-axis and the time rate of the corresponding field quantity, respectively. Therefore, the derivative $\partial \boldsymbol{v}(x_k, t)/\partial t$ is not the acceleration of the particle which at time t is at point $\{x_k\}$, because another particle will be at this very point at time $t + \Delta t$. In a waterfall, for example, the velocity at a fixed point is constant $[\partial \boldsymbol{v}(x_k, t)/\partial t = 0]$ but particles are accelerated by the action of the gravitational force.

We now obtain an expression for the acceleration in a Eulerian representation. Consider the fluid particle which is at the point $\{x_k\}$ at time t. The same particle will be at the point $\{x_k + \Delta x_k\}$ at time $t + \Delta t$. Hence, the ith component of acceleration is

$$\lim_{\Delta t \to 0} \frac{v_i(x_k + \Delta x_k, t + \Delta t) - v_i(x_k, t)}{\Delta t} = \lim_{\Delta t \to 0} \frac{v_i(x_k, t) + \dfrac{\partial v_i}{\partial x_k}\Delta x_k + \dfrac{\partial v_i}{\partial t}\Delta t - v_i(x_k, t)}{\Delta t}.$$

Taking into account that

$$\lim_{\Delta t \to 0} \frac{\Delta x_k}{\Delta t} = v_k,$$

we obtain

$$a_i = \frac{\partial v_i}{\partial t} + v_k \frac{\partial v_i}{\partial x_k} = \left(\frac{\partial}{\partial t} + v_k \frac{\partial}{\partial x_k}\right) v_i \tag{6.3}$$

or, in vector form,

$$a = \frac{\partial v}{\partial t} + (v \cdot \nabla)v = \left(\frac{\partial}{\partial t} + v \cdot \nabla\right)v. \tag{6.3a}$$

Analogously one can find the time rate of any other vector or scalar field quantity f connected with the given particle. Such a derivative is called a *convected* or a *material* derivative; it is also the *derivative along the trajectory.* It is denoted by df/dt (sometimes the notation Df/Dt is used, too). The partial derivative $\partial f/\partial t$ in the Eulerian representation is also called a *local* one. In our notation, the relationship between convected and local derivatives is

$$\frac{d}{dt} = \frac{\partial}{\partial t} + v \cdot \nabla. \tag{6.4}$$

Henceforth we will mostly use the Eulerian representation.

The convected derivative of an integral extended over volume V moving with the fluid and hence containing the same particle all the time is of importance for some purposes:

$$I = \int_V F(x_i, t)\, dV, \qquad \frac{dI}{dt} = \frac{d}{dt}\int_V F\, dV,$$

where F is an arbitrary vector (or scalar) function. We perform integration using the Lagrangian variables $x_i = x_i(\xi_k, t)$. In these variables the integration will be extended over a fixed volume V_0 (position of V at $t = 0$) at any time. Hence, the operations d/dt and the integration commute and

$$\frac{dI}{dt} = \frac{d}{dt}\int_{V_0} FJ\, d\xi_1\, d\xi_2\, d\xi_3 = \int_{V_0} \frac{d}{dt}(FJ)\, d\xi_1\, d\xi_2\, d\xi_3, \tag{6.5}$$

where J represents the Jacobian (6.1). Returning in (6.5) to the Eulerian variables (inverse transformation), we have

$$\frac{dI}{dt} = \int_V \frac{1}{J}\frac{d(FJ)}{dt}\, dV = \int_V \left(\frac{dF}{dt} + \frac{F}{J}\frac{dJ}{dt}\right) dV.$$

Using also (6.2, 4), we obtain finally

$$\frac{dI}{dt} = \frac{d}{dt}\int_V F\, dV = \int_V \left[\frac{\partial F}{\partial t} + (v \cdot \nabla)F + F(\nabla v)\right] dV. \tag{6.6}$$

6.2 System of Equations of Hydrodynamics

6.2.1 Equation of Continuity

One of the fundamental equations of fluid motion is the *equation of continuity* or the *conservation law of matter*. It follows from the evident fact that the mass of a liquid in a volume containing the same particles all the time is constant. This implies that:

$$\frac{d}{dt}\int_V \rho\, dV = 0. \tag{6.7}$$

Using (6.4, 6), we rewrite (6.7) as

$$\int_V \left(\frac{\partial \rho}{\partial t} + \boldsymbol{v}\nabla\rho + \rho\nabla\boldsymbol{v}\right) dV = 0. \tag{6.8}$$

Since (6.8) must be true for an arbitrary V, the integrand must vanish everywhere. As a result, we obtain the continuity equation in the Eulerian variables, which can be written in one of three forms:

$$\frac{\partial \rho}{\partial t} + \boldsymbol{v}\nabla\rho + \rho\nabla\boldsymbol{v} = 0 \quad \text{or} \quad \frac{\partial \rho}{\partial t} + \nabla(\rho\boldsymbol{v}) = 0 \quad \text{or} \quad \frac{d\rho}{dt} + \rho\nabla\boldsymbol{v} = 0. \tag{6.9}$$

The integral form of the continuity equation (6.8) can be transformed to show its physical meaning more clearly. For this purpose we use the Gauss divergence theorem to transform the volume integral of $\boldsymbol{v}\nabla\rho + \rho\nabla\boldsymbol{v} = \nabla(\rho\boldsymbol{v})$ into an integral extended over the bounding surface S:

$$\int_V \nabla(\rho\boldsymbol{v})\, dV = \int_S \rho\boldsymbol{v}\cdot\boldsymbol{n}\, dS,$$

where $\boldsymbol{n}$ is the outward normal on S. Then we obtain from (6.8),

$$\frac{\partial}{\partial t}\int_V \rho\, dV = -\int_S \rho\boldsymbol{v}\cdot\boldsymbol{n}\, dS; \tag{6.10}$$

thus the time rate of the fluid mass inside a fixed volume V is equal (with opposite sign) to the mass of the fluid flowing out of the volume across the bounding surface. Note that this principle and (6.10) can be chosen as the point of departure for obtaining (6.8) with the help of the inverse transformation.

6.2.2 The Euler Equation

The equation of fluid motion or the *Euler* equation can be obtained from the Newton's second law which states that the time rate of momentum of a volume of fluid is equal to the resultant force acting on this volume, or in mathematical form,

$$\frac{d}{dt}\int_V \rho \boldsymbol{v}\, dV = \boldsymbol{F} + \boldsymbol{F}_S, \tag{6.11}$$

where $\boldsymbol{F} = \int_V \rho \boldsymbol{f}\, dV$ is the external *body force* ($\boldsymbol{f}$ being the force per unit mass), $\boldsymbol{F}_S$ is the force exerted by the surrounding fluid on the volume V. This force is equal (with opposite sign) the force with which the volume acts on the surrounding medium or for an ideal fluid (no viscous forces), $\boldsymbol{F}_S = -\int_S p\boldsymbol{n}\, dS$, where p is the pressure, S the surface bounding V and $\boldsymbol{n}$ the outward normal to S. According to the vector divergence theorem we have

$$\boldsymbol{F}_S = -\int_S p\boldsymbol{n}\, dS = -\int_V \nabla p\, dV. \tag{6.12}$$

Then (6.11) may be written as

$$\int_V \left[\frac{d(\rho \boldsymbol{v})}{dt} + \rho \boldsymbol{v}(\nabla \boldsymbol{v})\right] dV = \int_V (-\nabla p + \rho \boldsymbol{f})\, dV. \tag{6.13}$$

Using the continuity equation (6.9), the integrand on the left-hand side may be transformed into

$$\frac{d(\rho \boldsymbol{v})}{dt} + \rho \boldsymbol{v}(\nabla \boldsymbol{v}) = \rho \frac{d\boldsymbol{v}}{dt} + \boldsymbol{v}\left(\frac{d\rho}{dt} + \rho \nabla \boldsymbol{v}\right) = \rho \frac{d\boldsymbol{v}}{dt} = \rho \frac{\partial \boldsymbol{v}}{\partial t} + \rho(\boldsymbol{v} \cdot \nabla)\boldsymbol{v}.$$

Since the volume V is arbitrary, the latter must be equal to the integrand on the right-hand side of (6.13). This yields Euler's equation of motion:

$$\frac{d\boldsymbol{v}}{dt} = \frac{\partial \boldsymbol{v}}{\partial t} + (\boldsymbol{v} \cdot \nabla)\boldsymbol{v} = -\frac{\nabla p}{\rho} + \boldsymbol{f}. \tag{6.14}$$

6.2.3 Completeness of the System of Equations

The last equation together with the continuity equation (6.9) comprises four scalar equations for five scalar quantities (density ρ, pressure p and the three velocity components v_i). Hence, the system is not yet complete. The *equation of state*

$$p = p(\rho, s) \tag{6.15}$$

as well as the equation for the entropy s must be added to it. As was said above, we assume the entropy of a given element of fluid to be constant so that

$$\frac{ds}{dt} = \frac{\partial s}{\partial t} + \boldsymbol{v}\nabla s = 0. \tag{6.16}$$

Of course, it does not mean that entropy must remain constant at a fixed point in space because different fluid elements with different entropy density may occur at this point at different times.

The entropy s can be excluded from our equations. Applying for this purpose the operation of a convected derivative to (6.15) and denoting

$$c^2 = (\partial p/\partial \rho)_s, \tag{6.17}$$

we obtain

$$\frac{dp}{dt} = c^2 \frac{d\rho}{dt}, \tag{6.18}$$

where the *velocity of sound*, $c = c(p, \rho)$, is the function of pressure and density (Chap. 12).

In the simplest case where the entropy density is constant throughout space, it is the equation of state

$$p = p(\rho) \tag{6.19}$$

which makes the system (6.9, 14) complete. The function $p(\rho)$ depends on the properties of the fluid under consideration. In particular, for an ideal (in the thermodynamical sense) gas, the equation of state is $pV^{\gamma} = \text{const}$, $\gamma = c_p/c_V$ which is the ratio of specific heat at constant pressure to that at constant volume. Since $V = 1/\rho$ is the unit mass volume, we have $p = p_0(\rho/\rho_0)^{\gamma}$.

The system of equations (6.9, 14) and (6.18 or 6.19) together with the appropriate boundary conditions determines the motion of an ideal fluid completely. At a boundary with a rigid body at rest, the normal component of the velocity v_n must be zero. Its tangential component in the ideal fluid can be arbitrary since the shear stresses are absent.

The gravitational force $\boldsymbol{f} = \boldsymbol{g}$ ($\boldsymbol{g}$ being the gravity acceleration directed to the earth's centre, $g = 9.81 \text{ ms}^{-2}$) is a typical example of a body force in Euler's equation (6.14). This equation may also include forces of attraction of the Sun and Moon (specifically in the theory of tides). In the dynamics of the ocean and the atmosphere, the noninertial character of the reference frame caused by Earth's rotation is very often important. Additional inertial forces must be included in (6.14) in this case: the *centrifugal* force $\boldsymbol{f}_{\text{cf}} = \Omega^2 \boldsymbol{r} = \nabla(\Omega^2 r^2/2)$, $\boldsymbol{\Omega}$ being the vector of the Earth's rotational frequency, and the *Coriolis force* $\boldsymbol{f}_{\text{c}} = -2\boldsymbol{\Omega} \times \boldsymbol{v}$. Note that all the forces mentioned above, except the Coriolis one, are *potential*, i.e. they can be represented as a gradient of some "potential" function u: $\boldsymbol{f} = -\nabla u$.

The case of an *incompressible* fluid has very important applications. The density of each fluid particle must remain constant: $d\rho/dt = \partial\rho/\partial t + \boldsymbol{v}\nabla\rho = 0$. It does not imply, however, that $dp/dt = 0$ which could be deduced from (6.18), since c is infinite in an incompressible fluid. The continuity equation (6.9) can now be written as $\nabla \boldsymbol{v} = 0$. Hence, for an incompressible fluid we have two equations

$$\frac{\partial \rho}{\partial t} + \boldsymbol{v}\nabla\rho = 0, \qquad \nabla\boldsymbol{v} = 0 \tag{6.20}$$

in addition to Euler's equation (6.14), and the system of equations remains complete and closed.

6.3 The Statics of Fluids

6.3.1 Basic Equations

Consider now a fluid at rest. Assuming $\boldsymbol{v} = 0$ in (6.14), we obtain the *hydrostatic equation*

$$\nabla p = \rho \boldsymbol{f}. \tag{6.21}$$

The case where the force $\boldsymbol{f}$ is potential is important: $\boldsymbol{f} = -\nabla u$. Then (6.21) becomes

$$\nabla p = -\rho \nabla u. \tag{6.22}$$

A solution to (6.22) does not always exist. Indeed, we have the gradient on the left-hand side. The right-hand side, however, can be represented as a gradient of some function only in the cases of specific dependence of ρ on the coordinates. In fact, by applying the curl operation to (6.22), we obtain $\nabla\rho \times \nabla u = 0$. Hence, the vector $\nabla\rho$ must be collinear with ∇u.

Consider now a fluid in the gravitational field when $u = gz$; the z-axis is assumed to be directed upward. Equation (6.22) becomes

$$\frac{\partial p}{\partial x} = \frac{\partial p}{\partial y} = 0, \qquad \frac{\partial p}{\partial z} = -\rho g. \tag{6.23}$$

This implies that the pressure depends only on z: $p = p(z)$. In the simplest case the density ρ is constant, then

$$p = -\rho g z + \text{const}.$$

If we have $p = p_0$ at some $z = z_0$ as the boundary condition, then const $= p_0 + \rho g z_0$ and

$$p = p_0 - \rho g(z - z_0). \tag{6.24}$$

In the atmosphere the change of density ρ with height must be taken into account. We consider the case of an isothermal atmosphere, assuming that the equation of state is that of an ideal gas, i.e.,

$$p = \frac{R}{\mu} \rho T, \tag{6.25}$$

where R is the universal gas constant and μ the molecular weight of the gas. Substituting p from (6.25) into (6.23), we obtain

$$\frac{RT}{\mu} \frac{d\rho}{dz} = -g\rho, \quad \ln \frac{\rho}{\rho_0} = -\frac{\mu g}{RT} z,$$

where $\rho_0 = \rho(0)$. The last equation yields

$$\rho = \rho_0 \exp(-\mu g z/RT). \tag{6.26}$$

This is the well-known *barometric formula*. The density decreases exponentially with height. The corresponding formula for a non-isothermal atmosphere $[T = T(z)]$ is obtained in Exercise 9.4.6.

Analogously, the so-called *barotropic* case can be examined for $p = p(\rho)$ [for example, $p = p_0(\rho/\rho_0)^\gamma$ for an isentropic atmosphere].

6.3.2 Hydrostatic Equilibrium. Väisälä Frequency

We can write down the equilibrium condition for a fluid with the general equation of state $p = p(\rho, s)$ in a gravitational field. Equation (6.23) now becomes

$$-\rho g = \frac{dp}{dz} = c^2 \frac{d\rho}{dz} + Y \frac{ds}{dz}, \tag{6.27}$$

where $Y = (\partial p/\partial s)_\rho$. Consider an element of fluid in equilibrium at the level z. Two forces act on it—the force of gravity $-g\rho(z)V_0$ and *Archimedes' force* which is the same in magnitude but directed upward. Suppose now that a vertical displacement ζ is given to element under investigation. Its volume changes due to the compressibility of the fluid and becomes $V_0 + \Delta V$, whereas the entropy remains constant (heat exchange neglected). The force acting on the displaced element is

$$F = -g\rho(z)V_0 + g\rho(z+\zeta)(V_0 + \Delta V),$$

i.e. again the sum of the weight and Archimedes' force. Expanding $\rho(z+\zeta)$ in a Taylor series and retaining only terms linear in ζ and ΔV, we obtain

$$F = g\rho(z)V_0\left(\frac{1}{\rho(z)} \frac{d\rho}{dz} \zeta + \frac{\Delta V}{V_0}\right).$$

The volume change ΔV can be found by applying the equation of state to the fluid element. Taking into account that $\Delta s = 0$, we again have, retaining only linear terms,

$$\Delta p = c^2 \Delta \rho = c^2 \Delta(m/V) = c^2 \rho(z) V_0 \Delta(V^{-1}) = -c^2 \rho(z) \Delta V/V_0 ,$$

where $m = \rho(z)V_0$, the mass of the element. Now we have $\Delta V/V_0 = -\Delta p/c^2\rho(z)$, but $\Delta p = \zeta\, dp/dz = -\rho(z)g\zeta$. Hence, $\Delta V/V_0 = g\zeta/c^2$ and we finally find for the force acting on the displaced parcel,

$$F = g\rho(z)V_0\left(\frac{1}{\rho}\frac{d\rho}{dz} + g/c^2\right)\zeta = -mN^2\zeta ,$$

where

$$N^2 = -g\left(\frac{1}{\rho}\frac{d\rho}{dz} + \frac{g}{c^2}\right) \tag{6.28}$$

is called the *Väisälä frequency*. It is of great importance in the theory of internal waves (Chap. 10). The element under consideration is in a stable equilibrium only if the force is opposite to the displacement, i.e., when

$$N^2 \geq 0 \quad \text{or} \quad \frac{1}{\rho}\frac{d\rho}{dz} + \frac{g}{c^2} \leq 0 . \tag{6.29}$$

Hence, the density gradient must be negative (ρ decreases with height) and sufficiently large in magnitude, otherwise the equilibrium is unstable and convective currents will arise.

Note also that the term $\rho^{-1}\, d\rho/dz$ in the expression for the Väisälä frequency (6.28) can be excluded. Indeed, we have, according to (6.27),

$$\frac{1}{\rho}\frac{d\rho}{dz} = -\frac{g}{c^2} - \frac{Y}{\rho c^2}\frac{ds}{dz}; \quad \text{hence},$$

$$N^2 = \frac{gY}{\rho c^2}\frac{ds}{dz} . \tag{6.30}$$

If the entropy is constant ($ds/dz = 0$), the equilibrium becomes neutral.

6.4 Bernoulli's Theorem and the Energy Conservation Law

6.4.1 Bernoulli's Theorem

The Bernoulli theorem holds for a steady flow ($\boldsymbol{v}$ does not depend on t) of an ideal "barotropic" fluid [equation of state $p = p(\rho)$]. To deduce this theorem we start with the Euler equation (6.14) with $\partial \boldsymbol{v}/\partial t = 0$ and the term $(\boldsymbol{v} \cdot \nabla)\boldsymbol{v}$ replaced by

$$(\boldsymbol{v} \cdot \nabla)\boldsymbol{v} = \nabla v^2/2 - \boldsymbol{v} \times \operatorname{curl} \boldsymbol{v}\,, \tag{6.31}$$

according to a known vector identity. Further, we introduce a new function w called *enthalpy:*

$$w = \varepsilon + pV = \varepsilon + p/\rho\,, \tag{6.32}$$

where ε is the *internal energy* per unit mass. In differential form we have

$$dw = d\varepsilon + dp/\rho - p\,d\rho/\rho^2\,. \tag{6.32a}$$

In the case of constant entropy ($ds = 0$), being of interest here, the internal energy differential $d\varepsilon = T\,ds - p\,dV$ will be

$$d\varepsilon = -p\,dV = p\,d\rho/\rho^2\,. \tag{6.33}$$

Hence, we have for the enthalpy differential,

$$dw = dp/\rho\,. \tag{6.34}$$

Now, assuming also that the external force is a potential one ($\boldsymbol{f} = -\nabla u$), we may write the Euler equation (6.14) as

$$\nabla(v^2/2 + w + u) = \boldsymbol{v} \times \operatorname{curl} \boldsymbol{v}\,. \tag{6.35}$$

The line with a tangent collinear with the velocity vector $\boldsymbol{v}$ at each point is called the *streamline.* Streamlines generally change their position in the course of time, but in the case of steady flow, they are fixed and coincide with the paths of the fluid particles. We choose some streamline and consider the projection of (6.35) on the tangent to this line at an arbitrary point. The vector product $\boldsymbol{v} \times \operatorname{curl} \boldsymbol{v}$ is orthogonal to $\boldsymbol{v}$, hence its projection is zero. The projection of the gradient is the derivative d/dl, dl being the element of the streamline. As a result, we have $d(v^2/2 + w + u)/dl = 0$ or

$$v^2/2 + w + u = C_1\,. \tag{6.36}$$

Hence, the quantity $v^2/2 + w + u$ is *invariant* along the *streamline.* This is the *Bernoulli theorem.*[5] The expression (6.36) is called the *Bernoulli integral.* The constant C_1 is different for different streamlines. In the particular case of

[5] Note that this quantity is also invariant along the line whose tangent is parallel to the vector curl $\boldsymbol{v}$ (vortex line). However, we limit ourselves to the simplest formulation of the Bernoulli theorem

potential flow (curl $\boldsymbol{v} = 0$), however, we have from (6.35)

$$\nabla(v^2/2 + w + u) = 0; \quad \text{hence},$$

$$v^2/2 + w + u = \text{const}, \tag{6.37}$$

where the constant is the same for any point.

In the case of incompressible fluid flow ($\rho = \text{const}$), we have from (6.34) $w = p/\rho$ up to within an arbitrary additive constant. In this case, (6.36) takes on the familiar form

$$v^2/2 + p/\rho + u = C_1. \tag{6.38}$$

6.4.2 Some Applications of Bernoulli's Theorem

We consider incompressible-fluid flow in a tube with a constriction (Fig. 6.1). Suppose we are interested in the question: where is the pressure higher, at the wide section A or at the narrow one B? The answer follows immediately from the Bernoulli theorem. In fact, since the velocity in the narrow section is larger (the mass rate at both sections is the same), the pressure at B is less, according to (6.38), if the potential u can be considered constant. If one makes holes at A and B and connects them by means of a thin glass tube partly filled with mercury, the difference in levels of the latter in the left and right parts of the connecting tube will be proportional to the pressure difference. Such a device, depicted schematically in Fig. 6.1, is called the *Venturi velocity meter*. Let us assign the subscripts 1 and 2 to the quantities p and v in the sections A and B, respectively. Then, in accordance with the Bernoulli theorem,

$$p_1/\rho + v_1^2/2 = p_2/\rho + v_2^2/2.$$

In addition, the relation $v_1 S_1 = v_2 S_2$ holds, S_1 and S_2 being the cross-sectional areas at A and B, respectively. From the last two equations, we obtain

$$v_1 = \sqrt{\frac{2(p_1 - p_2)}{\rho(S_1^2/S_2^2 - 1)}}.$$

Hence, if the pressure difference $p_1 - p_2$ is known, the velocity v_1 can be determined.

Consider a two-dimensional flow around two closely placed infinite cylinders

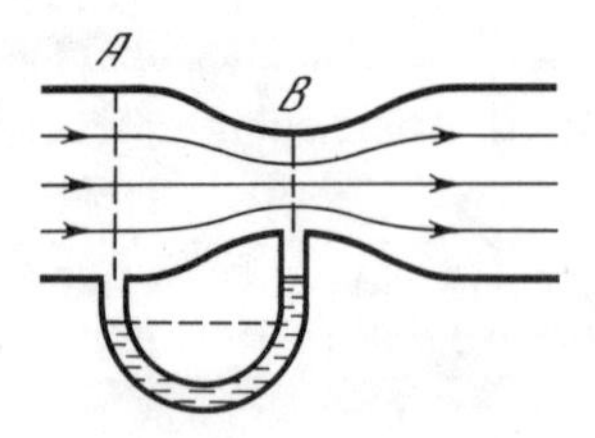

Fig. 6.1. A venturi velocity meter

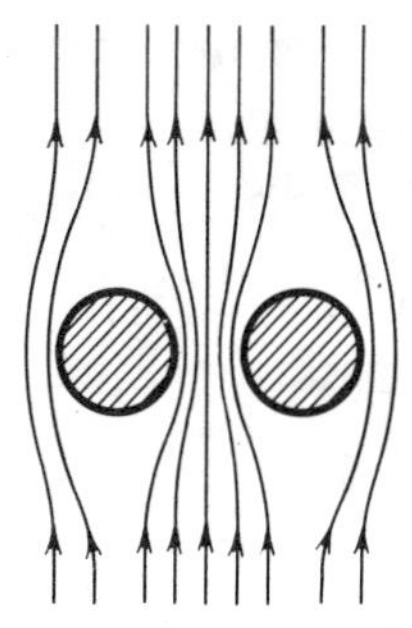

Fig. 6.2. An attraction force occurs between cylinders, closely placed in the flow

with axes perpendicular to the flow direction (Fig. 6.2). An increase in velocity takes place between the cylinders (streamlines here are situated more densely). It causes a pressure drop, according to the Bernoulli theorem, and hence an attraction force occurs between the cylinders. This phenomenon is well known in shiphandling because two ships moving closely in the same direction attract each other. Sometimes it could be the cause of a collision.

The elementary theory of liquid flow from a vessel through a hole is based on an application of the Bernoulli theorem. Consider some streamline connecting the free surface of the liquid in the vessel and the hole (dashed line in Fig. 6.3). Let the levels of the surface and the hole be z_1 and z_2, respectively. At $z = z_1$, the pressure is atmospheric $p = p_0$ and the velocity $v \simeq 0$ (the surface area is assumed large compared with the area of the hole). At the hole $z = z_2$, we have $v = v_2$ and also $p = p_0$ (we neglect the atmospheric pressure difference between z_1 and z_2). Equating the left-hand side of (6.38) at $z = z_1$ and $z = z_2$ and setting $u = gz$, yields

$$gz_2 + v_2^2/2 = gz_1 ,$$

and the well-known formula follows

$$v_2 = \sqrt{2g(z_1 - z_2)} . \tag{6.39}$$

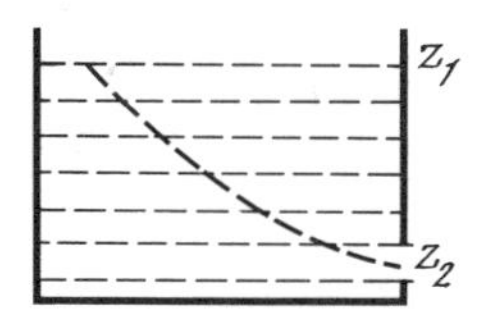

Fig. 6.3. Fluid outflow through the hole

6.4.3 The Bernoulli Theorem as a Consequence of the Energy-Conservation Law

The tube formed by all the streamlines passing through a small surface element normal to them is called a *flux tube*. Consider an element of a flux tube bounded by two cross sections whose areas are S_1 and S_2. The pressure, velocity, density and potential of an external force at these cross sections, i.e. p_1, v_1, ρ_1, u_1 and p_2, v_2, ρ_2, u_2, are supposed to be constant across S_1 and S_2. Let the flow be directed from S_1 towards S_2. The inflow of fluid mass across S_1 into the tube element during the time Δt is $\rho_1 S_1 v_1 \Delta t$. The outflow of mass across S_2 during the same period is $\rho_2 S_2 v_2 \Delta t$. In a steady flow these masses are equal:

$$\Delta M = \rho_1 S_1 v_1 \Delta t = \rho_2 S_2 v_2 \Delta t . \tag{6.40}$$

The pressure p_1 does the work $p_1 S_1 v_1 \Delta t$ on the inflowing fluid at S_1 and the outflowing fluid does the work $p_2 S_2 v_2 \Delta t$ on the external medium at S_2. The difference between these gives the energy increase, according to the energy conservation law,

$$p_1 S_1 v_1 \Delta t - p_2 S_2 v_2 \Delta t = \Delta M (E_2 - E_1) , \tag{6.41}$$

where E_1 and E_2 are the unit mass energy at S_1 and S_2, respectively. This energy is the sum of kinetic $v^2/2$, potential u, and internal ε energies:

$$E = v^2/2 + u + \varepsilon . \tag{6.42}$$

Substituting this into (6.41) and also taking into (6.40) account, we find

$$\frac{p_1}{\rho_1} - \frac{p_2}{\rho_2} = \frac{v_2^2}{2} + u_2 + \varepsilon_2 - \frac{v_1^2}{2} - u_1 - \varepsilon_1 .$$

Introducing the corresponding enthalpies $w_i = \varepsilon_i + p_i/\rho_i$, $i = 1, 2$, we obtain the Bernoulli theorem (6.36) which is now just a consequence of the *energy conservation law.*

6.4.4 Energy Conservation Law in the General Case of Unsteady Flow

We confine ourselves again to the case of an ideal fluid and potential external force. According to (6.42), the unit volume energy (energy density) is

$$\mathscr{E} = \rho E = \rho \frac{v^2}{2} + \rho u + \rho \varepsilon .$$

Assuming that u does not depend on time ($\partial u/\partial t = 0$), we obtain for the time rate of the energy density $\mathscr{E}$

$$\frac{\partial \mathscr{E}}{\partial t} = \left(\frac{v^2}{2} + u\right)\frac{\partial \rho}{\partial t} + \rho \boldsymbol{v} \cdot \frac{\partial \boldsymbol{v}}{\partial t} + \frac{\partial}{\partial t}(\rho\varepsilon) .$$

Using the continuity equation (6.9), the first term in the last equation can be written as

$$\left(\frac{v^2}{2}+u\right)\frac{\partial\rho}{\partial t}=-\left(\frac{v^2}{2}+u\right)\nabla(\rho\boldsymbol{v})\,.$$

The second term is transformed by using the Euler equation (6.14) with $\boldsymbol{f}=-\nabla u$:

$$\rho\boldsymbol{v}\cdot\frac{\partial\boldsymbol{v}}{\partial t}=-\rho\boldsymbol{v}\cdot\frac{\nabla p}{\rho}-\rho\boldsymbol{v}\cdot\nabla u-\rho\boldsymbol{v}\cdot(\boldsymbol{v}\cdot\nabla)\boldsymbol{v}\,.$$

Here, $\nabla p/\rho=\nabla w$ in accordance with (6.34). With (6.31) and the identity $\boldsymbol{v}\cdot(\boldsymbol{v}\times\operatorname{curl}\boldsymbol{v})\equiv 0$, we obtain

$$\rho\boldsymbol{v}\cdot(\boldsymbol{v}\cdot\nabla)\boldsymbol{v}=-\rho\boldsymbol{v}\cdot\nabla(v^2/2)\,.$$

By grouping respective terms, we find

$$\frac{\partial\mathscr{E}}{\partial t}=-\left(\frac{v^2}{2}+u\right)\nabla(\rho\boldsymbol{v})-\rho\boldsymbol{v}\cdot\nabla\left(\frac{v^2}{2}+u+w\right)+\frac{\partial(\rho\varepsilon)}{\partial t}\,.$$

The time rate of the internal energy density can be determined using (6.32, 33). We have $d(\rho\varepsilon)=\rho\,d\varepsilon+\varepsilon\,d\rho=p\rho^{-1}\,d\rho+\varepsilon\,d\rho=w\,d\rho$. This yields, with (6.9),

$$\frac{\partial(\rho\varepsilon)}{\partial t}=w\frac{\partial\rho}{\partial t}=-w\nabla(\rho\boldsymbol{v})\,, \tag{6.43}$$

and finally,

$$\frac{\partial\mathscr{E}}{\partial t}=\frac{\partial}{\partial t}\left(\rho\frac{v^2}{2}+\rho u+\rho\varepsilon\right)=-\nabla\left[\rho\boldsymbol{v}\left(\frac{v^2}{2}+u+w\right)\right]. \tag{6.44}$$

Integrate this relation over an arbitrary fixed volume V. The integral on the right-hand side can be transformed into the integral over the bounding surface S if one applies the Gauss divergence theorem to the vector $\boldsymbol{A}=\rho(V^2/2+u+w)\boldsymbol{v}$. Thus, we obtain the integral form of the energy conservation law for a moving fluid:

$$\frac{\partial}{\partial t}\int\limits_V\left(\rho\frac{v^2}{2}+\rho u+\rho\varepsilon\right)dV=-\int\limits_S\rho\left(\frac{v^2}{2}+u+w\right)v_n\,dS\,. \tag{6.45}$$

On the left-hand side we have the rate of change of the energy in volume V. On the right-hand side, there appears the *energy flux* across the bounding surface S, in accordance with the energy conservation law. Hence, the vector

$$\mathscr{N}\equiv\rho\left(\frac{v^2}{2}+u+w\right)\boldsymbol{v}=\rho\left(\frac{v^2}{2}+u+\varepsilon+\frac{p}{\rho}\right)\boldsymbol{v} \tag{6.46}$$

is the specific *energy flux* and the sum $v^2/2 + u + w$ is *the energy transported by the unit mass of fluid.* Note that w and not the internal energy ε is added here to the mechanical energy $v^2/2 + u$. This is because the fluid is not simply transporting the energy but also doing some work on the volume under consideration; for example, compressing it. It provides the additional energy flux $p\boldsymbol{v}$. Indeed, let us consider again isentropic flow in the flux tube. The energy flux across S_1 and S_2 must be the same as in the case of steady flow. This cannot be ΔME, the product of transported mass and energy density, since, according to (6.41), this quantity is generally not conserved. It follows, however, from (6.46) that the quantity $v^2/2 + u + w$ is conserved instead of the product $\Delta M(v^2/2 + u + w)$, since ΔM is also conserved.

Note that in the case of an incompressible fluid, $d\varepsilon = 0$ but $dw = dp/\rho$, according to (6.34). Since ε and w are determined to within additive constants, we can set $\varepsilon = 0$ and $w = p/\rho$ in this case.

Using (6.45), we can easily obtain the Bernoulli theorem. To this end choose the element of the flux tube between cross sections S_1 and S_2 as volume V and consider the case of a steady flow ($\partial/\partial t = 0$). Then it follows from (6.45) since $v_n = 0$ at the lateral walls of the tube, that

$$\rho_1 v_1 S_1(v_1^2/2 + u_1 + w_1) = \rho_2 v_2 S_2(v_2^2/2 + u_2 + w_2).$$

This immediately yields the Bernoulli theorem (6.36), since

$$\rho_1 v_1 S_1 = \rho_2 v_2 S_2$$

according to the mass conservation law.

6.5 Conservation of Momentum

6.5.1 The Specific Momentum Flux Tensor

By definition, $\rho\boldsymbol{v}$ is the *momentum* per unit volume. For the components of its time rate, using the equations of continuity (6.9) and motion (6.14), we have

$$\frac{\partial}{\partial t}(\rho v_i) = \rho\frac{\partial v_i}{\partial t} + v_i\frac{\partial \rho}{\partial t} = -\frac{\partial p}{\partial x_i} - \rho v_k\frac{\partial v_i}{\partial x_k} + f_i - v_i\frac{\partial(\rho v_k)}{\partial x_k}.$$

The sum of the second and fourth terms on the right-hand side is $\partial(\rho v_i v_k)/\partial x_k$. Writing also $\partial p/\partial x_i = \partial(\delta_{ik}p)/\partial x_k$, where δ_{ik} is the Kroneker symbol, we obtain

$$\frac{\partial}{\partial t}(\rho v_i) = -\frac{\partial}{\partial x_k}(p\,\delta_{ik} + \rho v_i v_k) + f_i. \tag{6.47}$$

Denoting

$$\Pi_{ik} = p\,\delta_{ik} + \rho v_i v_k, \tag{6.48}$$

we rewrite (6.47) as

$$\frac{\partial}{\partial t}(\rho v_i) = -\frac{\partial \Pi_{ik}}{\partial x_k} + f_i . \tag{6.49}$$

Now we integrate the last equation over an arbitrary fixed volume V:

$$\frac{\partial}{\partial t}\int_V \rho v_i \, dV = -\int_V \frac{\partial \Pi_{ik}}{\partial x_k} dV + \int_V f_i \, dV .$$

The first integral on the right-hand side can be transformed into an integral over the bounding surface S with the help of the Gauss divergence theorem (Theorem A.4). Then,

$$\frac{\partial}{\partial t}\int_V \rho v_i \, dV = -\int_S \Pi_{ik} n_k \, dS + \int_V f_i \, dV , \tag{6.50}$$

where $\{n_k\}$ are the unit vector components of the outward normal to S.

The last equation shows that the change of momentum in V is related to its flux across the bounding surface and to the action of an external force. If the latter is absent ($f_i = 0$), (6.50) yields the *law of momentum conservation:*

$$\frac{\partial}{\partial t}\int_V \rho v_i \, dV = -\int_S \Pi_{ik} n_k \, dS , \tag{6.51}$$

i.e., the time rate of the momentum in V is equal to its flux across the confining surface. Therefore, Π_{ik} is called the *tensor of the specific momentum flux.*

6.5.2 Euler's Theorem

In applications it is often important to know the force exerted by the moving fluid on the surrounding medium. Consider it for a steady flow with no external force. Equation (6.51) here assumes the following form:

$$\int_S \Pi_{ik} n_k \, dS = \int_S (p\,\delta_{ik} + \rho v_i v_k) n_k \, dS = 0 . \tag{6.52}$$

We choose S for the bounding surface of some small part of the flux tube. A sectional view of this part can be seen in Fig. 6.4. The flow velocity is $\boldsymbol{v} = -v\boldsymbol{n}$,

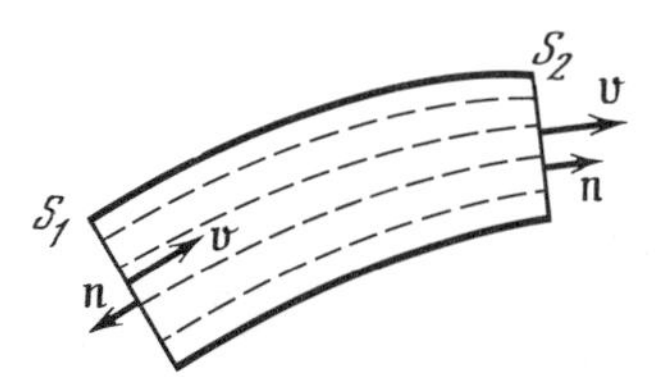

Fig. 6.4. Segment of the flux tube

$v = |\boldsymbol{v}|$ at the in let face of the part under consideration and $\boldsymbol{v} = v\boldsymbol{n}$ at the outlet. The integral

$$\int_{S_L} \rho v_i v_k n_k \, dS$$

over the lateral walls of the tube vanishes since $v_k n_k = \boldsymbol{v} \cdot \boldsymbol{n} = 0$ (the velocity is tangent to the walls). Thus (6.52) takes the form

$$\int_S p n_i \, dS = - \int_{S_1 + S_2} \rho v_i v_k n_k \, dS \, . \tag{6.53}$$

The left-hand side here is the ith component of the required force $\boldsymbol{F}$ acting on the surrounding medium. Since $v_k n_k = \boldsymbol{v} \cdot \boldsymbol{n} = -v$ at S_1 and $v_k n_k = \boldsymbol{v} \cdot \boldsymbol{n} = v$ at S_2, (6.53) becomes

$$F_i = - \int_{S_1} \rho v^2 n_i \, dS - \int_{S_2} \rho v^2 n_i \, dS \, . \tag{6.54}$$

Thus, we have deduced the *Euler theorem*, i.e., the force exerted by the flow in the flux tube upon the surrounding medium is the sum of the forces

$$- \int_{S_j} \rho v^2 \boldsymbol{n} \, dS \, , \quad j = 1, 2 \, ,$$

acting on the inlet and outlet faces of the tube.

Note that the vector

$$\mathscr{P}_1 = - \int_{S_1} \rho v^2 \boldsymbol{n} \, dS \tag{6.55}$$

is the *momentum flowing* into the chosen part of the flux tube across S_1 per unit time. The minus sign is because $\boldsymbol{v}$ and the outward normal $\boldsymbol{n}$ have opposite directions at S_1. Analogously,

$$\mathscr{P}_2 = \int_{S_2} \rho v^2 \boldsymbol{n} \, dS \tag{6.56}$$

is the *momentum outflow* across S_2 per unit time. Hence, (6.54) can be written as

$$\boldsymbol{F} = \mathscr{P}_1 - \mathscr{P}_2 \, , \tag{6.57}$$

i.e., the force $\boldsymbol{F}$ is equal to the momentum loss by the flux tube per unit time. Multiplying (6.57) by dt and changing the signs, we obtain the well-known classical form of the momentum theorem:

$$-\boldsymbol{F} \, dt = (\mathscr{P}_2 - \mathscr{P}_1) \, dt \, ,$$

the impulse of the force equals the momentum increment in the time interval dt. Here, $-\boldsymbol{F}$ is the force exerted by the surrounding fluid on the fluid in the flux tube, in accordance with the Newton's third law.

6.5.3 Some Applications of Euler's Theorem

Consider the force exerted by the flowing fluid on the tube walls. If the tube is cylindrical with a straight generator, we have the trivial case where $\mathscr{P}_2 = \mathscr{P}_1$ and the force is zero. The case is nontrivial when the tube is curved and possibly of variable cross section. Part of such a tube is shown in Fig. 6.4. The total force acting on the boundaries of the part under investigation is

$$\boldsymbol{F} = \int_S p\boldsymbol{n}\, dS = \boldsymbol{F}_{\mathrm{w}} + \int_{S_1} p\boldsymbol{n}\, dS + \int_{S_2} p\boldsymbol{n}\, dS\,.$$

The first term is the required force on the lateral walls of the tube, while the second and the third are forces on the cross sections S_1 and S_2, respectively.

We have for the force on the lateral walls, according to Euler's theorem (6.54),

$$\boldsymbol{F}_{\mathrm{w}} = -\int_{S_1} (p + \rho v^2)\boldsymbol{n}\, dS - \int_{S_2} (p + \rho v^2)\boldsymbol{n}\, dS\,. \tag{6.58}$$

Hence, this force is determined in terms of the flow parameters at the ends of the tube. In the case where the end faces are the planes of areas S_1, S_2, and the flow parameters p, ρ and $\boldsymbol{v}$ are constant across these faces, (6.58) becomes

$$\boldsymbol{F}_{\mathrm{w}} = -S_1(p_1 + \rho_1 v_1^2)\boldsymbol{n}_1 - S_2(p_2 + \rho_2 v_2^2)\boldsymbol{n}_2\,. \tag{6.58a}$$

Here, the subscripts 1 and 2 refer to the inlet and outlet ends of the tube, respectively.

Euler's theorem can also be applied to the determination of the so-called "*outflow coefficient*" of a jet in the case of liquid outflow from a vessel. The outflow velocity was determined above, see (6.39). However, the flow rate cannot be obtained just by multiplying this velocity with the hole area S. Indeed, all fluid particles approaching the hole have, according to (6.39), the same velocity v_2, but the directions of their motion are different. To determine the flow rate we use the fact (known from experiment) that at some small distance from the hole outside the vessel, the jet contracts and the streamlines become parallel. We denote the cross sectional area of the jet at this place by S'. The flow rate is $v_2 S'$. The ratio S'/S is called the *outflow coefficient of the jet*. It is not easy to calculate this coefficient in the case of a usual circular hole due to the complexity of streamlines near the hole. In one case, however, when a straight cylindrical tube is attached to the hole inside the vessel (*Borda mouthpiece*, Fig. 6.5), it can be found with the help of Euler's theorem.

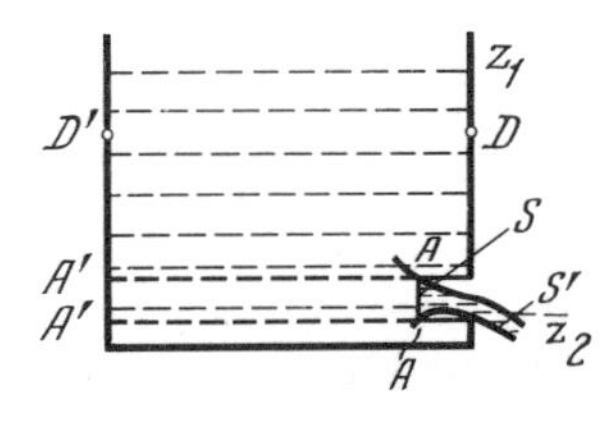

Fig. 6.5. Fluid outflow from the vessel with a Borda mouthpiece

The outflow of momentum from the jet in the time interval Δt is $\rho v_2^2 S' \Delta t$. To produce this outflow, the force F must exist, whose impulse is equal to this momentum outflow. The only force we have here is the pressure on the walls of the vessel. Since the inlet end of the Borda mouthpiece AA is far from the walls, the fluid at the walls is practically at rest, and this pressure may be calculated from hydrostatics. Note that the pressures at arbitrary points D and D′ are equal in magnitude and balance each other. An exception poses the area A′A′ opposite the hole. The hydrostatic pressure at this area is $\rho g(z_1 - z_2)S$ (the dimensions of the hole are supposed to be small compared with $z_1 - z_2$). This hydrostatic force produces the momentum flow rate through the hole:

$$\rho g(z_1 - z_2)S = \rho v_2^2 S' .$$

Substituting v_2 from (6.39), we obtain for the outflow coefficient $S'/S = 0.5$. An unbalanced hydrostatical force at the area A′A′ gives rise to the motion of the vessel in the direction opposite to the jet (jet propulsion).

Note that in the absence of a Borda mouthpiece, the fluid-particle velocities at the wall near the hole are not zero. Then, according to the Bernoulli theorem, the pressure in the neighbourhood of the hole must be lower than the hydrostatic one. Hence, the unbalanced force at the opposite side of the vessel as well as the outflow coefficient must be larger ($S'/S \simeq 0.61$ in this case, Exercise 7.5.8).

6.6 Vortex Flows of Ideal Fluids

The fluid flow is called *potential* or *irrotational* if $\operatorname{curl} \boldsymbol{v} = 0$ everywhere. If in some volume $\operatorname{curl} \boldsymbol{v} \neq 0$, the flow is called *vortex*. Let us consider the basic properties of the vortex flow.

6.6.1 The Circulation of Velocity

The integral

$$\Gamma = \oint_l \boldsymbol{v} \cdot d\boldsymbol{r} \tag{6.59}$$

extended over the closed contour l is called the *circulation of velocity* with respect to this contour. This integral can be transformed into a surface integral taken over an arbitrary open surface S confined by l. Indeed, according to the well-known Stoke's theorem, we have for an arbitrary vector $\boldsymbol{A}$

$$\oint \boldsymbol{A} \cdot d\boldsymbol{r} = \int_S \boldsymbol{n} \cdot \operatorname{curl} \boldsymbol{A} \, dS ,$$

where the positive direction of $\boldsymbol{n}$ is related to the positive sense of traversing l by the following right-hand rule: if the thumb points along the curve in the positive sense, then fingers will point towards the positive direction of $\boldsymbol{n}$. We assume also that S is a simply-connected surface, i.e. it can be contracted to a point. Now we obtain for the circulation,

$$\Gamma = \int_S \boldsymbol{n} \cdot \operatorname{curl} \boldsymbol{v}\, dS\,. \tag{6.60}$$

Obviously, it follows from this formula that in potential flow ($\operatorname{curl} \boldsymbol{v} = 0$), the circulation of velocity over an arbitrary simply connected contour is zero. This means in turn that in a potential flow there are no closed streamlines. Indeed, if such a line exists we could choose it as the contour l in (6.59) and obtain $\Gamma \neq 0$ since $\boldsymbol{v}$ and $d\boldsymbol{r}$ are parallel and $\boldsymbol{v} \cdot d\boldsymbol{r}$ is everywhere positive. The requirement for a contour to be simply connected is important here. If we choose l as a contour around a cylinder (such a contour cannot be contracted to a point while l is in a fluid all the time), the circulation can be different from zero even in a potential flow around a cylinder.

6.6.2 Kelvin's Circulation Theorem

The following theorem is rather important: *the circulation of velocity with respect to any closed "fluid" contour* (the contour consisting of the same fluid particles all the time) *remains unchanged* (Lord Kelvin, 1869). To prove it we obtain from (6.59)

$$\frac{d\Gamma}{dt} = \frac{d}{dt} \oint_l \boldsymbol{v} \cdot d\boldsymbol{r}\,,$$

where l is a closed fluid contour. To calculate the convected derivative we have to take into account the change of the contour's position in space, hence

$$\frac{d\Gamma}{dt} = \oint_l \left[\frac{d\boldsymbol{v}}{dt} \cdot d\boldsymbol{r} + \boldsymbol{v} \cdot d\left(\frac{d\boldsymbol{r}}{dt} \right) \right]. \tag{6.61}$$

In the second term here, $d\boldsymbol{r}/dt = \boldsymbol{v}$ and $\boldsymbol{v} \cdot d\boldsymbol{v} = d(v^2/2)$. To transform the first term we use the Euler equation (6.14), assuming $\boldsymbol{f} = -\nabla u$ (potential force) and taking into account that $\nabla w = \nabla p/\rho$ [isentropic change, see (6.34)]. Hence,

$$\frac{d\boldsymbol{v}}{dt} = -\nabla u - \nabla w = -\nabla(u + w)\,.$$

Now (6.61) assumes the form

$$\frac{d\Gamma}{dt} = \oint_l d\left(\frac{v^2}{2} - u - w \right) = 0\,.$$

It vanishes since it is an integral over a closed contour of an exact differential.

6.6.3 Helmholtz Theorems

There are some important theorems which were proved by Helmholtz before Kelvin (in 1858). Nevertheless, we will present them as corollaries of Kelvin's theorem. But before doing so we will introduce some new definitions and terms.

The vector $\boldsymbol{\omega} = \operatorname{curl} \boldsymbol{v}$ is called the *vorticity* or *vortex vector*. The *vortex line* is a curve whose tangent at each point has the direction of $\boldsymbol{\omega}$. A *vortex tube* consists of all vortex lines passing through a surface element normal to them. Now we can formulate the Helmholtz theorems as follows:

Theorem 1. The elements of an ideal fluid free of vorticity at some initial time, remains so indefinitely. To prove it we note that, according to Kelvin's theorem, the circulation

$$\Gamma = \oint_l \boldsymbol{v} \cdot d\boldsymbol{r}$$

with respect to a closed fluid contour remains unchanged. This integral can be transformed into the integral over the "fluid" open surface S bounded by l:

$$\Gamma = \int_S \boldsymbol{n} \cdot \operatorname{curl} \boldsymbol{v} \, dS ,$$

which must also remain unchanged. If at a certain instant $\operatorname{curl} \boldsymbol{v} = 0$ everywhere on S, then $\Gamma = 0$ for any l placed on S and it persists in being zero indefinitely. Indeed, if at some part of S, $\operatorname{curl} \boldsymbol{v}$ does not become zero, then we would choose the contour l in such a manner that $\boldsymbol{n} \cdot \operatorname{curl} \boldsymbol{v}$ does not change its sign inside it and hence $\Gamma \neq 0$ for this contour. However, this contradicts what was said above. Consequently, vortexes cannot be created in an ideal liquid moving in a potential field. Obviously, the inverse statement also holds, i.e. if $\operatorname{curl} \boldsymbol{v} \neq 0$ at some time, the vorticity cannot be destroyed since Γ remains unchanged.[6]

Theorem 2. A vortex line consists of the same fluid particles, i.e., it moves together with the fluid.

To prove this we take a vortex tube and consider the circulation Γ over an arbitrary, simply-connected closed contour l situated completely on the tube wall. In accordance with (6.60), this circulation can be represented as an integral over the surface S which is part of the tube wall. We have $\Gamma = 0$ initially, since $\boldsymbol{n} \cdot \operatorname{curl} \boldsymbol{v} = 0$ due to the definition of a vortex tube. It follows from Kelvin's theorem that $\Gamma = 0$ permanently for the contour l, i.e., $\boldsymbol{n} \cdot \operatorname{curl} \boldsymbol{v} = 0$ permanently for S. In other words, S must stay as part of the vortex tube's wall permanently. Since S (and l) can be chosen arbitrarily on the tube's wall, hence the same fluid particles will remain on the tube's wall indefinitely. We now make the cross sectional area of the tube tend to zero obtaining the vortex line. Consequently,

---

[6] In practice, it is a common fact that vortexes can appear and disappear (smoke rings, whirlpools, etc.). It is due to the viscosity of the fluid or its baroclinity [the thermal equation of state does not simply reduce to $p = p(\rho)$], or else because the external force is a nonpotential.

the latter also consists permanently of the same fluid particles.

Theorem 3. The flux of the vortex vector $\boldsymbol{\omega} = \operatorname{curl} \boldsymbol{v}$ through the vortex tube cross section, i.e., the quantity $\int_S \boldsymbol{n} \cdot \boldsymbol{\omega}\, dS$, remains unchanged along this tube.

To prove it, we note that the flux of vector $\boldsymbol{\omega}$ through any closed surface S_0 is zero. Indeed, we have, according to the Gauss divergence theorem,

$$\int_{S_0} \boldsymbol{n} \cdot \boldsymbol{\omega}\, dS = \int_V \operatorname{div} \boldsymbol{\omega}\, dV = 0$$

since $\operatorname{div} \boldsymbol{\omega} = \operatorname{div} \operatorname{curl} \boldsymbol{v} = 0$. Let S_0 be the bounding surface of part of the vortex tube: $S_0 = S_1 + S_2 + S'$, where S' is the wall's surface and S_1 and S_2 are the end faces of the part chosen. The flux of $\boldsymbol{\omega}$ through S' is zero due to the definition of the vortex tube ($\boldsymbol{n} \cdot \operatorname{curl} \boldsymbol{v} = 0$ at the walls), hence,

$$\int_{S_1} \boldsymbol{n} \cdot \boldsymbol{\omega}\, dS + \int_{S_2} \boldsymbol{n} \cdot \boldsymbol{\omega}\, dS = 0 .$$

The first term is negative since $\boldsymbol{n}$ is directed opposite to $\boldsymbol{\omega}$ at S_1. We see that the influx and outflux of the vector $\boldsymbol{\omega}$ are balanced, hence the theorem is proved.

When S_1 (as well as S_2) is sufficiently small and can be assumed constant across it, then the quantity ωS, called the *vortex strength*, remains unchanged along the tube. This means, in turn, that the vortex tube has neither a beginning nor an end in the fluid. It is closed, for otherwise its ends would be at infinity or at the walls of the vessel.

6.7 Exercises

6.7.1. Prove that

$$J^{-1}\, dJ/dt = \operatorname{div} \boldsymbol{v} \quad \text{with} \quad J = \frac{\partial(x_1, x_2, x_3)}{\partial(\xi_1, \xi_2, \xi_3)}$$

is the Jacobian of the transformation from Eulerian to Lagrangian coordinates.

Solution: In accordance with the rule for the differentiation of determinants, we obtain from (6.1) ($v_i \equiv dx_i/dt$ is the ith component of velocity):

$$\frac{dJ}{dt} = \frac{\partial(v_1, x_2, x_3)}{\partial(\xi_1, \xi_2, \xi_3)} + \frac{\partial(x_1, v_2, x_3)}{\partial(\xi_1, \xi_2, \xi_3)} + \frac{\partial(x_1, x_2, v_3)}{\partial(\xi_1, \xi_2, \xi_3)} .$$

For the row elements containing $v_i[x_k(\xi_m, t), t]$ in each of the three Jacobians we have

$$\frac{\partial v_i}{\partial \xi_j} = \frac{\partial v_i}{\partial x_k}\frac{\partial x_k}{\partial \xi_j} = \frac{\partial v_i}{\partial x_1}\frac{\partial x_1}{\partial \xi_j} + \frac{\partial v_i}{\partial x_2}\frac{\partial x_2}{\partial \xi_j} + \frac{\partial v_i}{\partial x_3}\frac{\partial x_3}{\partial \xi_j} .$$

Decompose now each Jacobian into the sum of three Jacobians, obtaining, for example,

$$\frac{\partial(v_1, x_2, x_3)}{\partial(\xi_1, \xi_2, \xi_3)} = \frac{\partial v_1}{\partial x_1}\frac{\partial(x_1, x_2, x_3)}{\partial(\xi_1, \xi_2, \xi_3)} + \frac{\partial v_1}{\partial x_2}\frac{\partial(x_2, x_2, x_3)}{\partial(\xi_1, \xi_2, \xi_3)} + \frac{\partial v_1}{\partial x_3}\frac{\partial(x_3, x_2, x_3)}{\partial(\xi_1, \xi_2, \xi_3)}$$
$$= \frac{\partial v_1}{\partial x_1} J$$

since the two other Jacobians are equal to zero (each has similar rows). Analogously,

$$\frac{\partial(x_1, v_2, x_3)}{\partial(\xi_1, \xi_2, \xi_3)} = \frac{\partial v_2}{\partial x_2} J, \qquad \frac{\partial(x_1, x_2, v_3)}{\partial(\xi_1, \xi_2, \xi_3)} = \frac{\partial v_3}{\partial x_3} J.$$

And finally,

$$dJ/dt = J\,\partial v_k/\partial x_k = J \operatorname{div} \boldsymbol{v}.$$

6.7.2. Derive the continuity equation in Lagrangian variables.

Solution: We have in accordance with the law of mass conservation

$$(d/dt)\int_V \rho(x_k, t)\,dV = 0,$$

where V is the arbitrary "fluid" volume (moving together with the fluid). In Lagrangian variables the left-hand side of the above equality becomes

$$\frac{d}{dt}\int_{V_0} \rho[x_k(\xi_l, t), t]J\,dV = \frac{d}{dt}\int_{V_0} \rho(\xi_l, t)J\,dV = \int_{V_0} \frac{d}{dt}[\rho(\xi_l, t)J]\,dV = 0.$$

Since V_0 is arbitrary, the integrand must be zero:

$$\frac{d}{dt}[\rho(\xi_1, \xi_2, \xi_3, t)J] = 0$$

which is the continuity equation in Lagrangian form.

6.7.3. Let the velocity field be given in Eulerian variables:

$$v_1 = v_2 = 0, \quad v_3(x_3, t) = f(x_3), \quad t \geq 0, \quad x_3 \geq 0.$$

Find the Lagrangian description of this motion. Consider the particular case where $f_3 = \sqrt{2gx_3}$.

Solution: Since $v_1 = v_2 = 0$, we have the dependence only on ξ_3 in Lagrangian variables $\{\xi_k\}$: $x_3 = x_3(\xi_3, t)$, $x_1 = \xi_1$, $x_2 = \xi_2$. Then $v_3(\xi_3, t) = \partial x_3(\xi_3, t)/\partial t = f[x_3(\xi_3, t)]$, or often integration (taking into account that $x_3 = \xi_3$ at $t = 0$),

$t = \int_{\xi_3}^{x_3} [dx_3/f(x_3)]$. If $f(x_3) = \sqrt{2gx_3}$, then $\sqrt{2}gt = 2(\sqrt{x_3} - \sqrt{\xi_3})$, $x_3 = \xi_3 + \sqrt{2g\xi_3}t + gt^2/2, v_3(\xi_3, t) = \sqrt{2g\xi_3} + gt, a_3(\xi_3, t) = g$. We have the motion with constant acceleration (the fluid motion is a waterfall).

6.7.4. A sphere of radius R_0 at $t = 0$ expands at $t > 0$ according to the law $R = R(t)[R(0) = R_0]$. Find this law if it is known that at $t > 0$ and $r > R(t)$, the velocity of the fluid particles is $v_r(r) \equiv v(r) = v_0 R_0^2/r^2$.

Solution: The velocity is specified by the Eulerian variable r. The Lagrangian variable $\varkappa$ is the particle coordinate at $t = 0$, i.e. $\varkappa = r(0)$. We have, according to the problem's condition,

$$v(\varkappa, t) = dr(\varkappa, t)/dt = R_0^2 v_0/r^2(\varkappa, t).$$

Integration of this equation under the condition $r(\varkappa, 0) = \varkappa$ yields

$$r(\varkappa, t) = (\varkappa^3 + 3R_0^2 v_0 t)^{1/3}, \qquad v(\varkappa, t) = R_0^2 v_0 (\varkappa^3 + 3R_0^2 v_0 t)^{-2/3}.$$

Hence, the desired law of the radius variation is

$$R(t) = r(R_0, t) = R_0(1 + 3v_0 t/R_0)^{1/3}.$$

6.7.5. Write the differential equation of streamlines if the Eulerian velocity $v_i(x_k, t)$ of the fluid motion is given.

Solution: Suppose that the streamlines are specified by the relation $\boldsymbol{r} = \boldsymbol{r}(s)$ in parametrical form. Since the tangent to the streamline is collinear with $\boldsymbol{v}$, then

$$d\boldsymbol{r}/ds = C\boldsymbol{v}, \qquad C = \text{const}.$$

Hence, the desired differential equation is

$$v_1^{-1}(dx_1/ds) = v_2^{-1}(dx_2/ds) = v_3^{-1}(dx_3/ds) = C.$$

6.7.6. Prove that in the case of steady flow as well as of the flow where $v_i(x_k, t) = \alpha_i(x_k)f(t)$, streamlines are the same as paths of fluid particles.

Solution: The paths of particle are defined, in general, by

$$(dx_i/dt) = v_i(x_k, t),$$

which can also be written in the form

$$v_1^{-1}(x_k, t)(dx_1/dt) = v_2^{-1}(x_k, t)(dx_2/dt) = v_3^{-1}(x_k, t)(dx_3/dt).$$

On the other hand, the streamlines are solutions of

$$v_1^{-1}(x_k, t)(dx_1/ds) = v_2^{-1}(x_k, t)(dx_2/ds) = v_3^{-1}(x_k, t)(dx_3/ds),$$

where the time t is a parameter. In the case of steady flow, $v_i = v_i(x_k)$ and both systems become equivalent. Analogously in the second case, the equation for streamlines

$$\alpha_1^{-1}(dx_1/ds) = \alpha_2^{-1}(dx_2/ds) = \alpha_3^{-1}(dx_3/ds)$$

and the equations for paths

$$\alpha_1^{-1}(dx_1/dt) = \alpha_2^{-1}(dx_2/dt) = \alpha_3^{-1}(dx_3/dt)$$

are equivalent.

6.7.7. Show that for a fluid rotating as a whole at an angular velocity $\boldsymbol{\Omega}$, the vortex vector is $\boldsymbol{\omega} = \text{curl } \boldsymbol{v} = 2\boldsymbol{\Omega}$.

Solution: We have for the components for $\boldsymbol{\omega}$:

$$\omega_1 = \frac{\partial v_3}{\partial x_2} - \frac{\partial v_2}{\partial x_3}, \qquad \omega_2 = \frac{\partial v_1}{\partial x_3} - \frac{\partial v_3}{\partial x_1}, \qquad \omega_3 = \frac{\partial v_2}{\partial x_1} - \frac{\partial v_1}{\partial x_2}.$$

Substituting here the velocity components when the fluid rotates as a whole (i.e., as a rigid body), yields

$$\boldsymbol{v} = \boldsymbol{\Omega} \times \boldsymbol{r} = \{\Omega_2 x_3 - \Omega_3 x_2, \Omega_3 x_1 - \Omega_1 x_3, \Omega_1 x_2 - \Omega_2 x_1\},$$

and we obtain

$$\omega_1 = 2\Omega_1, \qquad \omega_2 = 2\Omega_2, \qquad \omega_3 = 2\Omega_3, \qquad \boldsymbol{\omega} = 2\boldsymbol{\Omega}.$$

6.7.8. Write the differential equation for vortex lines for the general case and integrate it for the particular case considered in the previous problem.

Solution: Let $\boldsymbol{r} = \boldsymbol{r}(s)$ $[x_i = x_i(s)]$ be a parametrical form of the vortex line equation. Then, according to the definition of this line, $d\boldsymbol{r}/ds = C\boldsymbol{\omega}$ where $\boldsymbol{\omega} =$ curl $\boldsymbol{v}$, $C =$ const. Taking components, the last equation assumes the form

$$\omega_1^{-1}(dx_1/ds) = \omega_2^{-1}(dx_2/ds) = \omega_3^{-1}(dx_3/ds).$$

In the particular case considered in Exercise 6.7.7, $\boldsymbol{\omega} = 2\boldsymbol{\Omega}$ and the equation becomes

$$(2\Omega_1)^{-1}(dx_1/ds) = (2\Omega_2)^{-1}(dx_2/ds) = (2\Omega_3)^{-1}(dx_3/ds).$$

Its solution is the straight line parallel to $\boldsymbol{\Omega}$:

$$(x_1 - x_{10})/\Omega_1 = (x_2 - x_{20})/\Omega_2 = (x_3 - x_{30})/\Omega_3$$

which is an obvious result since the vortex vector $\boldsymbol{\omega} = 2\boldsymbol{\Omega}$ is constant.

6.7.9. For the case of an ideal gas in the gravitational field, find the pressure dependence on height if the temperature dependence is $T = T(z)$. Consider these particular cases:

a) $T = T_0 = \text{const}$; b) $T = T_0(1 - z/H), z < H$;
c) $T = T_0(1 - z^2/H^2)$, $z < H$.

Solution: The equation of state for an ideal gas is $p = RT(z)\rho/\mu$, where μ is the molecular weight. Consequently, $\rho = \mu p/RT(z)$ and $\rho_0 = \mu p_0/RT_0$ or $\mu/R = \rho_0 T_0/p_0$. Substitution of ρ into the hydrostatic equation (6.23) yields a differential equation which is easy to integrate:

$$\frac{dp}{p} = -g\frac{\rho_0 T_0}{p_0}\frac{dz}{T(z)}, \qquad p = p_0 \exp\left(-\frac{g\rho_0 T_0}{p_0}\int_0^z \frac{dz}{T(z)}\right).$$

a) when $T(z) = T_0$, we obtain $T_0 \int_0^z T^{-1}(z)\,dz = z$ and (6.26) for $p(z)$;

b) $T(z) = T_0(1 - z/H)$, $T_0 \int_0^z T^{-1}(z)\,dz = \int_0^z (1 - z/H)^{-1}\,dz = -H\ln(1 - z/H)$ and $p(z) = p_0(1 - z/H)^{\alpha_1}$, $\alpha_1 = g\rho_0 H/p_0$;

c) when $T(z) = T_0(1 - z^2/H^2)$, we have $T_0 \int_0^z T^{-1}(z)\,dz = (H/2)\ln|(1 + z/H)/(1 - z/H)|$ and $p(z) = p_0[(1 - z/H)/(1 + z/H)]^{\alpha_2}$, $\alpha_2 = \alpha_1/2 = g\rho_0 H/2p_0$.

6.7.10. Find the pressure as a function of height for a fluid with the equation of state $p = p_0(\rho/\rho_0)^q$ in the gravitational field.

Solution: Substitution of the given relation $\rho = \rho(p)$ into (6.23) yields

$$p = p_0\left(1 - \frac{q-1}{q}\frac{g\rho_0 z}{p_0}\right)^{q/(q-1)}.$$

6.7.11. A cylindrical vessel with a vertical axis filled by an ideal fluid rotates in the gravitational field about this axis at the angular velocity $\boldsymbol{\Omega}$. Determine the shape of the free surface of the fluid.

Solution: The Euler equation in a rotating coordinate system is

$$d\boldsymbol{v}/dt = -\nabla p/\rho - g\nabla z - \boldsymbol{a}_{\text{cf}},$$

where $\boldsymbol{a}_{\text{cf}}$ is centrifugal acceleration. For the case of equilibrium we have in cylindrical coordinates ($\boldsymbol{v} = 0$):

$$-\rho^{-1}(\partial p/\partial r) + \Omega^2 r = 0, \quad (\rho r)^{-1}(\partial p/\partial \varphi) = 0, \quad \rho^{-1}(\partial p/\partial z) = -g.$$

After integration this set of equations yields

$$p(r, z) = \rho\Omega^2 r^2/2 - g\rho z + p_0,$$

where $p_0 = p(0, 0)$. At the free surface $p(r, z) = p_a$ is the atmospheric pressure.

Hence,

$$z(r) = (\Omega^2 r^2/2g) + (p_0 - p_a)/g\rho\,.$$

6.7.12. Determine the pressure in the Earth as a function of the distance from its center assuming that the medium of the Earth is an incompressible liquid of density ρ. Neglect the earth rotation.

Solution: The hydrostatic equation for this case is

$$dp/dr = -\rho\, du/dr = -\rho g(r)\,.$$

According to Newton's gravitational law, the gravitational acceleration is $g(r) = \gamma M(r)/r^2$ where $M(r) = 4\pi r^3\rho/3$, is the mass of the liquid in the sphere of the radius r, $\gamma = 6.67 \times 10^{-8}\ \mathrm{cm}^{-3}\,\mathrm{s}^{-2}\,\mathrm{g}^{-1}$ is the gravitational constant. In particular, the well-known gravity acceleration $g_0 = 9.81\ \mathrm{ms}^{-2}$ at the Earth surface can be obtained, assuming the $r = R = 6400$ km is the earth's radius. Now we have

$$dp/dr = -\rho g_0 r/R\,, \qquad p(r) = p_a + \rho g_0(R^2 - r^2)/2R\,,$$

where p_a is the pressure at the Earth's surface. The maximum pressure at the Earth's center is $p(0) = p_a + \rho g_0 R/2$. The average density of the Earth can be expressed in terms of g_0, γ and R as $\rho = 3g_0/4\pi\gamma R$, which yields for $p(r)$ and $p(0)$:

$$p(r) = p_a + \frac{3g_0^2}{8\pi\gamma}\left(1 - \frac{r^2}{R^2}\right), \qquad p(0) = p_a + \frac{3}{8\pi}\frac{g_0^2}{\gamma} \simeq 10^6\ \mathrm{atm}\,.$$

6.7.13. The free surface of a liquid in a vessel is at a level z_0 above its bottom. At level z there is a hole. Determine the distance at which the outflowing liquid jet reaches the plane of the bottom ($z = 0$). Also determine z (with z_0 fixed) when this distance is maximum.

Solution: The horizontal outflow velocity is $v = \sqrt{2g(z_0 - z)}$, according to (6.39). In free fall in the gravitational field the fluid particles will reach the bottom's plane after the time $t_0 = \sqrt{2z/g}$. The horizontal distance covered during this time is $x(z) = vt_0 = 2\sqrt{z(z_0 - z)}$. It is maximum $x_{\max} = z_0$ when $z = z_0/2$.

6.7.14. Determine the dependence of the horizontal cross-sectional area of a vessel $S(z)$ on the vertical distance under the condition that the time rate of the surface level of the liquid out-flowing through the hole dz/dt is constant.

Solution: The outflow velocity is $v = \sqrt{2g(z - z_0)}$. Assuming that the outflow coefficient k is constant, we have for the rate of mass outflow

$$dm/dt = \rho k S_0 v = \rho k S_0 \sqrt{2g(z - z_0)}\,,$$

where S_0 is the hole's area. On the other hand, we have

$$dm/dt = \rho S(z)\, dz/dt\,.$$

Equating these quantities we obtain

$$S(z) = \frac{kS_0\sqrt{2g}}{dz/dt}\sqrt{z - z_0}\,.$$

6.7.15. An ideal gas is outflowing adiabatically through a small hole in the vessel. Determine the outflow velocity if the pressure is p_0 in the vessel and p outside it.

Solution: According to Bernoulli's theorem for a compressible gas, we have $w_0 + v_0^2/2 = w + v^2/2$, w being the enthalpy. Choosing the beginning of a streamline sufficiently far from the hole where $v_0 = 0$, we get for the outflow velocity $v^2 = 2(w_0 - w)$. Using further the enthalpy differential $dw = dp/\rho$ and the adiabatic equation $p = p_0(\rho/\rho_0)^\gamma$, $\gamma = c_p/c_v$, we find

$$w = \int \frac{dp}{\rho} = \frac{\gamma}{\gamma - 1}\frac{p_0}{\rho_0}\left(\frac{p}{p_0}\right)^{(\gamma-1)/\gamma} = \frac{\gamma}{\gamma - 1}\frac{p}{\rho}\,.$$

Hence,

$$v^2 = \frac{2\gamma}{\gamma - 1}\frac{p_0}{\rho_0}\left[1 - \left(\frac{p}{p_0}\right)^{(\gamma-1)/\gamma}\right] = \frac{2\gamma}{\gamma - 1}\left(\frac{p_0}{\rho_0} - \frac{p}{\rho}\right).$$

This velocity can also be expressed in terms of the sound velocity in the gas. We have

$$c^2 = \frac{dp}{d\rho} = c_0^2\left(\frac{p}{p_0}\right)^{(\gamma-1)/\gamma} = \gamma\frac{p}{\rho}, \qquad v^2 = \frac{2}{\gamma - 1}(c_0^2 - c^2),$$

where c_0 is the sound velocity in the gas at rest and c is that in the gas just leaving the vessel.

6.7.16. Find the rate of mass outflow in the previous exercise if the hole's area is S_0. Assume the outflow coefficient $k = 1$. Consider the case where $p = 0$ (outflow into the vacuum) and explain the result.

Solution: We have

$$\frac{dm}{dt} = S_0\rho v = S_0\rho_0\left(\frac{p}{p_0}\right)^{1/\gamma}\sqrt{\frac{2\gamma}{\gamma - 1}}\left(\frac{p_0}{\rho_0}\right)^{1/2}\left[1 - \left(\frac{p}{p_0}\right)^{(\gamma-1)/\gamma}\right]^{1/2}$$

$$= \frac{\rho_0 S_0 c_0}{\sqrt{(\gamma - 1)/2}}\sqrt{\left(\frac{p}{p_0}\right)^{2/\gamma} - \left(\frac{p}{p_0}\right)^{(\gamma+1)/\gamma}}\,.$$

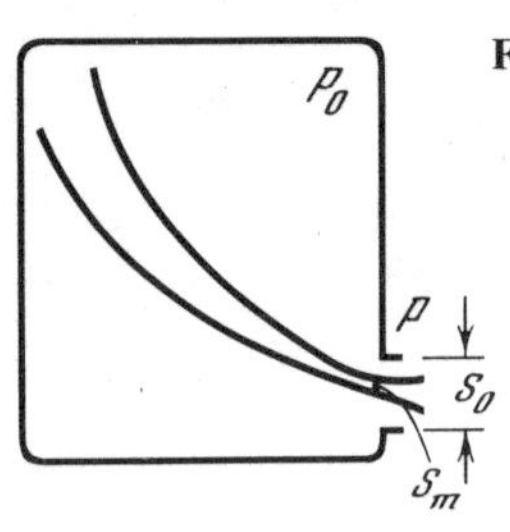

Fig. 6.6. Flux tube in the case of gas outflow from the vessel

In the case where $p = 0$, this yields $dm/dt = 0$, an obvious absurdity. To obtain the right result, consider the flux tube (Fig. 6.6) along which $w + v^2/2 = \text{const}$. Differentiating the latter expression along the flux tube we obtain

$$v(dv/dl) + dw/dl = v(dv/dl) + \rho^{-1} dp/dl = 0 .$$

On the other hand, the mass flux is constant along the tube: $\rho v S = \text{const}$. Hence,

$$vS(d\rho/dl) + \rho S(dv/dl) + \rho v(dS/dl) = 0 .$$

For the relative rate of change of the cross-sectional area of the tube, this yields

$$S^{-1}\frac{dS}{dl} = -v^{-1}\frac{dv}{dl} - \rho^{-1}\frac{d\rho}{dl} = -v^{-1}\frac{dv}{dl} - \rho^{-1}\frac{d\rho}{dp}\frac{dp}{dl}$$
$$= -v^{-1}\frac{dv}{dl}\left(1 - \frac{v^2}{c^2}\right).$$

The tube becomes thinner ($dS/dl < 0$) when $v < c$. When the outflow velocity approaches the sound velocity $v \to c$, then $dS/dl \to 0$ and the cross-sectional area becomes a minimum ($S|_{v=c} = S_{\mathrm{m}}$). After that when $v > c$ the tube broadens. Determine the gas parameters at the section S_{m} where $v = v_{\mathrm{m}}, c = c_{\mathrm{m}} = v_{\mathrm{m}}$. From the last formula in Exercise 6.7.15 we have $c_{\mathrm{m}}^2 = 2(c_0^2 - c_{\mathrm{m}}^2)/(\gamma - 1)$, which yields $c_{\mathrm{m}}^2 = 2c_0^2/(\gamma + 1)$. Since $c_{\mathrm{m}}^2 = c_0^2(p_{\mathrm{m}}/p_0)^{(\gamma-1)/\gamma}$, we have further

$$\frac{p_{\mathrm{m}}}{p_0} = \left(\frac{c_{\mathrm{m}}}{c_0}\right)^{2\gamma/(\gamma-1)} = \left(\frac{2}{\gamma+1}\right)^{\gamma/(\gamma-1)}$$

and, moreover, $\rho_{\mathrm{m}}/\rho_0 = [2/(\gamma + 1)]^{1/(\gamma-1)}$. It is reasonable to assume that a minimum cross section S_{m} occurs at the hole. Then the rate of mass change of the gas in the vessel is

$$\frac{dm}{dt} = \rho_{\mathrm{m}} v_{\mathrm{m}} S_0 = \left(\frac{2}{\gamma+1}\right)^{1/(\gamma-1)} \rho_0 c_0 \sqrt{\frac{2}{\gamma+1}} S_0 = \left(\frac{2}{\gamma+1}\right)^{(1+\gamma)/2(\gamma-1)} \rho_0 c_0 S_0 .$$

6.7.17. The same as Exercise 6.7.12 but take into account the Earth's rotation, find the form of the Earth's surface.

7. Potential Flow

It was shown above that vortexes cannot originate or disappear in an ideal (inviscid) fluid moving in the field of a potential force. Consequently, if the flow is potential initially (i.e., curl $\boldsymbol{v} = 0$ everywhere), it remains so indefinitely. Of course, one can constantly observe the creation and disappearance of the vortexes in a real fluid. Nevertheless, it is reasonable to first consider some problems of a potential flow since this very often appears to be a good approximation to real fluid flow; in addition, the equations describing this flow are much simpler.

7.1 Equations for a Potential Flow

7.1.1 Velocity Potential

Note first of all that the potential-flow velocity can be expressed in terms of one scalar function, the *velocity potential* $\varphi(x_i, t)$, assuming $\boldsymbol{v} = \nabla\varphi$. The condition for potential flow curl $\boldsymbol{v} = \operatorname{curl} \nabla\varphi = 0$ is satisfied automatically by this. The Euler equation (6.14) may be written down assuming a potential external force ($\boldsymbol{f} = -\nabla u$) and taking into account the vector identity (6.31) and the expression for the enthalpy differential (6.34):

$$\nabla(\partial\varphi/\partial t + v^2/2 + w + u) = 0\,.$$

Hence we immediately obtain

$$\frac{\partial\varphi}{\partial t} + \frac{v^2}{2} + w + u = F(t)\,. \tag{7.1}$$

The last equation can be considered a form of the *Bernoulli principle.* An arbitrary function of time $F(t)$ can be reduced to zero by the transform $\varphi' = \varphi + \int F(t)\,dt$, which does not influence $\boldsymbol{v}$. Now omitting the prime at φ', we rewrite (7.1) as

$$\frac{\partial\varphi}{\partial t} + \frac{v^2}{2} + w + u = 0\,. \tag{7.1a}$$

The simplest equations occur in the case of incompressible fluid flow where $d\rho/dt = 0$. Moreover, we assume in this chapter that the density ρ does not depend on spatial coordinates, too, i.e., $\rho = \text{const}$. In this case, we have $w = p/\rho$ up to an additive constant which can be included in φ. The continuity equation (6.9) also assumes the simple form: $\operatorname{div} \boldsymbol{v} = 0$ or

$$\Delta\varphi = 0, \tag{7.2}$$

when $\Delta = \operatorname{div} \operatorname{grad}$ is the Laplacian operator.

Thus, the problem of a potential flow of *inviscid incompressible* fluid reduces to the solution of one scalar equation (7.2) with the appropriate boundary conditions. The potential $\varphi(x_i, t)$ being known, the velocity is obtained from $\boldsymbol{v} = \nabla\varphi$ and the pressure, according to (7.1a), as

$$p = -\rho\left(\frac{v^2}{2} + u + \frac{\partial\varphi}{\partial t}\right). \tag{7.3}$$

One must be careful while applying this formula because the potential in it is defined to within an arbitrary function of time $F(t)$ which can influence p. This function can be determined if the pressure is prescribed in advance at some point.

7.1.2 Two-Dimensional Flow. Stream Function

The most complete theory is that of the potential *two-dimensional flow* of an *ideal incompressible* fluid. We can assume $x_3 = z$, $v_3 = v_z = 0$ in this case, $v_1 = v_x$ and $v_2 = v_y$ being independent of z. We introduce the new function $\psi(x, y, t)$ termed the *stream function* and defined in such a way that

$$v_x = \partial\psi/\partial y, \quad v_y = -\partial\psi/\partial x. \tag{7.4}$$

The continuity equation is now satisfied automatically: $\operatorname{div} \boldsymbol{v} = \partial v_x/\partial y + \partial v_y/\partial x = 0$. It can also be shown that the lines $\psi = \text{const}$ are the flow *streamlines*. Indeed, the tangent to a streamline is collinear with the velocity $\boldsymbol{v}$, by definition. Hence, the relation $dy/dx = v_y/v_x$ or $v_x\,dy - v_y\,dx = 0$ holds for the streamline $y = y(x)$. Taking (7.4) into account we obtain

$$\frac{\partial\psi}{\partial x}dx + \frac{\partial\psi}{\partial y}dy = d\psi = 0,$$

i.e., $\psi = \text{const}$ along the streamlines.

Using the stream function ψ, one can easily find the flow rate of the fluid across the plane curve connecting two points A and B. By definition, we have

$$\mathcal{N} = \int_A^B \boldsymbol{v} \cdot \boldsymbol{n}\,dl, \tag{7.5}$$

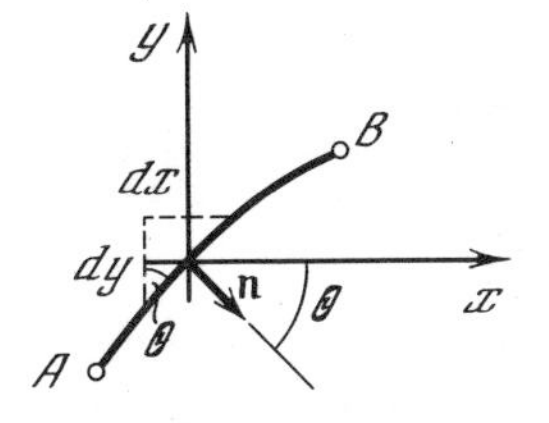

Fig. 7.1. To determine the flow rate of a fluid across the curve AB

where $\boldsymbol{n}$ is the unit normal to the curve (Fig. 7.1) and dl is an element of its length. In (7.5) we have

$$\boldsymbol{v}\cdot\boldsymbol{n} = v_x\cos(\boldsymbol{nx}) + v_y\cos(\boldsymbol{ny}),$$

$$dl\cos(\boldsymbol{nx}) = dl\cos\theta = dy,$$

$$dl\cos(\boldsymbol{ny}) = dl\cos(\theta+\pi/2) = -dl\sin\theta = -dx,$$

where $\boldsymbol{x}$ and $\boldsymbol{y}$ are the unit vectors along the corresponding axes. Now, taking also (7.4) into account, we can write

$$\mathscr{N} = \int_A^B \left(\frac{\partial\psi}{\partial y}dy + \frac{\partial\psi}{\partial x}dx\right) = \int_A^B d\psi = \psi_B - \psi_A. \tag{7.6}$$

Hence, the *flow rate* across a part of a curve is equal to the *difference of the values of the stream function* at the ends of the considered part.

We have not used yet the fact that the flow is potential. Therefore, the stream function can be introduced for vortex flow. The vortex vector $\boldsymbol{\omega} = \operatorname{curl}\boldsymbol{v} = \omega\nabla z$ can be related to the stream function since

$$\omega = \frac{\partial v_y}{\partial x} - \frac{\partial v_x}{\partial y} = -\left(\frac{\partial^2\psi}{\partial x^2} + \frac{\partial^2\psi}{\partial y^2}\right) = -\Delta\psi. \tag{7.7}$$

For potential flow ($\boldsymbol{\omega} = 0$), ψ obeys the equation

$$\Delta\psi = 0. \tag{7.8}$$

We see now that in the case of a *potential two-dimensional flow* of an *ideal incompressible* fluid, both the *velocity potential* $\varphi(x, y, t)$ and the *stream function* $\psi(x, y, t)$ *satisfy the Laplace equation.*

Using $\boldsymbol{v} = \nabla\varphi$ and (7.4), we can establish a relationship between φ and ψ:

$$\frac{\partial\varphi}{\partial x} = \frac{\partial\psi}{\partial y}, \quad \frac{\partial\varphi}{\partial y} = -\frac{\partial\psi}{\partial x}. \tag{7.9}$$

It follows from these equations that

$$\frac{\partial \varphi}{\partial x}\frac{\partial \psi}{\partial x} + \frac{\partial \varphi}{\partial y}\frac{\partial \psi}{\partial y} = \nabla\varphi \cdot \nabla\psi = 0\,, \tag{7.10}$$

i.e., the families of curves $\varphi = \text{const}$ and $\psi = \text{const}$ are mutually orthogonal.

The functions φ and ψ are, in a sense, equivalent. Up to now the flows have been considered with φ as the velocity potential and ψ as the stream function. However, with the same φ and ψ, we could imagine another, the so-called "*conjugate*" flow where ψ is the velocity potential and φ the stream function.

7.2 Applications of Analytical Functions to Problems of Hydrodynamics

7.2.1 The Complex Flow Potential

The properties of the functions φ and ψ allow us to apply the powerful theory of analytical functions of complex variables to the case of two-dimensional potential flow of an ideal incompressible fluid. Let us recall that the function of a complex variable assigns a complex number $F(z)$ to each point $z = x + \mathrm{i}y$ in the complex plane. The class of functions of a complex variable contains the very important subclass of analytical functions. The latter can be differentiated, i.e., the limit

$$F'(z) = \lim_{\Delta z \to 0} \frac{F(z + \Delta z) - F(z)}{\Delta z} \tag{7.11}$$

exists, which is called the derivative of $F(z)$.

Let

$$F(z) = \alpha(x, y) + \mathrm{i}\beta(x, y)\,, \tag{7.12}$$

where α and β are real functions. The condition for the differentiability of $F(z)$ is more restrictive than just the differentiability of its real and imaginary parts α and β. Namely, the so-called *Cauchy-Riemann conditions*

$$\partial\alpha/\partial x = \partial\beta/\partial y\,, \qquad \partial\beta/\partial x = -\partial\alpha/\partial y \tag{7.13}$$

must be satisfied. This fact is fundamental in the theory of functions of a complex variable. It is directly related to the existence of the limit (7.11) and to the independence of the latter of the direction of the increment Δz. We will illustrate this by calculating the limit (7.11) along two mutually perpendicular directions:

$\Delta z \equiv \Delta x$

$$F' = \lim_{\Delta x \to 0} \frac{\alpha(x+\Delta x, y) + \mathrm{i}\beta(x+\Delta x, y) - \alpha(x,y) - \mathrm{i}\beta(x,y)}{\Delta x} = \frac{\partial \alpha}{\partial x} + \mathrm{i}\frac{\partial \beta}{\partial x},$$

$\Delta z \equiv \mathrm{i}\,\Delta y$

$$F' = \lim_{\Delta y \to 0} \frac{\alpha(x, y+\Delta y) + \mathrm{i}\beta(x, y+\Delta y) - \alpha(x,y) - \mathrm{i}\beta(x,y)}{\mathrm{i}\,\Delta y} = \frac{\partial \beta}{\partial y} - \mathrm{i}\frac{\partial \alpha}{\partial y}.$$

Equating the real and imaginary parts of these expressions, we obtain the Cauchy-Riemann equations (7.13). Remember now that the velocity potential φ and the stream function ψ of a two-dimensional flow of ideal incompressible fluid satisfy exactly these equations, see (7.9). Hence, φ and ψ can be considered the real and imaginary parts of some analytical function $F(z)$ called the *complex flow potential.*

Using (7.13), one can prove that the real and imaginary parts of any analytical function $F(z)$ satisfy the Laplace equation. Indeed, differentiating the first of these equations with respect to x, the second with respect to y and adding the results we get

$$\frac{\partial^2 \alpha}{\partial x^2} + \frac{\partial^2 \alpha}{\partial y^2} = \Delta \alpha = 0\,.$$

In the same way one can prove that $\Delta \beta = 0$.

Hence, we have two conjugate flows for an arbitrary analytical function $F(z)$. The velocity potentials of these flows are $\varphi_1 = \mathrm{Re}\{F(z)\}$ [then $\psi_1 = \mathrm{Im}\{F(z)\}$] or $\varphi_2 = \mathrm{Im}\{F(z)\}$ [then $\psi_2 = \mathrm{Re}\{F(z)\}$]. Sometimes it is more convenient to take

$$\varphi_2 = \mathrm{Re}\{\pm \mathrm{i}F(z)\}\,, \qquad \psi_2 = \mathrm{Im}\{\pm \mathrm{i}F(z)\}\,.$$

7.2.2 Some Examples of Two-Dimensional Flows

Let us consider some simple functions $F(z)$ and the respective flows.

i) $F(z) = az$ (a is a real constant), i.e., $\varphi + \mathrm{i}\psi = ax + \mathrm{i}ay$ or $\varphi = ax$, $\psi = ay$. Hence, the flow is that with constant velocity $v_x = a$, $v_y = 0$. The streamlines ($\psi = \mathrm{const}$) are parallel to the x-axis. To obtain the conjugate flow, consider a new function $F(z)$ which differs from the previous one by the factor $(-\mathrm{i})$: $F(z) = -\mathrm{i}az = ay - \mathrm{i}ax$ or $\varphi = ay$, $\psi = -ax$. Here the velocity $v_x = 0$, $v_y = a$ is also constant and the streamlines ($\psi = \mathrm{const}$) are parallel to the y-axis.

ii) $F(z) = m \ln z\,, \quad z \neq 0, \quad m$ is real. (7.14)

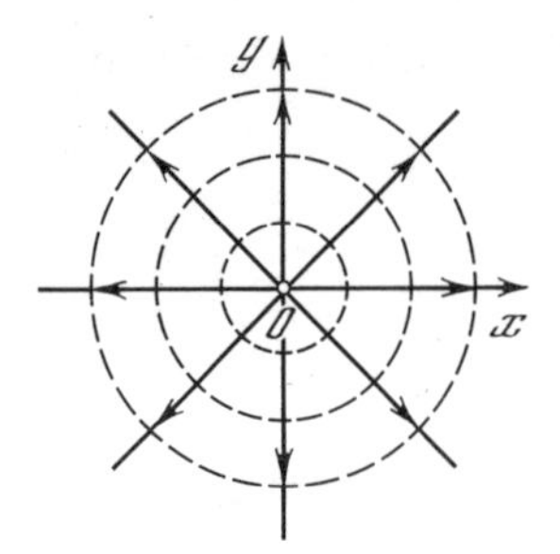

Fig. 7.2. Streamlines (full lines) and equipotential lines (dotted lines) for $F(z) = m \ln z$

Introducing the polar coordinates r and θ so that

$$x = r\cos\theta\,, \quad y = r\sin\theta\,, \quad z = x + \mathrm{i}y = r\exp(\mathrm{i}\theta)\,,$$

we transform (7.14) into

$$F(z) = m(\ln r + \mathrm{i}\theta)\,, \quad \varphi = m\ln r\,, \quad \psi = m\theta\,.$$

Hence, the radii $\theta = \text{const}$ are streamlines whereas the circles $r = \text{const}$ are the equipotential lines (Fig. 7.2). The particles move along the radii at the velocity

$$v_r = \partial\varphi/\partial r = m/r\,.$$

The right-hand side is positive if $m > 0$, i.e., we have a source at the point $r = 0$ in this case. The total flow of mass across an arbitrary circle $r = \text{const}$

$$2\pi r v_r = 2\pi m$$

is constant in accordance with the mass conservation law. Then the quantity $2\pi m$ is called the *strength of the source*. In the opposite case when $m < 0$, there is a sink at the point $r = 0$.

We obtain the complex flow potential for the conjugate flow by multiplying (7.14) by $-\mathrm{i}$:

$$F(z) = -\mathrm{i}\,m\ln z = m(\theta - \mathrm{i}\ln r)\,, \quad \varphi = m\theta\,, \quad \psi = -m\ln r\,. \tag{7.15}$$

The streamlines in this flow are concentric circles $r = \text{const}$ (dotted lines in Fig. 7.2), whereas the equipotential lines are the radii $\theta = \text{const}$ (full lines in Fig. 7.2). The velocity is

$$v_r = 0\,, \quad v_\theta = \frac{1}{r}\frac{\partial\varphi}{\partial\theta} = \frac{m}{r}\,.$$

Hence, (7.15) corresponds to the particles' rotation about the point $r = 0$ at a speed inversely proportional to the distance r from this point. This rotation is counterclockwise if $m > 0$ and clockwise if $m < 0$. Here we have a so-called *vortex source* with the circulation independent of r and equal to

$$\Gamma = 2\pi r v_\theta = 2\pi m \,. \tag{7.16}$$

Note that the existence of a nonvanishing circulation does not contradict Kelvin's theorem and the assumption about the potentiality of the flow. The point $r = 0$ is a singular one in this case and must be excluded from the region under consideration. Hence, the latter is not simply connected and Kelvin's theorem is inapplicable. The flow is potential everywhere except at the point $r = 0$.

7.2.3 Conformal Mapping

Conformal mapping is a well-developed branch of the theory of analytical functions. Its application to the theory of two-dimensional flow of an ideal incompressible fluid turns out to be very effective. Let us suppose that we have found a complex flow potential $F(z)$ for some domain satisfying the appropriate boundary conditions. By conformal mapping we now transform the complex plane, introducing the new variable ζ which is related to the variable z by $z = f(\zeta)$. The complex flow potential transforms as

$$F(z) = F[f(\zeta)] = \phi(\zeta)$$

and in the ζ-plane describes a new flow in a new domain. In such a way, using a suitable conformal mapping (transformation of $z \to \zeta$), one can obtain the solution of rather complicated problems from simpler ones. We will illustrate this method for wedgelike domains.

Assume an initial flow with constant velocity in the half space $y > 0$ with a rigid wall at $y = 0$ ($v_x = v_0, v_y = 0$, Fig. 7.3a). The complex flow potential $F(z) = v_0 z$ considered above describes this flow. Using the relation $z = R(\zeta/R)^\alpha$ ($\alpha, R > 0$), we transform the halfplane $y > 0$ into the wedge in the ζ-plane (Fig. 7.3b). The complex flow potential becomes

$$\phi(\zeta) = v_0 R(\zeta/R)^\alpha \,.$$

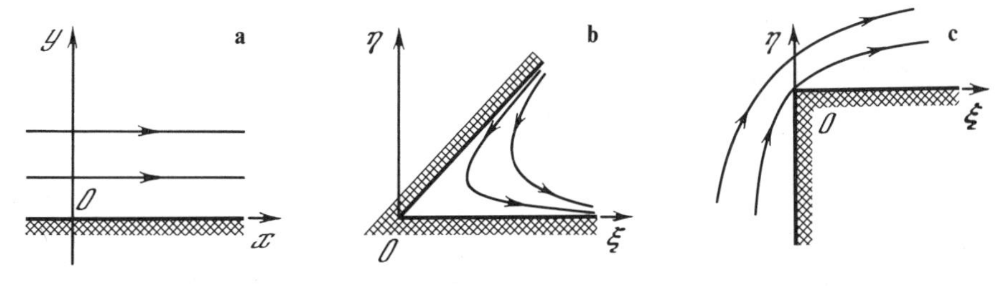

Fig. 7.3a–c. Conformal mapping of the half-plane $y > 0$ into the wedge with the help of $z = R(\zeta/R)^\alpha$

We introduce the polar coordinates r and θ, according to the relation $\zeta = r\exp(\mathrm{i}\theta)$, so that

$$\phi(\zeta) = v_0 R(r/R)^\alpha \exp(\mathrm{i}\alpha\theta) = v_0 R(r/R)^\alpha(\cos\alpha\theta + \mathrm{i}\sin\alpha\theta).$$

This yields for the velocity potential and the stream function

$$\varphi = v_0 R(r/R)^\alpha \cos\alpha\theta,$$

$$\psi = v_0 R(r/R)^\alpha \sin\alpha\theta$$

and for the velocity components

$$\begin{aligned} v_r &= \partial\varphi/\partial r = v_0\alpha(r/R)^{\alpha-1}\cos\alpha\theta, \\ v_\theta &= r^{-1}\,\partial\varphi/\partial\theta = -v_0\alpha(r/R)^{\alpha-1}\sin\alpha\theta. \end{aligned} \tag{7.17}$$

We observe that v_θ is zero at radii $\theta = 0$ and $\theta = \pi/\alpha$, where rigid walls can be placed without influencing the flow. Hence, we have obtained the solution of the problem of flow in a wedge with vertex angle π/α. The case $\alpha = 4$ is shown in Fig. 7.3b.

It follows from (7.17) that v_r, $v_\theta \to 0$ if $r \to 0$ and $\alpha > 1$, i.e., the velocity is zero at the vertex of the wedge. If $\alpha < 1$ we have flow around the wedge. The case $\alpha = 2/3$ is shown in Fig. 7.3c. It follows from (7.17) that v_r, $v_\theta \to \infty$ if $r \to 0$ in this case.

The complex flow potential $F(z) = \varphi + \mathrm{i}\psi$ can be regarded as an analytical function implementing a conformal mapping of some domain in the z-plane (the so-called "*flow domain*") onto the F-plane. It is useful in some cases to consider the conformal mapping of a flow domain onto the *complex velocity* plane

$$w(z) = F'(z) = \frac{\partial\varphi}{\partial x} + \mathrm{i}\frac{\partial\psi}{\partial x} = v_x - \mathrm{i}v_y. \tag{7.18}$$

Finally, there is the possibility for conformal mapping of the F-plane onto the w-plane. This last conformal mapping is governed by the transformation $w(F) = w[z(F)]$, where $z(F)$ is the transformation inverse to $F(z)$.

Suppose that the transformation $w(F)$ is known. The complex flow potential $F(z)$ can then be found by integration of the obvious relation

$$dF/dz = w(F). \tag{7.19}$$

The use of this procedure for solving some hydrodynamic problems is illustrated in Exercises 7.5.7, 8.

7.3 Steady Flow Around a Cylinder

7.3.1 Application of Conformal Mapping

Let us consider the flow around a circular cylinder. This case is important by itself and because some other more sophisticated cases can be treated by conformal mapping using the flow around a cylinder as the initial flow.

Consider a circle $r = R$ (R: radius of the cylinder) in the plane $z = x + \mathrm{i}y = r\exp(\mathrm{i}\theta)$ (Fig. 7.4a). We seek the solution of the Laplace equation $\Delta\varphi = 0$ in the domain outside the circle, under the condition that the normal velocity at the boundary $r = R$ is zero and at infinity ($r \to \infty$), the undisturbed flow has the velocity components $v_x = -v_0$, $v_y = 0$. In terms of the potential these conditions are

$$\left.\frac{\partial\varphi}{\partial r}\right|_{r=R} = 0\,, \quad \left.\frac{\partial\varphi}{\partial x}\right|_{r\to\infty} = -v_0\,, \quad \left.\frac{\partial\varphi}{\partial y}\right|_{r\to\infty} = 0\,. \tag{7.20}$$

The potential ϕ and the stream function ψ ($\Delta\psi = 0$) together give the complex flow potential $F(z) = \varphi(x, y) + \mathrm{i}\psi(x, y)$. We now implement a conformal mapping of the z-plane such that the exterior domain of the circle $r = R$ transforms onto the whole ζ-planes except the cut between -1 and $+1$ (Fig. 7.4b). The well-known *Joukovski transformation* can be used for this purpose:

$$\zeta = \xi + \mathrm{i}\eta = \frac{1}{2}\left(\frac{z}{R} + \frac{R}{z}\right). \tag{7.21}$$

In this way, the circle $z = R\exp(\mathrm{i}\theta)$ is transformed into the straight cut between $\zeta = -1$ and $\zeta = 1$ traversed twice. Instead of $F(z)$, we now have

$$\phi(\zeta) = \phi\left[\frac{1}{2}\left(\frac{z}{R} + \frac{R}{z}\right)\right] = \varphi_1(\xi, \eta) + \mathrm{i}\psi_1(\xi, \eta)\,.$$

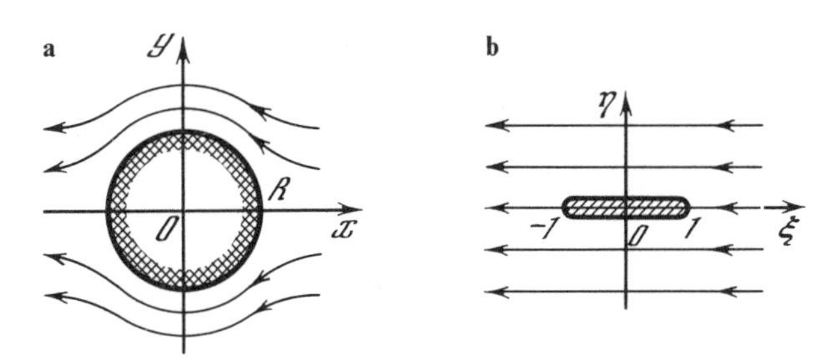

Fig. 7.4a, b. Conformal mapping of the circular cylinder (a) on the plane $\xi = \xi + \mathrm{i}\eta$ (b)

The condition $(\partial\varphi/\partial r)_{r=R} = 0$ now becomes $(\partial\varphi_1/\partial\eta)_{\eta=0} = 0$ since the angle between the intersecting lines (the right angle between the boundary and its normal in our case) does not change under conformal mapping. Further, one has at $r \to \infty$,

$$\zeta = \xi + \mathrm{i}\eta \to z/2R = (x/2R) + \mathrm{i}y/2R$$

which yields $\varphi(x, y) \to \varphi_1(x/2R, y/2R)$ at $r \to \infty$. Hence,

$$\left.\frac{\partial\varphi}{\partial x}\right|_{r\to\infty} = -v_0 \to \left.\frac{1}{2R}\frac{\partial\varphi_1}{\partial\xi}\right|_{\varkappa\to\infty}, \quad \left.\frac{\partial\varphi}{\partial y}\right|_{\varkappa\to\infty} = 0 = \left.\frac{1}{2R}\frac{\partial\varphi_1}{\partial\eta}\right|_{\varkappa\to\infty}$$

$$\text{or } (\partial\varphi_1/\partial\xi)_{\varkappa\to\infty} = -2Rv_0\,, \quad (\partial\varphi_1/\partial\eta)_{\varkappa\to\infty} = 0\,,$$

where

$$\varkappa = \sqrt{\xi^2 + \eta^2}\,.$$

As a result, we have the problem of flow past an infinitely thin strip under the condition that the fluid velocity is $-2Rv_0$ in front of the strip. The solution of this problem is the flow with the velocity independent of the space variables:

$$v_\xi = -2Rv_0\,, \quad v_\eta = 0\,.$$

The complex flow potential is, for this case (Sect. 7.2),

$$\phi(\zeta) = -2Rv_0\zeta\,.$$

Returning to the variable z with the help of (7.21), we obtain the complex flow potential of the initial problem

$$F(z) = -2Rv_0\zeta(z) = -Rv_0\left(\frac{z}{R} + \frac{R}{z}\right) = -v_0\left(r\mathrm{e}^{\mathrm{i}\theta} + \frac{R^2}{r}\mathrm{e}^{-\mathrm{i}\theta}\right). \tag{7.22}$$

Hence, the potential is

$$\varphi = \mathrm{Re}\{F(z)\} = -v_0(r + R^2/r)\cos\theta \tag{7.23}$$

and the velocity components are

$$\begin{aligned} v_r &= \partial\varphi/\partial r = -v_0(1 - R^2/r^2)\cos\theta\,, \\ v_\theta &= r^{-1}\,\partial\varphi/\partial\theta = v_0(1 + R^2/r^2)\sin\theta\,. \end{aligned} \tag{7.23a}$$

The boundary conditions (7.20) for φ are satisfied. In fact, we have $(\partial\varphi/\partial r)_{r=R} = 0$ and $\varphi \to -v_0 r\cos\theta = -v_0 x$ at $r \to \infty$; hence, $(\partial\varphi/\partial x)_{r\to\infty} = -v_0$, $(\partial\varphi/\partial y)_{r\to\infty} = 0$.

For the velocity components at the surface of the cylinder ($r = R$), we have

$$v_r = 0\,, \quad v_\theta = 2v_0\sin\theta\,. \tag{7.24}$$

The maximum velocity is at the midsection ($\theta = \pm\pi/2$) where

$$v = v_\theta = 2v_0 .$$

7.3.2 The Pressure Coefficient

The pressure at the cylindrical surface can be determined by the Bernoulli theorem

$$p_0 + \rho v_0^2/2 = p + \rho v^2/2 , \tag{7.25}$$

where p_0 and v_0 are the pressure and velocity at infinity and p and v are the same quantities at the surface. Substitution of v from (7.24) into (7.25) yields for the pressure

$$p = p_0 + \rho(v_0^2 - v^2)/2 = p_0 + \rho v_0^2(1 - 4\sin^2\theta)/2 . \tag{7.26}$$

The pressure is $p_0 + \rho v_0^2/2$ at the stagnation point ($\theta = 0$) where the flow is divided. Here it is greater than at infinity. The same pressure occurs at the symmetric point at the rear of the cylinder ($\theta = \pi$). As we approach the midsection ($\theta = \pm\pi/2$), the pressure drops monotonically to the value $p_0 - 3\rho v_0^2/2$ which is less than p_0. The quantity

$$K = \frac{p - p_0}{\rho v_0^2/2} = 1 - 4\sin^2\theta \tag{7.27}$$

is termed the *pressure coefficient*. It does not depend on the radius of the cylinder, the fluid density or the flow velocity. This fact is a manifestation of the principle of the so-called *hydrodynamic similarity* which will be discussed below. Owing to this principle, the foregoing theory could be experimentally proved by determination of the pressure-coefficient dependence on θ for a circular cylinder with arbitrary fixed R, v_0 and ρ.

Experiments with real fluids do not confirm the law (7.27), however. The coefficient K is equal to unity at the stagnation point ($\theta = 0$) again but does not drop as low as $K = -3$ near the midsection, as follows from (7.27), and does not increase up to unity at the rear critical point ($\theta = \pi$). This is because the flow of the type considered above where the streamlines follow the boundary, as in Fig. 7.4, is impossible for real fluids. It turns out that the streamlines leave the cylinder surface in the region $\pi/2 < \theta < \pi$, and vortexes are generated there. We will discuss this phenomenon later. All this does not mean, however, that the theory of an ideal fluid flow is useless, the cylinder is of a rather poor streamline profile. For a good streamline airfoil (such as an airplane wing, for example), this theory is closer to reality.

7.3.3 The Paradox of d'Alembert and Euler

We have seen above that the pressure distribution (7.26) in the flow around the circular cylinder is symmetric with respect to the midsection $\theta = \pi/2$. Hence, the total force on the cylinder is zero. This is the particular case of a more general statement: there is no net force (a couple of forces is possible) on a body of arbitrary but smooth shape in ideal fluid flow. Its proof for the general three-dimensional case is based on Euler's theorem. Suppose first that a body is placed into a cylindrical tube with rigid walls and a generator parallel to the streamlines at infinity (Fig. 7.5). The lateral surface of the tube S' as well as the

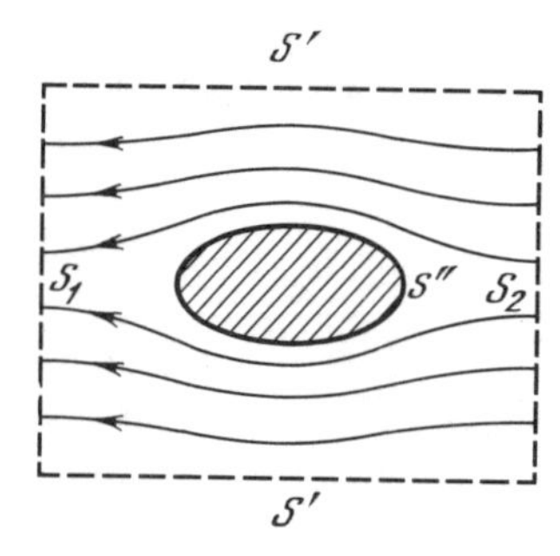

Fig. 7.5. Proof of the paradox of d'Alembert and Euler

body's surface S'' coincide with certain streamlines. Consider two cross sections of the tube S_1 and S_2 on both sides of the body, sufficiently far from it, where the flow velocity can be assumed undisturbed and equal to $-v_0$. The force on the fluid between the surfaces $S' + S_1 + S_2$ and S'' is equal to the momentum flux difference between the cross sections S_1 and S_2, according to Euler's theorem. The areas of S_1 and S_2 as well as the velocities at these cross sections are equal, hence the total force on the fluid is zero. To determine the force on the body, consider the component of the force on the fluid along the flow direction. Such a component is zero at S', since the pressure on the wall is perpendicular to $\mathbf{v}_0$. Three other components are the force F on the fluid by the body and pressure forces at the cross section. Their sum must be zero: $F + p_1 S_1 - p_2 S_2 = 0$. Now, since $S_1 = S_2$, $p_1 = p_2 = p_0$, we obtain $F = 0$. Hence, the force exerted by the fluid on the body must also be zero, according to Newton's third law. Displacement of the tube's walls to infinity changes nothing in our reasoning. Hence, the force on the body will also be zero in unlimited flow.

We can superimpose a homogeneous flow with velocity v_0 on the flow considered above . Then the flow velocity will be zero at infinity and we obtain the case of a body moving in a fluid at rest. It follows from the aforesaid that a body moving in an ideal fluid at constant velocity experiences no drag.

7.3.4 The Flow Around a Cylinder with Circulation

In the case considered above, the force perpendicular to the flow velocity was also zero [expression (7.26) is symmetric with respect to the plane $\theta = 0$]. We have quite another case, however, if the circulation flow around a cylinder is superimposed on the flow considered. The complex flow potential will now be the sum of that for symmetric flow (7.22) and that for circulation flow (7.15). Hence we have, taking into (7.16) account,

$$F(z) = -v_0(z + R^2/z) + (\Gamma/2\pi i)\ln z\,. \tag{7.28}$$

Obtaining φ and ψ from this expression, one can determine all characteristics of the flow. Naturally, the normal component of the velocity v_2 on the cylinder surface vanishes. The tangential component v_θ was found above separately for both flows, see (7.16, 24). The total velocity is their sum:

$$v\,|_{r=R} = v_\theta|_{r=R} = 2V_0 \sin\theta + \Gamma/2\pi R\,. \tag{7.29}$$

Substituting this quantity as v into the Bernoulli theorem (7.25), we obtain for the pressure distribution over the cylinder surface

$$\begin{aligned} p &= p_0 + \rho[v_0^2 - (2v_0 \sin\theta + \Gamma/2\pi R)^2]/2 \\ &= p_0 + \rho[v_0^2(1 - 4\sin^2\theta) - (2\Gamma v_0/\pi R)\sin\theta - (\Gamma/2\pi R)^2]/2\,. \end{aligned} \tag{7.30}$$

The pressure p is symmetric with respect to the midsection ($\theta = \pi/2$), therefore the cylinder again experiences no drag. However, the pressure is not symmetric with respect to the plane $\theta = 0$ due to the presence of the term $-(2\Gamma v_0/\pi R)\sin\theta$ in (7.30). Hence, the pressure is different on the upper and lower halves of the cylinder. To understand the reason for this, mentally superimpose the counter-clockwise circulation flow (Γ positive) upon the symmetrical flow shown in Fig. 7.4a. The velocities of the flow will be of the same sign above the cylinder and of opposite sign below it. Hence, the resultant velocity will be greater above the cylinder. The pressure according to the Bernoulli theorem will be greater below the cylinder, hence a force in the y-direction will arise (*hydrodynamic lift*). We will show that this force per unit length of the cylinder is

$$F_y = -2 \int_{-\pi/2}^{\pi/2} pR \sin\theta \, d\theta\,.$$

Indeed, the area element at the surface of the cylinder is $R\,d\theta$. The projection of the force per unit area on the y-axis is $-p\sin\theta$ and we obtain above formula for F_y if we also take the symmetry of p about the midsection ($\theta = \pi/2$) into account and perform the integration in the limits $(-\pi/2, \pi/2)$ instead of $(0, 2\pi)$. Substituting p from (7.30) (taking into account only the antisymmetric term) into the integral for F_y, we obtain

$$F_y = \frac{2\Gamma v_0 \rho}{\pi} \int_{-\pi/2}^{\pi/2} \sin^2\theta \, d\theta = \rho v_0 \Gamma . \tag{7.31}$$

The lift is proportional to the velocity of the symmetric flow v_0 as well as to the circulation Γ. If the latter changes its sign, the "lift" also changes its direction.

Note that the expression for the lift $F_y = \rho v_0 \Gamma$ holds for any cylindrical body (not only of circular cross section). This is the Kutta-Joukovski theorem.

The so-called rotor propulsion device which was proposed for ships use the force given by (7.31). The idea is that if a cylinder rotating about the vertical axis is installed on the ship, it gives rise to a force perpendicular to the wind direction. However, such a device is not used in practice yet.

7.4 Irrotational Flow Around a Sphere

7.4.1 The Flow Potential and the Particle Velocity

In the general case of three-dimensional flow, the conformal mapping method cannot be used and we have to solve the Laplacian equation directly under corresponding boundary conditions. We will demonstrate this approach in the case of flow around a sphere.[7]

Suppose that a fluid is at rest at infinity and a sphere moves with a constant velocity $\boldsymbol{v}_0$. We again introduce the velocity potential φ ($\boldsymbol{v} = \nabla\varphi$) which must satisfy the Laplace equation (7.2). The boundary condition is the equality of the normal velocity of the spherical particle $\boldsymbol{v}_0 \cdot \boldsymbol{r}/r$ to that of the fluid $\boldsymbol{v} \cdot \boldsymbol{r}/r$. A spherical system of coordinates is used here with the origin at the center of the sphere (Fig. 7.6). Our problem is formulated mathematically as

$$\begin{aligned} &\Delta\varphi = 0 , \quad \boldsymbol{v}|_{r\to\infty} = 0 , \\ &\left.\frac{\boldsymbol{v}\cdot\boldsymbol{r}}{r}\right|_{r=R} = \left.\frac{\boldsymbol{v}_0\cdot\boldsymbol{r}}{r}\right|_R , \quad r = \sqrt{x^2+y^2+z^2} . \end{aligned} \tag{7.32}$$

The well-known fundamental solution of the Laplace equation satisfying our condition at infinity is $\varphi = A/r$, where A is constant. Such a solution describes, for instance, the gravitational potential of a point mass or the electrical

[7] Since this problem has cylindrical symmetry it can be also reduced to a two-dimensional one. We prefer, however, the way used below.

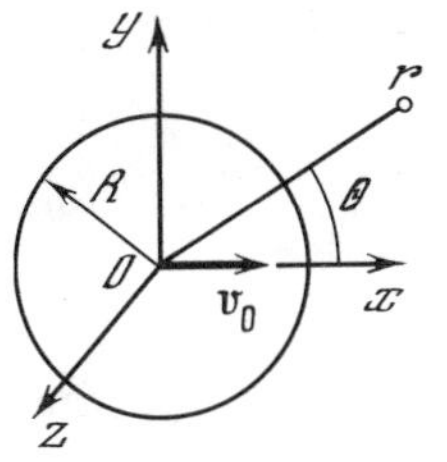

Fig. 7.6. Geometry in the case of the flow around a sphere

potential of a point charge. It cannot serve as the solution to our problem, however, since it is centrally symmetric whereas our problem has no such symmetry due to the existence of flow with velocity $\boldsymbol{v}_0$. Let us try, therefore, another simple solution of the Laplace equation, i.e., the potential of the dipole $\boldsymbol{B}\nabla(r^{-1})$, $\boldsymbol{B}$ being the constant vector called the dipole moment. It satisfies the Laplace equation since the operations $\partial/\partial x$ and Δ commute. The condition at infinity is also satisfied since $\nabla(r^{-1}) \sim r^{-2}$ at $r \to \infty$.

The sphere's velocity $\boldsymbol{v}_0$ enters the boundary condition at $r = R$ linearly and we look, therefore, for a solution in the form

$$\varphi = A\boldsymbol{v}_0\nabla(r^{-1}) = -A\boldsymbol{v}_0 \cdot \boldsymbol{r}/r^3$$

which yields for the particle velocity

$$\boldsymbol{v} = -A\boldsymbol{v}_0/r^3 + 3A(\boldsymbol{v}_0 \cdot \boldsymbol{r})\boldsymbol{r}/r^5 .$$

Substitution of this expression into the boundary condition at the surface of the sphere gives

$$-Av_{0r}/R^3 + 3Av_{0r}/R^3 = v_{0r}$$

whence $A = R^3/2$. Now we have the required solution for flow around the sphere:

$$\varphi = -R^3(\boldsymbol{v}_0 \cdot \boldsymbol{r})/2r^3 , \qquad \boldsymbol{v} = R^3\boldsymbol{v}_0/2r^3 + 3R^3(\boldsymbol{v}_0 \cdot \boldsymbol{r})\boldsymbol{r}/2r^5 . \tag{7.33}$$

If we superimpose the homogeneous flow of the velocity $-\boldsymbol{v}_0$ upon this flow, we obtain the solution of the problem of the flow around a sphere at rest when the flow velocity at infinity is $-\boldsymbol{v}_0$. For the velocity in this new problem we obtain, according to (7.33),

$$\boldsymbol{v} = -\boldsymbol{v}_0 - R^3\boldsymbol{v}_0/2r^3 + 3R^3(\boldsymbol{v}_0 \cdot \boldsymbol{r})\boldsymbol{r}/2r^5 \tag{7.34}$$

or in components,

$$\begin{aligned} v_r = \boldsymbol{v}\boldsymbol{r}/r &= -v_0 \cos\theta - (R^3 v_0 \cos\theta)/2r^3 \\ &\quad + 3(R^3 v_0 \cos\theta)/2r^3 = -v_0 \cos\theta(1 - R^3/r^3) , \\ v_\theta &= v_0 \sin\theta(1 + R^3/2r^3) . \end{aligned} \tag{7.35}$$

In the last expression, we have taken into account that $v_{0\theta} = v_0 \sin\theta$.

At the surface of the sphere ($r = R$) we have $v_r = 0$ and

$$v_\theta = v = 3(v_0 \sin\theta)/2\,, \tag{7.36}$$

i.e., $v = 0$ at both critical points $\theta = 0$ and $\theta = \pi$. The pressure distribution over the sphere can be determined with the help of the Bernoulli theorem. Substitution of v from (7.36) into (7.25) yields

$$p = p_0 + \rho v_0^2[1 - (9/4)\sin^2\theta]\,. \tag{7.37}$$

The pressure is symmetric about the midsection plane ($\theta = \pi/2$) and minimum at it. The pressure coefficient analogous to (7.27) is

$$K = \frac{p - p_0}{\rho v_0^2/2} = 1 - \frac{9}{4}\sin^2\theta\,. \tag{7.38}$$

7.4.2 The Induced Mass

Since the pressure distribution is symmetric about the plane $\theta = \pi/2$, the total force exerted by the flow on the sphere is zero which is also in accordance with the d'Alambert-Euler paradox. However, it is true only if the sphere's velocity is constant ($v_0 = \text{const}$). If an acceleration of the sphere takes place, additional *induced inertia* arises due to the flow's action.

To determine the additional "*induced mass*" of the sphere, we suppose that there is a force F providing it with a constant acceleration a. If the sphere were initially at rest, then at the time $T = v_0/a$, its velocity is v_0 and the distance covered $s = aT^2/2 = v_0^2/2a$. Hence, the work done by the force is $Fs = Fv_0^2/2a$. It must be equal to the kinetic energy generated in the system "*sphere plus fluid*":

$$Fv_0^2/2a = mv_0^2/2 + \int_V (\rho v^2/2)\,dV;$$

the integration is over the exterior of the sphere ($r > R$). This equation can be written as

$$F = ma + Ma\,, \quad \text{where}$$

$$M = (\rho/v_0^2)\int_{r>R} v^2\,dV \tag{7.39}$$

is the *induced mass*. To determine M for our case, we have from (7.33)

$$v_r = (v_0 R^3 \cos\theta)/r^3\,, \qquad v_\theta = (v_0 R^3 \sin\theta)/2r^3\,,$$

$$v^2 = v_0^2(R^6/r^6)(1 + 3\cos^2\theta)/4$$

and

$$\int_{r>R} v^2\,dV = \tfrac{2}{3}\pi R^3 v_0^2 .$$

Now we obtain for M,

$$M = \tfrac{2}{3}\pi R^3 \rho = \tfrac{1}{2}\cdot\tfrac{4}{3}\pi R^3 \rho = \tfrac{1}{2} V_{\mathrm{S}}\rho \qquad (7.40)$$

where V_{S} is the volume of the sphere. Hence, the induced mass is equal to half the mass of the fluid displaced by the sphere.

7.5 Exercises

7.5.1. The fluid rotates about the vertical axis, the angular velocity being dependent on the radius $\Omega(r)$. Determine the function $\Omega(r)$ assuming that the motion is potential.

Solution: The particle velocity is

$$\boldsymbol{v} = \{u, v\} = \boldsymbol{\Omega} \times \boldsymbol{R} = \{-\Omega y, \Omega x\},$$

where $\boldsymbol{R} = \{x, y, z\} = \{\boldsymbol{r}, z\}$. This yields for $\boldsymbol{\omega} = \operatorname{curl} \boldsymbol{v}$:

$$\boldsymbol{\omega} = \omega \nabla z, \qquad \omega = \partial v/\partial x - \partial u/\partial y = \Omega + x\Omega' x/r + \Omega + y\Omega' y/r = 2\Omega + r\Omega' .$$

For potential flow $\omega = 0$ or $\Omega' \equiv d\Omega/dr = -2\Omega/r$. The solution of this equation is $\Omega(r) = \Omega_0 a^2/r^2$ where $\Omega_0 = \Omega(a)$; a is arbitrary.

7.5.2. Find the shape of the free surface of a fluid in a gravitational field under the same conditions as in the previous exercise.

Solution: The gravity potential is gz (z is positive upward). Introducing the centrifugal inertial force $\rho\Omega^2\boldsymbol{r} = \rho\Omega_0^2(a/r)^4\boldsymbol{r}$ with its potential per unit mass $\Omega_0^2 a^4/2r^2$, write the Bernoulli equation (6.38) as

$$\frac{p}{\rho} + \frac{\Omega_0^2 a^4}{2r^2} + gz = gz_0 = \text{const} .$$

The constant z_0 is the same for all the fluid since the motion is potential (Sect. 6.4.1). For the free surface ($p = 0$), we now obtain:

$$z = z_0 - \Omega_0^2 a^4/2gr^2 .$$

Note that $z \to -\infty$ as $r \to 0$.

7.5.3. Find the shape of the free surface of a fluid rotating about the vertical axis in the gravitational field assuming that the angular velocity is Ω_0 if $r \leq a$ and $\Omega_0(a/r)^2$ if $r \geq a$.

Solution: The shape of the free surface at $r < a$ is (Exercise 6.7.11)

$$z(r) = H + \Omega_0^2 r^2/2g\,, \qquad H = (p_0 - p_a)/g\rho\,.$$

For $r > a$, it was found in the previous exercise. The continuity condition for $z(r)$ at $r = a$ yields the relation $z_0 = H + (\Omega_0 a)^2/g$. Hence, we have for the free surface

$$z(r) = H + \frac{\Omega_0^2 a^2}{2g} \cdot \begin{matrix} r^2/a^2\,, & r \leq a\,, \\ 2 - a^2/r^2\,, & r \geq a\,. \end{matrix}$$

Note that dz/dr is also continuous at $r = a$.

7.5.4. Find the complex potential $F(z)$ for an incompressible fluid flow around an elliptical cylinder. The flow velocity at infinity is $v_x = -v_0$, $v_y = 0$; the x-direction is one of the principal axes of the ellipse (Fig. 7.7).

Solution: First we find the conformal mapping of the exterior of the ellipse in the z-plane on the exterior of the circle $|\zeta| > R$ in the ζ-plane. For this purpose we write the ellipse's equation in polar coordinates: $x = a\cos\theta$, $y = b\sin\theta$, $z = x + iy = a\cos\theta + ib\sin\theta$, or

$$z = \frac{a+b}{2}\exp(i\theta) + \frac{a-b}{2}\exp(-i\theta) = \frac{a+b}{2R}R\exp(i\theta) + \frac{a-b}{2R}\frac{R^2}{R\exp(i\theta)}\,.$$

Now one can see that the desired conformal mapping is given by the generalized Joukovski transformation (7.21):

$$z = f(\zeta) = \frac{a+b}{2R}\left(\zeta + \frac{a-b}{a+b}\frac{R^2}{\zeta}\right).$$

We assume $R = (a+b)/2$, then $z = \zeta + (a^2 - b^2)/4\zeta$ and $z \to \zeta$ at infinity. The inverse transformation is $\zeta = (z + \sqrt{z^2 - a^2 + b^2})/2$. Now in the plane $\zeta =$

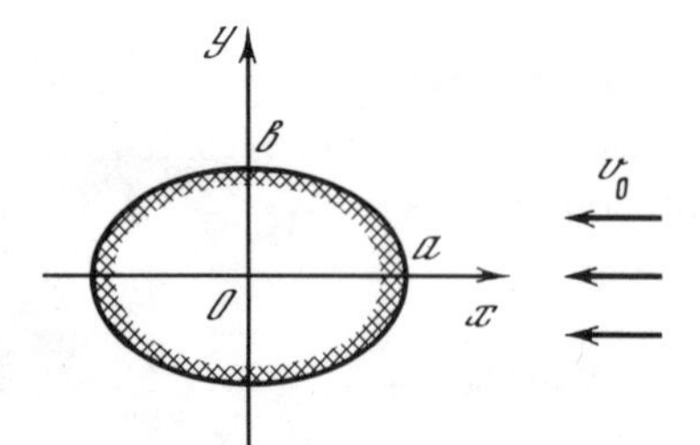

Fig. 7.7. The flow around an elliptical cylinder

$\xi + \mathrm{i}\eta$, we have the problem of flow around a circular cylinder of radius $R = (a+b)/2$ under the conditions at infinity:

$$\left.\frac{\partial\phi(\xi,\eta)}{\partial\xi}\right|_{|\zeta|\to\infty} = \left.\frac{\partial\phi(x,y)}{\partial x}\right|_{|z|\to\infty} = -v_0\,, \qquad \left.\frac{\partial\phi}{\partial\eta}\right|_{|\zeta|\to\infty} = 0\,.$$

The complex flow potential for this problem is, according to (7.22), $\phi(\zeta) = -v_0[\zeta + (a+b)^2/4\zeta]$. Using the inverse transformation $\zeta(z)$, we finally obtain the complex flow potential for the flow around the elliptical cylinder:

$$F(z) = \phi[\zeta(z)] = -\frac{v_0}{2}\left(z + \sqrt{z^2 - a^2 + b^2} + \frac{(a+b)^2}{z + \sqrt{z^2 - a^2 + b^2}}\right)$$

$$= -\frac{v_0}{2}(a+b)\left(\frac{z + \sqrt{z^2 - a^2 + b^2}}{a+b} + \frac{z - \sqrt{z^2 - a^2 + b^2}}{a-b}\right).$$

7.5.5. The same Exercise 7.5.4, but at infinity the flow makes an angle α with the x-axis.

Solution: Consider first the case of a circular cylinder. The rotation of the coordinate axes corresponds to the conformal transformation $\zeta = z\exp(-\mathrm{i}\alpha)$. Taking this into account, we obtain from (7.22)

$$F_c(z) = -v_0[z\exp(-\mathrm{i}\alpha) + R^2 z^{-1}\exp(\mathrm{i}\alpha)]\,.$$

In the case of an elliptical cylinder, we use the transformation (Exercise 7.5.4) $\zeta = (z + \sqrt{z^2 - a^2 + b^2})/2$, i.e., conformal mapping of the ellipse exterior on the exterior of the circle of radius $R = (a+b)/2$. Using $F_c(z)$ instead of (7.22), we finally obtain

$$F_e(z) = -\frac{v_0}{2}(a+b)\left[\frac{z + \sqrt{z^2 - a^2 + b^2}}{a+b}\exp(-\mathrm{i}\alpha) + \frac{z - \sqrt{z^2 - a^2 + b^2}}{a-b}\exp(\mathrm{i}\alpha)\right].$$

7.5.6. Find the complex potential of the flow around an infinitely thin strip of the width $2a$. At infinity the flow velocity is v_0 in magnitude and makes an angle α with the strip's plane (Fig. 7.8). Consider the particular cases $\alpha = 0$ and $\alpha = \pi/2$.

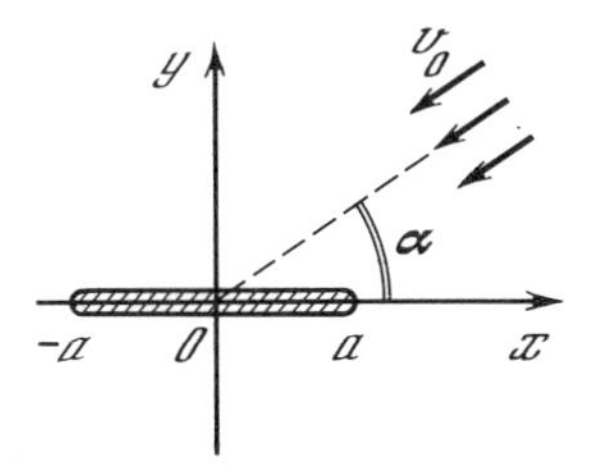

Fig. 7.8. The flow around a strip

Solution: The strip is the particular case of an elliptical cylinder when $b = 0$. Using $F_e(z)$, from Exercise 7.5.5 we derive

$$F(z) = -v_0(z\cos\alpha - \mathrm{i}\sqrt{z^2 - a^2}\sin\alpha) = v_{0x}z - \mathrm{i}v_{0y}\sqrt{z^2 - a^2}\,.$$

$F(z) = -v_0 z$ when $\alpha = 0$, i.e., the flow remains undisturbed by the strip in this case (which is only natural). When $\alpha = \pi/2$ (the flow at infinity is orthogonal to the strip), $F(z) = \mathrm{i}v_0\sqrt{z^2 - a^2}$. Note that the complex velocity $F'(z)$ is infinite at the strip's edges ($z = \pm a$).

7.5.7. An incompressible fluid flows out of the halfspace $x < 0$ through the slot LL′ of width $2D$. We are faced with a two-dimensional problem (Fig. 7.9a), AB and A′ B′ being the jet's boundaries into the free halfspace $x > 0$. Let p_2 be the pressure in the latter and p_1 be that at infinity in the halfspace $x < 0$. Find the conformal mapping of the flow in the plane $z = x + \mathrm{i}y$ onto the new plane of the complex flow potential $F = \varphi + \mathrm{i}\psi$ as well as on the new plane of the complex velocity $w = u - \mathrm{i}v$.

Solution: Note that the problem has mirror symmetry about the x-axis and that this axis is one of the streamlines ($\psi|_{y=0} = \text{const}$). Without loss of generality, we can assume that $\psi|_{y=0} = 0$. LAB is also a streamline. At $x \to \infty$ the flow becomes homogeneous ($u = \partial\psi/\partial y = v_0$, $v = -\partial\psi/\partial x = 0$), hence $\psi|_{x\to\infty} = v_0 y$ and $\psi = v_0 d$ at the streamline LAB. By analogy, $\psi = -v_0 d$ at the streamline L′ A′ B′. This implies that in the plane of the complex potential $F = \varphi + \mathrm{i}\psi$, the flow domain is a strip (Fig. 7.9b).

The velocity v_0 for $x \to \infty$ can be found with the help of the Bernoulli theorem:

$$\frac{p_1}{\rho} = \frac{p_2}{\rho} + \frac{v_0^2}{2}, \qquad v_0 = \sqrt{2(p_1 - p_2)/2}\,.$$

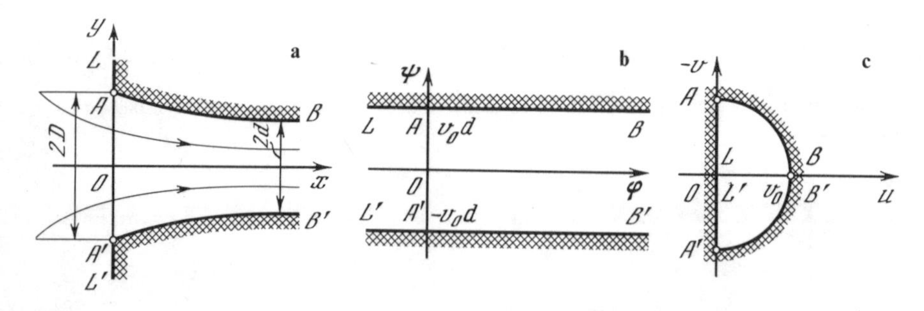

Fig. 7.9a–c. Mapping of the flow region (the fluid outflow through the slot) (**a**) onto the z-plane, (**b**) onto the plane of the complex flow potential, and (**c**) onto the complex velocity plane

It follows from this theorem also that on AB and A′ B′ the velocity has constant magnitude $|v_0|$ and changes from $\boldsymbol{v} = \{0, -v_0\}$ to $\boldsymbol{v} = \{v_0, 0\}$ on AB and from $\boldsymbol{v} = \{0, v_0\}$ to $\boldsymbol{v} = \{v_0, 0\}$ on A′ B′. On LA we have $\boldsymbol{v} = \{0, v\}$; v changes monotonically from $v = 0$ (as $y \to \infty$) to $v = -v_0$ (at $y = 0$). Also taking the symmetry into account, we conclude from the above that in the plane of complex velocity, the flow domain is the interior of the half circle (Fig. 7.9c).

7.5.8. Using the results of the previous exercise, find the complex flow potential and the outflow coefficient for the fluid flowing out through a slot.

Solution: First we find the transformation $w = w(F)$ (notation as in Exercise 7.5.7). For this purpose we transfer each of the regions in Figs. 7.9b, 7.9c by conformal mappings $\zeta = \zeta(F) = \zeta(w)$ into the halfplane $\zeta > 0$ of the variable $\zeta = \xi + \mathrm{i}\eta$ (Fig. 7.10). It is easy to verify that the first mapping is given by the function $\zeta = \exp(-\pi F/2v_0 d)$. The second one is the Joukovski transformation (7.21) with a rotation by $\pi/2$: $\zeta = -\mathrm{i}(w/\mathrm{i}v_0 + \mathrm{i}v_0/w)/2$. The inverse transformation is $w = v_0(\sqrt{\zeta^2 + 1} - \zeta)$. Consequently, the mapping of the strip in the F-plane onto the half circle in the w-plane is

$$w(F) = v_0\left[\sqrt{1 + \exp(-\pi F/v_0 d)} - \exp(-\pi F/2v_0 d)\right].$$

Now, since $w = F'$, the desired transformation $F(z)$ can be found from

$$v_0\, dz = dF/(\sqrt{1 + \alpha^2} - \alpha) = (\alpha + \sqrt{1 + \alpha^2})\, dF\,, \qquad \alpha = \exp(-\pi F/2v_0 d)\,.$$

Integration yields the inverse transformation $z = z(F)$:

$$z = C - \frac{2d}{\pi}\left[\frac{\pi F}{2v_0 d} + \alpha + \sqrt{1 + \alpha^2} + \ln(\sqrt{1 + \alpha^2} - 1)\right].$$

The complex constant $C = C' + \mathrm{i}C''$ can be determined from the conditions on the streamlines. Consider, for example, the streamline LAB (Fig. 7.9a) where $\psi = v_0 d$. Hence, $F = \varphi + \mathrm{i}v_0 d$ $(-\infty < \varphi < \infty)$. Substituting this result into the

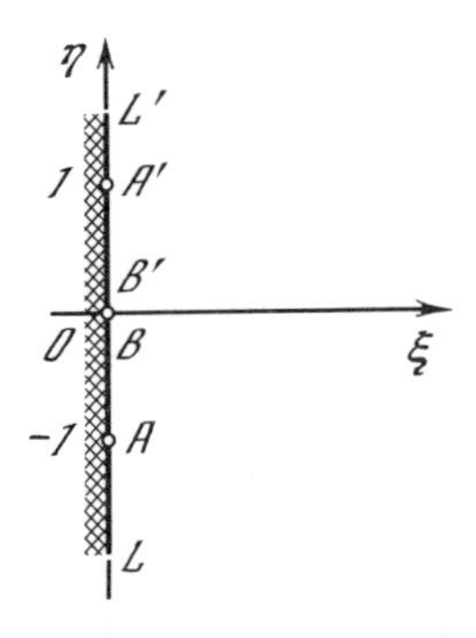

Fig. 7.10. The half-plane $\zeta = \xi + \mathrm{i}\eta$

expression for $z(F)$ we obtain

$$\alpha = \exp(-\mathrm{i}\pi/2 - \gamma), \quad \gamma = \pi\varphi/2v_0 d \quad \text{and}$$

$$z|_{\psi = v_0 d} = C - \frac{2d}{\pi}\left[\gamma + \mathrm{i}\frac{\pi}{2} - \mathrm{i}e^{-\gamma} + \sqrt{1 - e^{-2\gamma}} + \ln(\sqrt{1 + e^{-\mathrm{i}\pi - 2\gamma}} - 1)\right]. \tag{7.41}$$

If $\varphi \to \infty$, then $z \to x + id$, $\gamma \to \infty$,

$$\ln[\sqrt{1 + \exp(-\mathrm{i}\pi - 2\gamma)} - 1] \simeq -\ln 2 - \mathrm{i}\pi - 2\gamma .$$

Hence,

$$x + id = C' + \mathrm{i}C'' + \varphi/v_0 - 2d(1 - \ln 2)/\pi + id .$$

Consequently, $C'' \equiv 0$. On the segment LA (corresponding to $\varphi < 0$) of this streamline, we have $z = \mathrm{i}y$, $\gamma < 0$, $\exp(-\gamma) > 1$; hence,

$$C' - \varphi/v_0 - 2d\ln|\sqrt{1 - \exp(-2\gamma)} - 1|/\pi = 0 .$$

But $\sqrt{1 - \exp(-2\gamma)} = \mathrm{i}\sqrt{\exp(-2\gamma) - 1}$, $|\mathrm{i}\sqrt{\exp(-2\gamma) - 1} - 1| = \exp(-\gamma)$; hence, $C' \equiv 0$ also.

The outflow coefficient which is $k = d/D$ by definition, can be derived from the condition $z(\mathrm{i}v_0 d) = \mathrm{i}D$ which yields $D = (\pi + 2)d/\pi$, $k = \pi/(2 + \pi) = 0.61$.

7.5.9. Find the induced mass per unit length of a circular cylinder moving with acceleration in a direction perpendicular to its axis.

Solution: By analogy to the case of a sphere (Sect. 7.7.2), first determine the kinetic energy of the fluid disturbed by the cylinder moving with velocity v_0. Adding the term $v_0 z$ to the complex flow potential (7.22), we obtain the complex flow potential for the case of the cylinder moving with velocity v_0:

$$F(z) = -v_0 R^2/z = -v_0 R^2 \exp(-\mathrm{i}\theta)/r , \qquad \varphi = -v_0(R^2/r)\cos\theta .$$

For the particle velocity we find

$$v_r = \partial\varphi/\partial r = v_0(R^2/r^2)\cos\theta ,$$
$$v_\theta = r^{-1}\partial\varphi/\partial\theta = v_0(R^2/r^2)\sin\theta ,$$
$$v^2 = v_0^2(R/r)^4 .$$

and for the kinetic energy,

$$\mathscr{E}_k = \frac{1}{2}\int_0^{2\pi}\int_0^{\infty} \rho v^2 r\,dr\,d\theta = \rho\pi R^2 v_0^2/2 = M v_0^2/2 ,$$

where the induced mass $M = \rho\pi R^2$ is equal to the mass of the displaced fluid.

7.5.10. Determine the induced mass of a sphere using the dynamical equation.

Solution: The Laplace equation holds at any fixed moment even when the sphere moves with acceleration $\boldsymbol{a}$. Then, assuming $\boldsymbol{v}_0 = \boldsymbol{a}t$, we obtain from (7.33)

$$\varphi = -\frac{1}{2}\left(\frac{R}{r}\right)^3 t\boldsymbol{a}\cdot\boldsymbol{r}, \qquad \boldsymbol{v} = -\frac{1}{2}\left(\frac{R}{r}\right)^3 \boldsymbol{a}t + \frac{3}{2}\left(\frac{R}{r}\right)^3 \frac{\boldsymbol{a}\cdot\boldsymbol{r}}{r^2}\boldsymbol{r}t\,.$$

The pressure on the surface of the sphere can be found with the help of (7.1) noting that $u = 0$, $w = p/\rho$ in this equation. But as $r \to \infty$, we have $\varphi \to 0$, $v \to 0$, $p \to p_0$ and obtain

$$p|_{r=R} = p_0 - \rho\left.\frac{v^2}{2}\right|_{r=R} - \rho\left.\frac{\partial\varphi}{\partial t}\right|_{r=R} = p_0 - \frac{\rho}{8}a^2t^2(1 + 3\cos^2\theta) + \frac{\rho}{2}aR\cos\theta\,.$$

The force resisting the sphere's movement is given by the integral

$$F_x = -\int\limits_S p\frac{x}{r}\,dS = -R^2\int\limits_0^{2\pi}\int\limits_0^{\pi}\Big[p_0 - \frac{\rho}{8}a^2t^2(1 + 3\cos^2\theta) + \frac{\rho}{2}aR\cos\theta\Big]\cos\theta\sin\theta\,d\theta\,d\varphi\,.$$

The first two terms in the square brackets are symmetric about the midsection's plane $\theta = \pi/2$ and yield zero after integration. Hence,

$$F_x = -2\pi R^2\int\limits_0^{\pi}\frac{\rho aR}{2}\cos^2\theta\sin\theta\,d\theta = \frac{2}{3}\pi R^3\rho a = \frac{1}{2}V_S\rho a = Ma\,,$$

where $M = V_S\rho/2$, i.e., the same as in (7.37).

7.5.11. A sphere of mass m and radius R is in equilibrium at the level z_0 in an incompressible stratified fluid of density $\rho = \rho(z)$ $[m = (4/3)\pi R^3\rho(z_0)]$. Determine the period of small oscillations of the sphere when it is displaced in the vertical direction.

Solution: The restoring force

$$F = mN^2(z_0)\zeta = -mg\rho^{-1}(d\rho/dz)\zeta|_{z_0}$$

acts on the sphere where ζ is the vertical displacement and $N(z_0)$ is the Väisälä frequency [see (6.28) where $c = \infty$]. Taking into account the induced mass of the sphere $m/2$, its equation of motion is

$$(3/2)\,m\ddot{\zeta} + mN^2(z_0)\zeta = 0$$

which yields for the frequency and the period of the oscillations

$$\omega^2 = 2N^2(z_0)/3\,, \qquad T = \sqrt{6}\pi/N(z_0)\,.$$

7.5.12. Determine the pressure force inside a weightless spherical shell stretching symmetrically without stresses in an incompressible fluid. Acceleration of the shell's points is a. Find the induced mass of the sphere in this case.

Solution: In the time Δt, the velocity of the shell points is $v = a\,\Delta t$. A central-symmetric solution of the Laplace equation satisfying the boundary condition $v|_R = (\partial\varphi/\partial r)_R = a\,\Delta t$ is $\varphi = -a\,\Delta t R^2/r$. The fluid-particle velocity is $v(r) = \partial\varphi/\partial r = a\,\Delta t R^2/r^2$, whereas the kinetic energy of the fluid is

$$\mathscr{E}_k = 2\pi\rho\int_R^\infty v^2r^2\,dr = 2\pi\rho(a\,\Delta t)^2R^4\int_R^\infty r^{-2}\,dr = 2\pi R^3\rho(a\,\Delta t)^2\,.$$

This energy must be equal to the work $A = F\,\Delta R = Fa(\Delta t)^2/2$, $F = 4\pi R^2 p$ being the force on the shell. This yields

$$F = 4\pi R^3\rho a = Ma\,, \qquad M = 4\pi R^3\rho\,.$$

Hence, the induced mass M is equal to three times the mass of the displaced fluid.

7.5.13. Prove that an infinite force (per unit length) is needed for expanding a cylindrical shell with constant velocity in an incompressible fluid.

Hint: Show that the energy of a layer of unit length along the cylinder axis in this case of cylindrical symmetry is infinite.

7.5.14. A sphere of mass m and radius R is dropped into the ocean of depth H. What time is needed for the sphere to reach the bottom of the ocean? The water is assumed as an ideal fluid.

8. Flows of Viscous Fluids

It was shown above that the flow of an ideal fluid in the field of a potential force produces no vortexes,—i.e., a potential flow remains potential at any time. In real flows, however, we regularly observe a production and destruction of vortexes. This is because the real fluid is a viscous one. The assumption that there is a slip at the boundary between a flowing fluid and a solid (as in the case of an ideal fluid) also turns out to be incorrect for a real fluid. All components of the velocity vanish on the surface of a body at rest, according to experiments. That is why the dust accumulates on the surfaces of bodies even when flow exists past these bodies (the dust on blades of a fan, for example).

8.1 Equations of Flow of Viscous Fluid

8.1.1 Newtonian Viscosity and Viscous Stresses

Let us first consider the viscosity or friction force in the following simple situation. Imagine two infinite parallel plates at a distance h. Let one of these plates (for example, the upper one, Fig. 8.1) move relative to the other at a constant velocity v_0. The force which must be applied to this plate to ensure its moment we denote as F. According to experiment, this force per unit area of the plate is directly proportional to v_0 and inversely proportional to h:

$$F/S = \eta v_0/h\,. \tag{8.1}$$

S is the area of the plate. The proportionality factor η is called *viscosity* (sometimes *Newtonian viscosity*). The so-called *kinematic viscosity* $\nu = \eta/\rho$ is often used in equations, namely the ratio of viscosity to fluid density. For

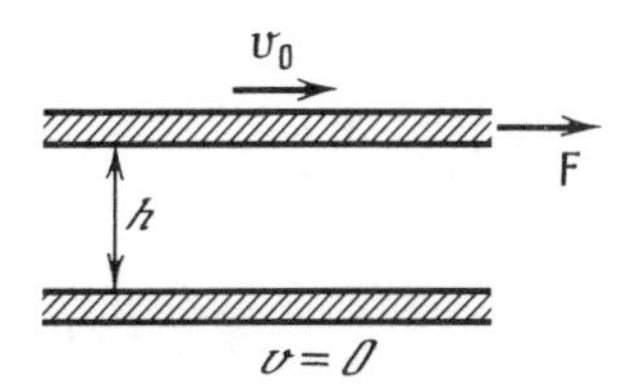

Fig. 8.1. Geometry for Couette flow

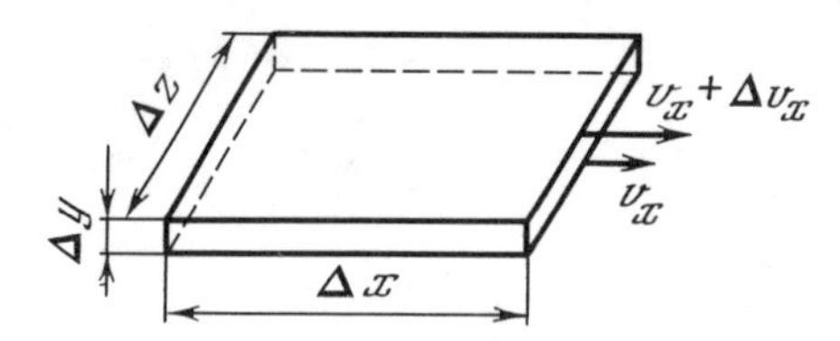

Fig. 8.2. To determine the tensor of viscous stresses

example, for water and air, η (g cm^{-1} s^{-1}) and ν (cm^2 s^{-1}) at the temperature of 20°C are:

$$\text{water:} \quad \eta = 0.01\,, \qquad \nu = 0.01$$

$$\text{air:} \quad \eta = 1.8 \times 10^{-4}\,, \qquad \nu = 0.15\,.$$

Viscosity depends on temperature. In liquids it decreases rapidly whereas in gases it increases slowly as the temperature increases.

We will write down a relationship analogous to (8.1) but for a general flow. Choose the x-axis parallel to the velocity vector at some point. Consider a small element of fluid at this point in the form of a rectangular parallelepiped whose dimensions are Δx, Δy and Δz (Fig. 8.2). Assume that $\Delta y \ll \Delta x$, $\Delta y \ll \Delta z$ and denote by $\Delta S = \Delta x\,\Delta z$ the area of the faces perpendicular to the y-axis. The particle velocity at the lower face of the volume under investigation is v_x and at the upper one $v_x + \Delta v_x$. A shear force ΔF must exist, according to (8.1), supporting this "shear current" such that

$$\Delta F/\Delta S = \eta\, \Delta v_x/\Delta y\,.$$

Assuming $\Delta y \to 0$ and $\Delta S \to 0$, we obtain one of the components of the so-called *viscous stress tensor:*

$$\sigma_{xy} = \lim_{\Delta S \to 0} (\Delta F/\Delta S) = \eta(\partial v_x/\partial y)\,. \tag{8.2}$$

In this chapter we will also formulate the general theory of viscous fluid flow.

8.1.2 The Navier-Stokes Equation

It was shown in Sect. 6.5 that in the case of an ideal fluid with no external force, the Euler equation can be written as a momentum conservation law

$$\partial(\rho v_i)/\partial t = -\partial \Pi_{ik}/\partial x_k\,, \tag{8.3}$$

where $\Pi_{ik} = p\,\delta_{ik} + \rho v_i v_k$ is the specific momentum flux tensor. Equation (8.3) can be generalized by taking into account the additional terms in Π_{ik} due to viscous stresses and assume

$$\Pi_{ik} = p\,\delta_{ik} + \rho v_i v_k - \sigma_{ik}\,, \tag{8.4}$$

where σ_{ik} is the *viscous stress tensor*. Its dependence on the particle velocity can be determined by quite general considerations. It was shown above that viscous forces arise only when different fluid particles move at different velocities. Hence, σ_{ik} must depend on velocity gradients, see (8.2), and not on the velocity itself. Assuming that these gradients are small, we expand σ_{ik} into a series retaining only the linear terms in the gradients. Moreover, σ_{ik} must be zero if the fluid rotates as a rigid body. This means that σ_{ik} depends only on the following combinations of velocity gradients:

$$(\partial v_i/\partial x_k) + (\partial v_k/\partial x_i), \qquad \partial v_m/\partial x_m, \tag{8.5}$$

which are zero in the case of a rigid motion. Indeed, if the fluid rotates as a whole with the angular velocity $\boldsymbol{\Omega} = \{\Omega_1, \Omega_2, \Omega_3\}$, we have $\boldsymbol{v} = \boldsymbol{\Omega} \times \boldsymbol{r}$, $\boldsymbol{r}$ being the position vector of the fluid particle (the origin is assumed to be on the axis of rotation). It immediately follows from this that $(\partial v_m/\partial x_m) = \operatorname{div} \boldsymbol{v} = 0$. We also have, for example,

$$v_1 = \Omega_2 x_3 - \Omega_3 x_2, \quad v_2 = \Omega_3 x_1 - \Omega_1 x_3, \quad (\partial v_1/\partial x_2) + (\partial v_2/\partial x_1) = 0 \text{ etc.}$$

The most general expression for the second-order tensor satisfying the conditions stated above is

$$\sigma_{ik} = a\left(\frac{\partial v_i}{\partial x_k} + \frac{\partial v_k}{\partial x_i}\right) + b\frac{\partial v_m}{\partial x_m}\delta_{ik}, \tag{8.6}$$

where a and b are constants independent of velocity or its gradients. The last term of (8.6) which is proportional to the rate of volume change is nonzero only in compressible fluids and important, for example, in the theory of sound waves in viscous fluids.

Now the equation of fluid motion can be obtained if we substitute (8.4) into (8.3) and take (8.6) into account. If the coefficients a and b are assumed to be independent of the coordinates, this yields

$$\rho\frac{\partial v_i}{\partial t} + v_i\frac{\partial \rho}{\partial t} = -\frac{\partial p}{\partial x_i} - \rho v_k\frac{\partial v_i}{\partial x_k} - v_i\frac{\partial}{\partial x_k}(\rho v_k) + a\frac{\partial^2 v_i}{\partial x_k^2}$$
$$+ a\frac{\partial^2 v_k}{\partial x_i \partial x_k} + b\frac{\partial^2 v_k}{\partial x_i \partial x_k}.$$

From this relation, using the continuity equation (6.9), we obtain the *Navier-Stokes equation:*

$$\rho\left(\frac{\partial v_i}{\partial t} + v_k\frac{\partial v_i}{\partial x_k}\right) = -\frac{\partial p}{\partial x_i} + a\,\Delta v_i + (a+b)\frac{\partial^2 v_k}{\partial x_i \partial x_k}, \tag{8.7}$$

or in vector form,

$$\rho \frac{d\boldsymbol{v}}{dt} \equiv \rho \left[\frac{\partial \boldsymbol{v}}{\partial t} + (\boldsymbol{v} \cdot \nabla)\boldsymbol{v} \right] = -\nabla p + a \Delta \boldsymbol{v} + (a+b)\nabla(\nabla \boldsymbol{v}) . \tag{8.8}$$

In the case of an incompressible fluid ($\nabla \boldsymbol{v} = 0$), the Navier-Stokes equation assumes the form

$$\rho \left[\frac{\partial \boldsymbol{v}}{\partial t} + (\boldsymbol{v} \cdot \nabla)\boldsymbol{v} \right] = -\nabla p + a \Delta \boldsymbol{v} . \tag{8.9}$$

8.1.3 The Viscous Force

The term $\partial \sigma_{ik}/\partial x_k$ on the right-hand side of (8.3) or (what is the same) the terms proportional to a or b in (8.7) may be considered the ith component of viscous force per unit volume:

$$f_i = a \frac{\partial^2 v_i}{\partial x_k^2} + (a+b) \frac{\partial^2 v_k}{\partial x_i \, \partial x_k} = a \Delta v_i + (a+b) \frac{\partial}{\partial x_i} (\operatorname{div} \boldsymbol{v}) . \tag{8.10}$$

Let us now find the force exerted by the viscous fluid on a body in the case of a steady flow. Imagine a closed surface S' surrounding the body (Fig. 8.3). In the case of a steady flow, (8.3) becomes $\partial \Pi_{ik}/\partial x_k = 0$. Integrating this equation over the volume between S' and the surface S of the body and using the Gauss divergence theorem for the tensor Π_{ik}, yields

$$-\int_S \Pi_{ik} n_k \, dS + \int_{S'} \Pi_{ik} n_k \, dS = 0 , \tag{8.11}$$

where $\boldsymbol{n} = \{n_1, n_2, n_3\}$ is the outward normal to S or S'.

Taking (8.4) into account and the fact that on S the flow velocity vanishes (adhesion of the viscous fluid to the solid surface), we write the first term in (8.11) as

$$F_i = -\int_S p n_i \, dS + \int_S \sigma_{ik} n_k \, dS . \tag{8.12}$$

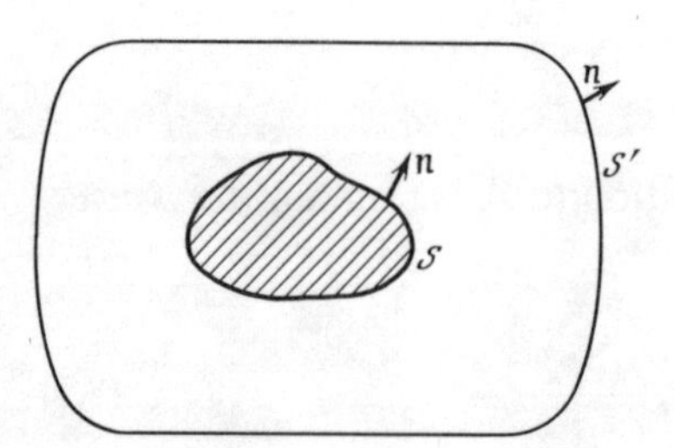

Fig. 8.3. To determine of the viscous force exerted by fluid on a body

This is the force exerted by the fluid on the body. The first term here is the resultant pressure force on the body and the second term

$$F'_i = \int_S \sigma_{ik} n_k \, dS \tag{8.13}$$

is the force due to viscosity. Its density (i.e., the force per unit area of the surface of the body) is

$$f'_i = \sigma_{ik} n_k . \tag{8.14}$$

Now, taking (8.11) into account, we can also write the force acting on the body as

$$F_i = \int_{S'} (-pn_i - \rho v_i v_k n_k + \sigma_{ik} n_k) \, dS , \tag{8.15}$$

i.e., this force can be calculated if the characteristics of the flow on an arbitrary surface S' are known. This approach may turn out to be much more convenient when compared with calculations by (8.12) because the surface of the body S as well as the flow near it may be rather complicated. On the other hand, the structure of the flow at large distances from the body is much simpler and can be found, for example, by linearization of the solution of the equation of motion with respect to disturbances from an unperturbed state (without a body).

8.2 Some Examples of Viscous Fluid Flow

8.2.1 Couette Flow

The simple flow considered in Sect. 8.1.1 (Fig. 8.1) is called the *plane Couette flow*. We now apply the Navier-Stokes equation to this flow. Let the x-axis be parallel to the velocity vector $\boldsymbol{v}_0$ and the y-axis perpendicular to the plates. The origin is on the lower fixed plane. Because of the symmetry, the velocity and pressure may depend only on y: $v_x = v = v(y)$, $p = p(y)$. Besides, $\partial/\partial t = 0$ since the flow is steady. Hence, we have for the components of (8.9),

$$\partial p/\partial y = 0 , \qquad \partial^2 v/\partial y^2 = 0 ,$$

which yields after integration: $p = p_0 = \text{const}$, $v = Ay + B$. The constants A and B are determined by the boundary conditions $v = 0$ at $y = 0$ and $v = v_0$ at $y = h$. We thus obtain $B = 0$, $A = v_0/h$; hence,

$$v = (v_0/h)y .$$

We now determine the force f per unit area exerted by the fluid on the lower plate. The pressure gives no tangential component and to calculate this force we can use (8.14) where σ_{ik} is given by the first term in (8.6) (incompressible fluid). By supposing that the subscripts 1 and 2 correspond to x and y, respectively, we obtain

$$f|_{y=0} = a\,\partial v/\partial y = av_0/h\,. \tag{8.16}$$

Obviously, this force has to be equal to the force (8.1) applied to the upper plate. Hence we have $a = \eta$. Now the Navier-Stokes equation for an incompressible fluid can be written as

$$\rho\left[\frac{\partial \boldsymbol{v}}{\partial t} + (\boldsymbol{v}\cdot\nabla)\boldsymbol{v}\right] = -\nabla p + \eta\,\Delta\boldsymbol{v}\,. \tag{8.17}$$

The important quantity in fluid dynamics is the *drag coefficient* C_{D} which is the ratio of the force f to the so-called *velocity head* $\rho v_{\mathrm{m}}^2/2$:

$$C_{\mathrm{D}} = 2f/\rho v_{\mathrm{m}}^2\,, \tag{8.18}$$

where v_{m} is the mean (or some characteristic) flow velocity. In the case of Couette flow, $v_{\mathrm{m}} = v_0/2$, $f = 2\eta v_{\mathrm{m}}/h$ and

$$C_{\mathrm{D}} = 4\nu/v_{\mathrm{m}}h = 4/\mathrm{Re}\,, \tag{8.19}$$

where $\mathrm{Re} = v_{\mathrm{m}}h/\nu$ is the so-called *Reynolds number* and $\nu = \eta/\rho$ is kinematic viscosity.

8.2.2 Plane Poiseuille Flow

Let us suppose that we have two fixed parallel plates a distance h apart. Consider the steady fluid flow between the plates supported by an external pressure gradient along the plates. Again let the x and y-axes be parallel and perpendicular to the plates, respectively. The fluid velocity can depend only on y in this case $[v_x = v(y)]$. The component of (8.17) along the y-axis is

$$\partial p/\partial y = 0\,,$$

i.e., the pressure is independent of y and depends only on x. The component of the same equation along the x-axis is

$$\partial^2 v/\partial y^2 = \eta^{-1}\,\partial p/\partial x\,. \tag{8.20}$$

Since the left-hand side of (8.20) depends only on y while its right-hand side depends only on x, it follows that both sides are equal to a constant. Integration of (8.20) yields

$$\partial v/\partial y = A - by/\eta\,, \qquad v = B + Ay - by^2/2\eta\,, \qquad b = -\partial p/\partial x\,. \tag{8.21}$$

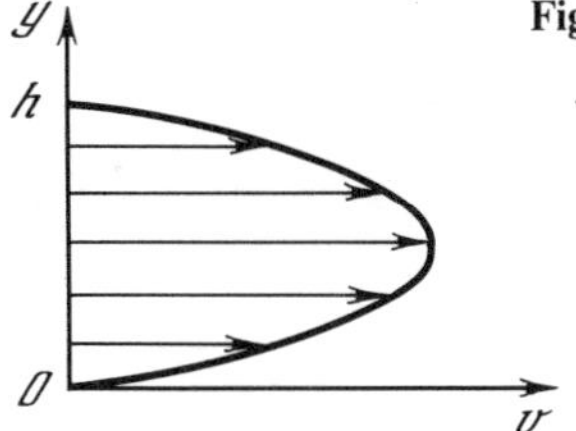

Fig. 8.4. Particle-velocity curve in Poisseuille flow

Using the boundary conditions $v = 0$ at $y = 0, h$, we find for the integration constants $B = 0$, $A = bh/2\eta$; hence,

$$v = by(h - y)/2\eta = -(\partial p/\partial x)y(h - y)/2\eta\,. \tag{8.22}$$

The velocity profile across the flow is parabolic (Fig. 8.4). We now determine the force per unit area exerted by the fluid on each plate. Quite analogously with the Couette flow considered above, see (8.16), we have

$$f_x = \eta \left.\frac{\partial v}{\partial y}\right|_{y=0} = \frac{bh}{2} = -\frac{\partial p}{\partial x}\frac{h}{2}\,. \tag{8.23}$$

Here, $f_x > 0$, $dp/dx < 0$, i.e., the force direction is the same as the flow direction whereas the pressure decreases as x increases.

The mean flow velocity is

$$v_{\mathrm{m}} = \frac{1}{h}\int_0^h v\,dy = \frac{bh^2}{12\eta} = -\frac{h^2}{12\eta}\frac{dp}{dx} \tag{8.24}$$

and the drag coefficient, according to (8.18),

$$C_{\mathrm{D}} = 12/\mathrm{Re}\,, \qquad \mathrm{Re} = v_{\mathrm{m}}h/\nu\,. \tag{8.25}$$

Again C_{D} depends on the Reynolds number Re only.

8.2.3 Poiseuille Flow in a Cylindrical Pipe

Let the radius of a pipe be R and the x-axis coincide with the axis of the cylinder. The only nonzero component of the velocity is $v_x = v(r)$. The latter depends only on the distance $r = \sqrt{y^2 + z^2}$ from the axis. The components of the Navier-Stokes equation (8.17) along the axes are

$$\partial p/\partial x + \eta\,\Delta v = 0\,, \qquad \partial p/\partial y = 0\,, \qquad \partial p/\partial z = 0\,. \tag{8.26}$$

It follows from the last two equations that $p = p(x)$. Since $v = v(r)$, we obtain from the first equation $-\partial p/\partial x = b$ which is constant.

The Laplace operator for the axisymmetric case is

$$\Delta = r^{-1}\frac{d}{dr}\left(r\frac{d}{dr}\right).$$

Now, (8.26) assumes the form

$$\frac{d}{dr}\left(r\frac{dv}{dr}\right) = -\frac{br}{\eta}. \tag{8.27}$$

The integration over r yields

$$r\frac{dv}{dr} = -\frac{br^2}{2\eta} + A, \qquad \frac{dv}{dr} = -\frac{br}{2\eta} + \frac{A}{r}.$$

The requirement that dv/dr must be finite everywhere including $r = 0$ yields $A = 0$, and the next integration gives

$$v = B - br^2/4\eta.$$

We have $v = 0$ at the pipe boundary $r = R$, hence $B = bR^2/4\eta$ and finally

$$v = \frac{b}{4\eta}(R^2 - r^2) = -\frac{1}{4\eta}\frac{dp}{dx}(R^2 - r^2). \tag{8.28}$$

The velocity distribution is a paraboloid of revolution. The mean velocity is

$$v_{\mathrm{m}} = \frac{1}{\pi R^2}\int_0^R\int_0^{2\pi} vr\,dr\,d\varphi = \frac{b}{2R^2\eta}\int_0^R (R^2 - r^2)r\,dr = \frac{bR^2}{8\eta}.$$

Now we can easily obtain the well-known *Poiseuille formula* for the flow rate of the fluid through the cylindrical tube:

$$Q = \pi R^2 v_{\mathrm{m}} = \frac{\pi b R^4}{8\eta} = \frac{\pi}{8\eta}\left|\frac{dp}{dx}\right| R^4. \tag{8.29}$$

The force per unit area of the tube boundary is

$$f_x = -\eta\left.\frac{dv}{dr}\right|_{r=R} = \frac{1}{2}bR = -\frac{1}{2}\frac{dp}{dx}R.$$

The force per unit length of the tube $2\pi R f_x = -\pi R^2\, dp/dx$ is balanced naturally by the pressure gradient. The drag coefficient depends again only on the Reynolds number:

$$C_{\mathrm{D}} = \frac{2f_x}{\rho v_{\mathrm{m}}^2} = \frac{8\nu}{v_{\mathrm{m}}R} = \frac{16}{\mathrm{Re}}, \qquad \mathrm{Re} = \frac{v_{\mathrm{m}}D}{\nu}, \tag{8.30}$$

where $D = 2R$ is the tube diameter.

The flows considered above occur when the flow velocities are sufficiently small (the Reynolds number is small). Formula (8.29) is in good agreement with experiment in these cases.

8.2.4 Viscous Fluid Flow Around a Sphere

Let us consider a fixed sphere of radius R at rest and assume that the steady flow of an incompressible fluid is homogeneous at infinity. The direction of the x-axis is chosen along the direction of $\boldsymbol{v}_0$ at infinity (Fig. 8.5). We now neglect the quadratic (inertial) term $\rho(\boldsymbol{v}\cdot\nabla)\boldsymbol{v}$ in the Navier-Stokes equation. This is possible if the Reynolds number is small. In fact, the characteristic scale of the flow is the sphere's diameter D; hence the order of magnitude of this term is $\rho v_0^2/D$, while the order of the viscous term $\eta\,\Delta\boldsymbol{v}$ is $\eta v_0/D^2$. Hence, the inertial term is small compared with the viscous one if $v_0 D/\nu = \mathrm{Re} \ll 1$.

Equation (8.17) without the inertial term and the incompressibility condition ($\nabla\boldsymbol{v} = 0$) constitute a closed system of equations for our problem:

$$\nabla p = \eta\,\Delta\boldsymbol{v}\,, \qquad \nabla\boldsymbol{v} = 0\,. \tag{8.31}$$

The flow velocity must be zero on the surface of the sphere and $\boldsymbol{v}_0$ at infinity:

$$\boldsymbol{v}\big|_{r=R} = 0\,, \qquad v_x\big|_{r\to\infty} = v_0\,, \qquad v_y\big|_{r\to\infty} = v_z\big|_{r\to\infty} = 0\,, \tag{8.32}$$

where $r = \sqrt{x^2+y^2+z^2}$.

We look for the solution of (8.31) under the boundary conditions (8.32) as a sum of a potential part ($\nabla\varphi$) and an additive term $\boldsymbol{w}$:

$$\boldsymbol{v} = \nabla\varphi + \boldsymbol{w}(x,y,z)\,. \tag{8.33}$$

Substitution of this expression into the second of (8.31) yields

$$\Delta\varphi + \nabla\boldsymbol{w} = 0\,. \tag{8.34}$$

Suppose that φ satisfies the condition $\nabla\varphi\big|_{r\to\infty} = \boldsymbol{v}_0$. Then we can assume:

$$\varphi = v_0 x + C_1(\partial r/\partial x) + C_2\,\partial r^{-1}/\partial x\,, \tag{8.35}$$

where C_1 and C_2 are, for the time being, arbitrary constants. Taking into account that in spherical coordinates $\Delta(r^{-1}) = 0$, $\Delta = r^{-2}(d/dr)(r^2\,d/dr)$, $\Delta r = 2/r$,

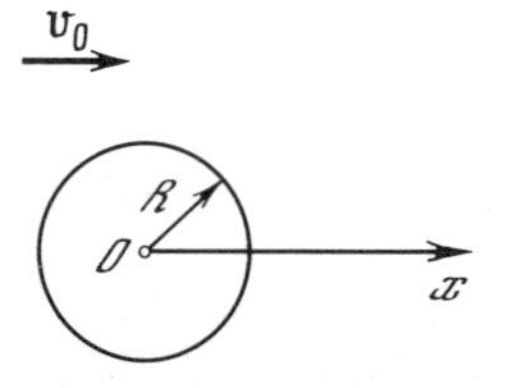

Fig. 8.5. Illustrating viscous fluid flow around a sphere

from (8.35)

$$\Delta\varphi = 2C_1 \frac{\partial}{\partial x}\left(\frac{1}{r}\right). \tag{8.36}$$

Note that (8.34) is satisfied with (8.36), if $\boldsymbol{w} = -(2C_1/r)\nabla x$.

Substituting this expression and (8.35) into (8.33), we obtain for the velocity

$$\begin{aligned}\boldsymbol{v} &= \nabla[v_0 x + C_1(x/r) - C_2 x/r^3] - 2C_1 r^{-1}\nabla x \\ &= (v_0 - C_1 r^{-1} - C_2 r^{-3})\nabla x + x(-C_1 r^{-3} + 3C_2 r^{-5})\boldsymbol{r}.\end{aligned} \tag{8.37}$$

This expression satisfies the required condition $\boldsymbol{v} = v_0 \nabla x$ as $r \to \infty$. To satisfy the condition on the sphere's surface ($\boldsymbol{v}|_{r=R} = 0$), we assume $3C_2 = C_1 R^2$. Then the second brackets on the right-hand side of (8.37) vanishes at $r = R$. The same is true for the first term in parentheses if we assume additionally that

$$C_1 = \tfrac{3}{4} v_0 R. \tag{8.38}$$

Then (8.37) becomes

$$\boldsymbol{v} = \left(1 - \frac{3}{4}\frac{R}{r} - \frac{1}{4}\frac{R^3}{r^3}\right)\boldsymbol{v}_0 - \frac{3}{4}\frac{v_0 R x}{r^2}\left(1 - \frac{R^2}{r^2}\right)\frac{\boldsymbol{r}}{r}, \tag{8.39}$$

or for the components

$$\begin{aligned}v_x &= v_0\left(1 - \frac{3}{4}\frac{R}{r} - \frac{1}{4}\frac{R^3}{r^3}\right) - \frac{3}{4} v_0 \frac{R x^2}{r^3}\left(1 - \frac{R^2}{r^2}\right), \\ v_y &= -\frac{3}{4} v_0 \frac{R x y}{r^3}\left(1 - \frac{R^2}{r^2}\right), \quad v_z = -\frac{3}{4} v_0 \frac{R x z}{r^3}\left(1 - \frac{R^2}{r^2}\right).\end{aligned} \tag{8.40}$$

8.2.5 Stokes' Formula for Drag

In order to find the pressure p from (8.31), we substitute $\boldsymbol{v}$ from (8.33) into it and take into account that $w_y = w_z = 0$, $\Delta w_x \sim \Delta(1/r) = 0$, we obtain

$$\nabla p = \eta\, \Delta(\nabla\varphi) = \nabla(\eta\, \Delta\varphi)$$

or, according to (8.36, 38),

$$\nabla p = \nabla\left[2\eta C_1 \frac{\partial}{\partial x}\left(\frac{1}{r}\right)\right] = \nabla\left(-\frac{3}{2}\eta v_0 \frac{R x}{r^3}\right).$$

The solution of this equation under the condition that the pressure vanishes at infinity is

$$p = -\tfrac{3}{2}\eta v_0 R x / r^3. \tag{8.41}$$

Let us now determine the force $\boldsymbol{F}$ exerted by the flow on the sphere. Obviously, $F_y = F_z = 0$ due to the symmetry of the flow about the x-axis. The viscous force per unit area of the sphere surface acting in the x-direction is, according to (8.29),

$$f'_1 = \sigma_{1k} n_k = \eta \left(\frac{\partial v_1}{\partial x_k} + \frac{\partial v_k}{\partial x_1} \right) n_k = \eta \left(\frac{\partial v_1}{\partial x_k} + \frac{\partial v_k}{\partial x_1} \right) \frac{x_k}{r} .$$

Here, as always, $x_1 = x$, $x_2 = y$, $x_3 = z$. The first term in the parentheses gives rise to the force

$$\left(\frac{\partial v_1}{\partial x_k} \frac{x_k}{r} \right)_{r=R} = \left(\frac{\partial v_1}{\partial r} \frac{\partial r}{\partial x_k} \frac{x_k}{r} \right)_{r=R} = \left(\frac{\partial v_1}{\partial r} \frac{x_k^2}{r^2} \right)_{r=R} = \left. \frac{\partial v_1}{\partial r} \right|_{r=R} .$$

In performing the differentiation we have disregarded the dependence of $v_1 \equiv v_x$ on $x \equiv x_1$ given by the last term in (8.40) for v_x. It is permissible because this term vanishes after differentiation with respect to x for $r = R$. The second term in f'_1 becomes

$$\left(\frac{\partial v_k}{\partial x_1} \frac{x_k}{r} \right)_{r=R} = \left(\frac{\partial v_k}{\partial r} \frac{x_1 x_k}{r^2} \right)_{r=R} = \left(\frac{\partial v_k}{\partial x_k} \frac{x_1}{r} \right)_{r=R} = 0$$

since the fluid is assumed incompressible ($\partial v_k / \partial x_k = 0$). Now taking (8.40) into account we obtain

$$f'_1 = f'_x = \eta \left. \frac{\partial v_x}{\partial r} \right|_{r=R} = \frac{3}{2} \frac{\eta v_0}{R} \left(1 - \frac{x^2}{R^2} \right) . \tag{8.42}$$

Besides this tangential force, there is also a force due to pressure acting on the sphere. The x-component of this force per unit area is, according to (8.41),

$$-p \left. \frac{x}{R} \right|_{r=R} = \frac{3}{2} \frac{\eta v_0}{R} \left. \frac{x^2}{R^2} \right|_{r=R} .$$

The resultant force per unit area of the sphere's surface is the sum of the latter and (8.42):

$$f_x = \frac{3}{2} \frac{\eta v_0}{R} . \tag{8.43}$$

The force exerted by flow on the whole sphere is equal to f_x multiplied by the sphere's surface area $4\pi R^2$:

$$F_x = 6\pi \eta R v_0 . \tag{8.44}$$

This is the well-known *Stokes' formula* for drag when the sphere moves slowly relative to the fluid. Note that in the case of an ideal (inviscid) fluid, this force is zero.

The drag coefficient (8.18) is

$$C_D = \frac{2F_x}{\pi R^2 \rho v_0^2} = \frac{24}{\mathrm{Re}}, \qquad \mathrm{Re} = \frac{v_0 D}{\nu}, \qquad D = 2R. \tag{8.45}$$

In the case of free fall of the sphere in the gravitational field, the constant velocity v_∞ is reached as $t \to \infty$. This velocity can be determined from the condition that the resistance force F_x is equal to the weight of the sphere $(4/3)\pi g \rho_1 R^3$ (ρ_1 is an average density of the sphere) minus the Archimedes force $(4/3)\pi g \rho R^3$. This yields for the limiting velocity

$$v_\infty = \frac{2}{9}\frac{gR^2}{\nu}\left(\frac{\rho_1}{\rho} - 1\right). \tag{8.46}$$

It has already been pointed out that the theory developed above is valid only if the Reynolds number is small ($\mathrm{Re} \ll 1$), i.e., in the case of a small v_∞, a large viscosity ν or small dimensions of the sphere (D is small). This theory can be improved, however, by taking into account the nonlinear (inertial) term in the Navier-Stokes equation. Then, the formula for the drag coefficient becomes $C_D \simeq (24/\mathrm{Re})(1 + 3\mathrm{Re}/16)$, provided the second and higher powers of Re are neglected. The last formula shows that the linear theory is valid within an accuracy of 10% if $\mathrm{Re} < 1/2$. In the case of free-falling spherical particles, we obtain from $\mathrm{Re} = 2Rv_\infty/\nu < 1/2$ with (8.46):

$$R < \frac{1}{2}\left(\frac{9\nu^2}{g}\frac{\rho}{|\rho_1 - \rho|}\right)^{1/3}, \qquad v_\infty < \frac{1}{2}\left(\frac{\nu g}{9}\frac{|\rho_1 - \rho|}{\rho}\right)^{1/3}. \tag{8.47}$$

In the case of a sphere of water in air (drops of rain), for example, the theory is valid if $R < 3 \times 10^{-3}$ cm ($v_\infty < 12$ cm s^{-1}).

8.3 Boundary Layer

The solution of $\Delta\varphi = 0$, $\boldsymbol{v} = \nabla\varphi$ for the potential flow of an ideal incompressible fluid was considered in Chap. 7. This solution also satisfies the Navier-Stokes equation (8.17) since the viscous term $\eta\,\Delta\boldsymbol{v}$ becomes zero in this case $[\Delta\boldsymbol{v} = \Delta(\nabla\varphi) = \nabla(\Delta\varphi) = 0]$. Hence, the solution for an ideal fluid can also be used to some extent for a real one. A drastic difference may only occur near some body present in the fluid. Indeed, the solutions for a real fluid and for an ideal one must satisfy different boundary conditions on the body's surface—the tangential velocity is *prohibited* in a real fluid and is not *specified* in an ideal one. However, in many important cases this difference is essential only in a thin layer near the surface called the *boundary layer*.

8.3.1 Viscous Waves

As a very simple example of a boundary layer, we consider viscous waves near an infinite plate oscillating in its own plane. Assume that the latter coincides with the xy-plane and we have a viscous homogeneous incompressible fluid in the halfspace $z > 0$. Let the oscillations be harmonic and in the x-direction,

$$v_x|_{z=0} = v_0 \exp(-\mathrm{i}\omega t), \quad v_y|_{z=0} = v_z|_{z=0} = 0. \tag{8.48}$$

In this case the ideal fluid would remain at rest due to slip at the plate. The viscous fluid adheres to the plate and (8.48) must be regarded the boundary conditions at $z = 0$ for the flow under consideration.

We turn to the Navier-Stokes' equation (8.17). Obviously, $v_y = 0$ due to the symmetry of the problem and for the same reason, v_x and v_z can depend only on z. Since the fluid is incompressible we have $\operatorname{div} \boldsymbol{v} = 0$ or $\partial v_z/\partial z = 0$, $v_z = \text{const}$. However, const $= 0$ since $v_z = 0$ at $z = 0$. The nonlinear term $\rho(\boldsymbol{v} \cdot \nabla)\boldsymbol{v} = \rho v_x(\partial v_x/\partial x)\nabla x$ in (8.17) is also zero. The z-component of this equation will be just $\partial p/\partial z = 0$ or $p = \text{const}$ and its x-component becomes ($v_x \equiv v$):

$$\partial v/\partial t = \nu\, \partial^2 v/\partial z^2, \quad \nu = \eta/\rho. \tag{8.49}$$

This is a linear equation of the same type as the equation of heat conduction. We look for its solution under the condition (8.48) assuming

$$v = v_0 \exp[\mathrm{i}(kz - \omega t)]. \tag{8.50}$$

Substitution of this expression into (8.49) yields

$$-\mathrm{i}\omega = -\nu k^2, \quad k = \sqrt{\mathrm{i}\omega/\nu} = \pm(1 + \mathrm{i})\sqrt{\omega/2\nu}. \tag{8.51}$$

Only the "+" sign is appropriate here, otherwise v in (8.50) grows infinitely as $z \to \infty$. Now we obtain from (8.50) the expression for *the viscous wave:*

$$v = v_0 \exp(-\sqrt{\omega/2\nu} z) \exp[\mathrm{i}(\sqrt{\omega/2\nu} z - \omega t)]. \tag{8.52}$$

The amplitude of this wave decreases e times at the distance

$$\delta = \sqrt{2\nu/\omega} \tag{8.53}$$

from the plate. The lower the viscosity, the smaller is δ, the thickness of the layer in which significant oscillations of the fluid are observed. At distances $z \gg \delta$, we have $\boldsymbol{v} = 0$ as in an ideal fluid.

8.3.2 The Boundary Layer. Qualitative Considerations

Let us consider a fluid flow around a body. One can ask the question: does the flow of viscous fluid approach that of an ideal one if the viscosity η approaches zero? It seems that the answer must be yes since the ratio of the viscous term to the inertial one in the Navier-Stokes equation is of the order of magnitude

$$\left|\frac{\eta \Delta \boldsymbol{v}}{\rho(\boldsymbol{v}\cdot\nabla)\boldsymbol{v}}\right| \sim \frac{\nu}{L v_0} = \frac{1}{\mathrm{Re}}, \qquad \mathrm{Re} = \frac{v_0 L}{\nu}, \tag{8.54}$$

where v_0 is the characteristic velocity of the flow and L the characteristic dimension of the body. One has $\mathrm{Re} \to \infty$ when $\nu \to 0$ and the viscous term becomes infinitely small compared with the inertial one. This reasoning, however, is right everywhere except in a thin boundary layer adjacent to the surface of the body. The flow velocity varies from zero on the surface to those characteristic of ideal fluid flow at a small distance from the surface. The less the viscosity, the smaller this distance (the thickness of the boundary layer), and the greater the velocity gradient normal to the surface. As a result, the viscous term $\nu \Delta \boldsymbol{v}$ in (8.17) having the order of magnitude of $\nu v_0/h^2$ (h: *the thickness of the boundary layer*) remains finite and comparable with other terms when $\nu \to 0$, $h \to 0$.

A formal reason for introducing a boundary layer when the Reynolds number of the flow is large is that we reduce the order of (8.17) dropping the viscous term. It immediately leads to a reduction in the number of boundary conditions we need for a solution. As a result, we cannot meet the zero condition for the tangential velocity at the body's surface and have to allow slip on this surface. But this contradicts experiment since real fluid adheres to the surface. This contradiction can be removed by the introduction of the concept of a boundary layer across which the velocity varies from zero to that corresponding to the flow of an ideal fluid.

The viscous wave considered in the previous subsection is the simplest case of a boundary layer. The width of the zone of existence of this wave, δ, decreases with decreasing viscosity. Outside this zone, the fluid remains at rest as in the case of an ideal fluid. However, the case considered there (an oscillating infinite plate) is also a special one because the nonlinear (inertial) term in the Navier-Stokes equation (8.17) is zero.

Now we consider another case which (though being idealistic, too) yields results of more general importance. This is the case of the two-dimensional steady incompressible fluid flow past a semi-infinite thin plate (Fig. 8.6). Let the latter lie in the halfplane $z = 0$ occupying the region $-\infty < y < \infty$, $x \geq 0$. At large distances from the plate ($x \to -\infty$) the flow is in the positive x-direction.

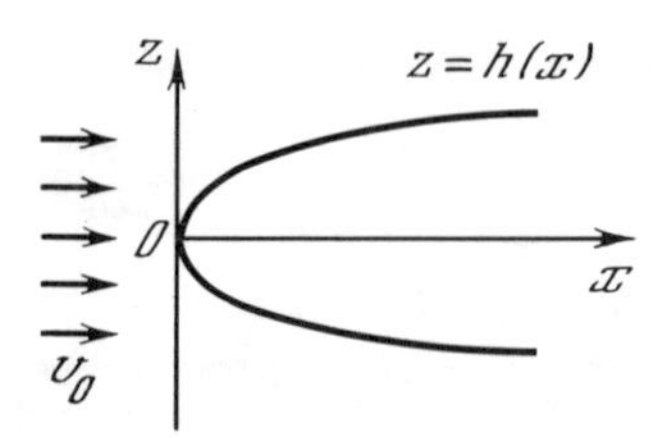

Fig. 8.6. Boundary layer in the flow past a semi-infinite thin plate

Now the problem which we have to solve may be formulated as follows, see (8.17),

$$\frac{\partial \boldsymbol{v}}{\partial t} + (\boldsymbol{v} \cdot \nabla)\boldsymbol{v} = -\frac{\nabla p}{\rho} + \nu \Delta \boldsymbol{v}, \qquad \nabla \boldsymbol{v} = 0,$$
$$\boldsymbol{v}\Big|_{\substack{z=0\\x\geqslant 0}} = 0, \qquad v_x|_{x\to-\infty} = v_0, \qquad v_z|_{x\to-\infty} = v_y|_{x\to-\infty} = 0. \tag{8.55}$$

The only characteristic scale of the flow-velocity variation along the plate is the distance of the reference point x from the edge of plate. The characteristic scale in the z-direction is obviously the thickness of the boundary layer itself. Assuming that the inertial and viscous terms in the Navier-Stokes equation have the same order of magnitude in the boundary layer, we can estimate the thickness of the latter $h(x)$. Assuming also that

$$\left|\frac{\partial^2 \boldsymbol{v}}{\partial z^2}\right| \sim \frac{v}{h^2} \gg \left|\frac{\partial^2 \boldsymbol{v}}{\partial x^2}\right| \sim \frac{v}{x^2} \quad \text{or} \quad h \ll x,$$

we obtain for the viscous term

$$\nu|\Delta \boldsymbol{v}| \sim \nu|\partial^2 \boldsymbol{v}/\partial z^2| \sim \nu v/h^2.$$

Since the flow is mainly along the plate ($\boldsymbol{v} \simeq v_x \nabla x$, $v \simeq v_x$), we have for the inertial term

$$|(\boldsymbol{v} \cdot \nabla)\boldsymbol{v}| \sim |v_x \,\partial v_x/\partial x| \sim v^2/x.$$

Now equating these estimates and noting that $v \simeq v_x \sim v_0$, the velocity in unperturbed flow, we obtain

$$\frac{\nu v_0}{h^2(x)} \sim \frac{v_0^2}{x}, \qquad h(x) \sim \left(\frac{\nu x}{v_0}\right)^{1/2} = x\left(\frac{\nu}{v_0 x}\right)^{1/2} = \frac{x}{\sqrt{\mathrm{Re}}}, \tag{8.56}$$

where $\mathrm{Re} = v_0 x/\nu \gg 1$ is the Reynolds number assumed large, otherwise the concept of a boundary layer is inconsistent.[8] Note that *the thickness of the boundary layer is proportional to the square root of the distance from the plate's edge.*

[8] It is possible to define the Reynolds number for the flow past a plate in another way:

$$(\mathrm{Re})_h = v_0 h(x)/\nu \sim x/h(x) \sim \sqrt{\mathrm{Re}},$$

but it makes no difference in our reasoning.

The tangential viscous force on the plate or the skin-friction resistance per unit area is, see (8.14),

$$|f| = \eta\left|\frac{\partial v_x}{\partial z} + \frac{\partial v_z}{\partial x}\right| \simeq \eta\left|\frac{\partial v_x}{\partial z}\right| \sim \frac{\eta v_0}{h(x)} = \rho\,\frac{\nu v_0}{h(x)} = \rho\left(\frac{\nu v_0^3}{x}\right)^{1/2}. \tag{8.57}$$

Correspondingly, we have for the drag coefficient,

$$C_{\mathrm{D}} = \frac{2f}{\rho v_0^2} \sim \left(\frac{\nu}{v_0 x}\right)^{1/2} = \frac{1}{\sqrt{\mathrm{Re}}}, \tag{8.58}$$

8.3.3 Prandl's Equation for a Boundary Layer

We consider the same case as in the previous subsection but in a more quantitative way. Equations (8.55) can be written in components as

$$\begin{aligned} &\frac{\partial v_x}{\partial t} + v_x\frac{\partial v_x}{\partial x} + v_z\frac{\partial v_x}{\partial z} = -\frac{1}{\rho}\frac{\partial p}{\partial x} + \nu\left(\frac{\partial^2 v_x}{\partial x^2} + \frac{\partial^2 v_x}{\partial z^2}\right),\\ &\frac{\partial v_z}{\partial t} + v_x\frac{\partial v_z}{\partial x} + v_z\frac{\partial v_z}{\partial z} = -\frac{1}{\rho}\frac{\partial p}{\partial z} + \nu\left(\frac{\partial^2 v_z}{\partial x^2} + \frac{\partial^2 v_z}{\partial z^2}\right),\\ &\frac{\partial v_x}{\partial x} + \frac{\partial v_z}{\partial z} = 0. \end{aligned} \tag{8.59}$$

To simplify these equations we estimate the order of magnitude of different terms using the results of the qualitative discussion above. Within the boundary layer $0 \le z \le h(x)$ we have

$$v_x \sim v_0,\ \partial/\partial x \sim x^{-1},\ \partial/\partial z \sim h^{-1},\ h \sim x/\sqrt{\mathrm{Re}} \ll x,\ \mathrm{Re} = v_0 x/\nu \gg 1.$$

Integrating the last equation in (8.59) over z from zero to z_1 $[0 \le z_1 \le h(x)]$ and taking into account that $v_z|_{z=0} = 0$ yields

$$v_z|_{z=z_1} = -\int_0^{z_1} (\partial v_x/\partial x)\,dz.$$

Since $\partial v_x/\partial x \sim v_0/x$, we obtain for the order of magnitude $v_z \sim v_0 h/x \ll v_x$.

For the order of magnitude of the terms in the first two equations in (8.59), we have

$$v_x(\partial v_x/\partial x) \sim v_0^2/x, \quad v_x(\partial v_z/\partial x) \sim v_0^2 h/x^2,\ \text{etc.}$$

This procedure leads to the following results:

$$(\partial p/\partial x) \sim v_0^2/x, \quad (\partial p/\partial z) \sim v_0^2 h/x^2 \ll \partial p/\partial x.$$

Hence we set $\partial p/\partial z = 0$, $p = p(x)$, i.e., the pressure can be assumed constant across the boundary layer. However, just outside the boundary layer, the pressure can be found under the assumption that the fluid is ideal. According to what was said above, the same pressure must be retained across the boundary layer. It permits us to regard $p(x)$ as a known function.

In the first equation in (8.59), the term $\partial^2 v_x/\partial x^2 \sim v_0/x^2$ can be neglected, compared with the term $\partial^2 v_x/\partial z^2 \sim v_0/h^2$. As a result, we have a system of two equations called *Prandtl's system for a boundary layer:*

$$\begin{aligned} &\frac{\partial v_x}{\partial t} + v_x \frac{\partial v_x}{\partial x} + v_z \frac{\partial v_x}{\partial z} = -\frac{1}{\rho}\frac{dp}{dx} + \nu \frac{\partial^2 v_x}{\partial z^2}, \\ &\frac{\partial v_x}{\partial x} + \frac{\partial v_z}{\partial z} = 0. \end{aligned} \tag{8.60}$$

This system with the boundary conditions $\boldsymbol{v} = 0$ on the surface of solid bodies and $v_x = v_0$, $v_z = 0$ at infinity completely determines the flow. Note that in the boundary layer the flow is not potential:

$$(\mathrm{curl}\,\boldsymbol{v})_y = (\partial v_z/\partial x) - (\partial v_x/\partial z) \simeq -\partial v_x/\partial z \neq 0.$$

Another integral form of Prandtl's equation is useful in some cases as first suggested by Karman. To obtain it we rewrite the inertial term in (8.60) using the second equation in (8.60):

$$v_x \frac{\partial v_x}{\partial x} + v_z \frac{\partial v_x}{\partial z} = v_x \frac{\partial v_x}{\partial x} + \frac{\partial}{\partial z}(v_x v_z) - v_x \frac{\partial v_z}{\partial z} = \frac{\partial (v_x)^2}{\partial x} + \frac{\partial (v_x v_z)}{\partial z}.$$

With this identity, we integrate the first equation in (8.60) over z between the limits $(0, h)$:

$$\int_0^h \frac{\partial v_x}{\partial t}\,dz + \int_0^h \frac{\partial (v_x)^2}{\partial x}\,dz + v_x v_z\bigg|_0^h = -\frac{h}{\rho}\frac{dp}{dx} + \nu \frac{\partial v_x}{\partial z}\bigg|_0^h. \tag{8.61}$$

We denote $v_x|_{z=h} = v_0(x)$ and transform different terms in the last equation in the following way:

$$\int_0^h \frac{\partial v_x}{\partial t}\,dz = \frac{\partial}{\partial t}\int_0^h v_x\,dz - v_0 \frac{\partial h}{\partial t}, \qquad \int_0^h \frac{\partial v_x^2}{\partial x}\,dz = \frac{\partial}{\partial x}\int_0^h v_x^2\,dz - v_0^2 \frac{\partial h}{\partial x},$$

$$\begin{aligned} v_x v_z\bigg|_0^h &= v_0 v_z\bigg|_{z=h} = v_0 \int_0^h \frac{\partial v_z}{\partial z}\,dz = -v_0 \int_0^h \frac{\partial v_x}{\partial x}\,dz \\ &= -v_0 \frac{\partial}{\partial x}\int_0^h v_x\,dz + v_0^2 \frac{\partial h}{\partial x}. \end{aligned}$$

Assuming also that $\partial v_x/\partial z|_{z=h} = 0$, we finally obtain the required *integral form of Prandtl's equations:*

$$\frac{\partial}{\partial t}\int_0^h v_x\,dz - v_0\frac{\partial h}{\partial t} + \frac{\partial}{\partial x}\int_0^h v_x^2\,dz - v_0\frac{\partial}{\partial x}\int_0^h v_x\,dz = -\frac{h}{\rho}\frac{dp}{dx} - \nu\left.\frac{\partial v_x}{\partial z}\right|_{z=0}. \tag{8.62}$$

8.3.4 Approximate Theory of a Boundary Layer in a Simple Case

We again consider the steady flow past a semi-infinite thin plate. In (8.62), $dp/dx = 0$ in this case since p is determined by solving the equations for an ideal fluid, and the thin plate does not disturb the ideal fluid flow. We assume for the function $v_x(z)$ in the boundary layer the simple form

$$v_x = v_0 \sin\xi\,, \qquad \xi = \pi z/2h\,, \qquad 0 \le z \le h(x)\,. \tag{8.63}$$

It satisfies our conditions $[v_x(0) = 0,\ v_x'(h) = 0]$ and differs very little from that in the exact theory, as well as in the experiment. Substituting (8.63) into the integral equation (8.62) and noting that $\partial/\partial t = 0$ in the case under investigation, we find

$$\int_0^h v_x^2\,dz = v_0^2\frac{2h}{\pi}\int_0^{\pi/2}\sin^2\xi\,d\xi = \frac{h}{2}v_0^2\,, \qquad \int_0^h v_x\,dz = v_0\frac{2h}{\pi}\int_0^{\pi/2}\sin\xi\,d\xi = \frac{2h}{\pi}v_0\,.$$

Thus, we obtain the ordinary differential equation for

$$h\frac{dh}{dx} = \frac{\nu}{v_0}\frac{\pi^2}{4-\pi}. \tag{8.64}$$

The solution of this equation under the condition $h(0) = 0$ is

$$h(x) = \sqrt{2\pi^2/(4-\pi)}\sqrt{\nu x/v_0}\,. \tag{8.65}$$

We see that (8.56) obtained by qualitative considerations gives the right order of magnitude for the thickness of the boundary layer.

The force per unit area exerted by the fluid on the plate is

$$f_x = \eta\left.\frac{\partial v_x}{\partial z}\right|_{z=0} = \eta v_0\frac{\pi}{2h} = \frac{1}{2}\rho v_0^2\sqrt{\frac{4-\pi}{2}}\sqrt{\frac{\nu}{v_0 x}}\,.$$

If the plate is not semi-infinite but has a finite extent L along the flow, we have for the drag per unit width

$$F = \int_0^L f_x\,dx = \frac{1}{2}\rho v_0^2\sqrt{\frac{4-\pi}{2}}\,2L\sqrt{\frac{\nu}{v_0 L}} = \sqrt{\frac{4-\pi}{2}}\frac{\rho v_0^2 L}{\sqrt{\mathrm{Re}}} = 1.31\rho\frac{v_0^2}{2}\frac{L}{\sqrt{\mathrm{Re}}}$$

which yields for the drag coefficient

$$C_{\mathrm{D}} = \frac{2F}{\rho v_0^2 L} = \frac{1.31}{\sqrt{\mathrm{Re}}}, \qquad \mathrm{Re} = \frac{v_0 L}{\nu}.$$

The exact solution of Prandtl's equation gives the same result for C_{D} except that the coefficient is 1.328 (instead of 1.31).

8.4 Exercises

8.4.1. The constants a and b in (8.6) are assumed independent. Find the relationship between them under the assumption that the mean pressure p_{m} in the fluid equals the pressure p in the hydrodynamic equations.

Solution: The mean pressure is determined in terms of the stress tensor $\sigma'_{ik} = -p\,\delta_{ik} + \sigma_{ik}$ in the following way:

$$p_{\mathrm{m}} = -\sigma'_{kk}/3 = p - \sigma_{kk}/3\,,$$

where σ_{ik} is the viscous stress tensor. The equality $p_{\mathrm{m}} = p$ holds if $\sigma_{kk} = 0$ or, taking (8.6) into account,

$$2a(\partial v_k/\partial x_k) + 3b(\partial v_k/\partial x_k) = 0\,, \qquad b = -2a/3\,.$$

Note that the assumption $p_{\mathrm{m}} = p$ cannot be deducted from general-physics laws. To take into account the possible difference between p_{m} and p, a new coefficient (the "second viscosity") is sometimes introduced so that $p_{\mathrm{m}} = p - \mu \nabla \boldsymbol{v}$. Then,

$$b = -\tfrac{2}{3}a + \mu = -\tfrac{2}{3}\eta + \mu\,.$$

It was shown experimentally that nonzero second viscosity exists only in some special cases. For ordinary classical fluids, $\mu = 0$.

8.4.2. Using the Navier-Stokes equation (8.8), derive an equation for the vorticity $\boldsymbol{\omega} = \operatorname{curl} \boldsymbol{v}$.

Solution: Using the vector identity (6.31) and noting that $a = \eta$, one can write (8.8) in the form

$$\frac{\partial \boldsymbol{v}}{\partial t} + \nabla\left(\frac{v^2}{2}\right) + \boldsymbol{\omega} \times \boldsymbol{v} = -\frac{\nabla p}{\rho} + \nu \Delta \boldsymbol{v} + (\nu + b/\rho)\nabla(\nabla \boldsymbol{v}),\ \nu = \eta/\rho\,.$$

Apply the operation curl to this equation and use the well-known vector identities

$$\operatorname{curl}(\nabla\varphi) = 0\,, \quad \operatorname{div}\operatorname{curl}\boldsymbol{a} = 0\,, \quad \operatorname{curl}(\varphi\boldsymbol{a}) = \varphi\operatorname{curl}\boldsymbol{a} - \boldsymbol{a} \times \nabla\varphi\,,$$

$$\operatorname{curl}(\boldsymbol{a} \times \boldsymbol{b}) = (\boldsymbol{b}\cdot\nabla)\boldsymbol{a} - (\boldsymbol{a}\cdot\nabla)\boldsymbol{b} + \boldsymbol{a}(\nabla\boldsymbol{b}) - \boldsymbol{b}(\nabla\boldsymbol{a})\,.$$

This yields the required equation

$$\frac{\partial \boldsymbol{\omega}}{\partial t} + (\boldsymbol{v} \cdot \nabla)\boldsymbol{\omega} - (\boldsymbol{\omega} \cdot \nabla)\boldsymbol{v} + \boldsymbol{\omega}(\nabla \boldsymbol{v}) = -\frac{\nabla p \times \nabla \rho}{\rho^2} + \nu \Delta \boldsymbol{\omega}.$$

If we use the identity $\Delta \boldsymbol{\omega} = \nabla(\nabla \boldsymbol{\omega}) - \operatorname{curl}\operatorname{curl}\boldsymbol{\omega}$, the last equation can be written in another form:

$$\frac{\partial \boldsymbol{\omega}}{\partial t} + (\boldsymbol{v} \cdot \nabla)\boldsymbol{\omega} - (\boldsymbol{\omega} \cdot \nabla)\boldsymbol{v} + \boldsymbol{\omega}(\nabla \boldsymbol{v}) = -\frac{\nabla p \times \nabla \rho}{\rho^2} - \nu \operatorname{curl}\operatorname{curl}\boldsymbol{\omega}.$$

8.4.3. Using the solution of Exercise 8.4.2, determine the rate of change of the circulation of velocity in a viscous fluid.

Solution: Integrate the last equation in Exercise 8.4.2 over any open "fluid" surface S (including the same fluid particles all the time):

$$\int_S \left[\frac{d\boldsymbol{\omega}}{dt} - (\boldsymbol{\omega} \cdot \nabla)\boldsymbol{v} + \boldsymbol{\omega}(\nabla \boldsymbol{v})\right] \cdot \boldsymbol{n}\, dS$$

$$= -\int_S \frac{\nabla p \times \nabla \rho}{\rho^2} \cdot \boldsymbol{n}\, dS - \nu \int_S (\operatorname{curl}\operatorname{curl}\boldsymbol{\omega}) \cdot \boldsymbol{n}\, dS,$$

where $d\boldsymbol{\omega}/dt = \partial\boldsymbol{\omega}/\partial t + (\boldsymbol{v} \cdot \nabla)\boldsymbol{\omega}$. The integral on the left-hand side can be transformed using the formula analogous to (6.6):

$$\int_S \left[\frac{d\boldsymbol{\omega}}{dt} - (\boldsymbol{\omega} \cdot \nabla)\boldsymbol{v} + \boldsymbol{\omega}(\nabla \boldsymbol{v})\right] \cdot \boldsymbol{n}\, dS = \frac{d}{dt}\int_S \boldsymbol{\omega} \cdot \boldsymbol{n}\, dS.$$

Using Stokes' formula we obtain further $\int_S (\operatorname{curl}\operatorname{curl}\boldsymbol{\omega}) \cdot \boldsymbol{n}\, dS = \oint_l (\operatorname{curl}\boldsymbol{\omega})\, d\boldsymbol{r}$,

$$\int_S \boldsymbol{\omega} \cdot \boldsymbol{n}\, dS = \int_S (\operatorname{curl}\boldsymbol{v}) \cdot \boldsymbol{n}\, dS = \oint_l \boldsymbol{v}\, d\boldsymbol{r} = \Gamma,$$

where l is the closed contour supporting S. As a result, we obtain for the rate of change of the circulation

$$\frac{d\Gamma}{dt} = -\int_S \frac{\nabla p \times \nabla \rho}{\rho^2} \cdot \boldsymbol{n}\, dS - \nu \oint_l (\operatorname{curl}\boldsymbol{\omega}) \cdot d\boldsymbol{r}.$$

If $\nu = 0$ and the equation of state has the form $p = p(\rho)$ (barotropic fluid), then $\nabla p \times \nabla \rho = (dp/d\rho)\nabla\rho \times \nabla\rho = 0$ and the last equation becomes the mathematical form of Thompson's theorem for an ideal fluid. We see that the change of circulation (generation or disappearance of vortexes) is due to the viscosity or a non-barotropic equation of state.

8.4.4. Assuming the particle-velocity field in the fluid is known, find the expression for the Laplacian of the pressure (Δp).

Hint: Apply the operator div to the Navier-Stokes equation (8.17). The answer is

$$\Delta p = -\rho \nabla [(\boldsymbol{v} \cdot \nabla) \boldsymbol{v}] = -\rho \frac{\partial^2 v_i v_k}{\partial x_i \partial x_k}.$$

8.4.5. Determine the velocity distribution and drag for Couette's flow between two coaxial circular cylindrical tubes: one of these tubes moves along the generator relative to the other at the velocity v_0.

Solution: Let the radius of the tube at rest be R_1, that of the moving one R_2. The fluid particle's velocity is along the generator and depends only on the distance r from the axis: $v_x = v(r)$. The inertial term $(\boldsymbol{v} \cdot \nabla)\boldsymbol{v}$ in the Navier-Stokes equation is zero. The radial and axial components of this equation are, respectively:

$$\frac{\partial p}{\partial r} = 0, \quad \frac{\partial p}{\partial x} = \eta \, \Delta v = \eta r^{-1} \frac{d}{dr}\left(r \frac{dv}{dr} \right).$$

However, p cannot depend on x due to the symmetry of the problem, hence $\partial p / \partial x = 0$, $p = \text{const}$ and the equation for v after integration is

$$v(r) = A \ln(r/R_1) + B.$$

Using the boundary conditions $v(R_1) = 0$, $v(R_2) = v_0$, we find for the integration constants

$$B = 0, \quad A = v_0 / \ln \alpha, \quad \alpha = R_2 / R_1.$$

Hence,

$$v(r) = v_0 \ln(r/R_1) \ln^{-1} \alpha.$$

The skin friction force on the jth cylinder ($j = 1, 2$) per unit area is

$$f_{xj} = (-1)^{j-1} \eta (\partial v / \partial r)_{r = R_j} = (-1)^{j-1} \eta v_0 / R_j \ln \alpha.$$

The drag per unit length of the cylinder has the same magnitude for both cylinders and naturally has opposite direction:

$$F_x = 2\pi R_j f_{xj} = (-1)^{j-1} 2\pi \eta v_0 / \ln \alpha.$$

8.4.6. Determine the velocity distribution in the Poiseuille flow between coaxial cylinders. Find the flow rate.

Solution: Components of the Navier-Stokes equation will be the same as in the previous exercise but now we have to assume that $\partial p/\partial x = \text{const}$. Integrating the equation for $v(r)$ twice, we obtain

$$v(r) = \frac{1}{4\eta}\frac{dp}{dx}r^2 + A\ln\frac{r}{R_1} + B.$$

Using the boundary conditions $v(R_1) = v(R_2) = 0$, we find

$$B = -\frac{1}{4\eta}\frac{dp}{dx}R_1^2, \qquad A = -\frac{1}{4\eta}\frac{dp}{dx}(R_2^2 - R_1^2)\ln^{-1}\alpha, \qquad \alpha = \frac{R_2}{R_1},$$

$$v = -\frac{R_1^2}{4\eta}\frac{dp}{dx}\left[1 - \frac{r^2}{R_1^2} + (\alpha^2 - 1)\frac{\ln(r/R_1)}{\ln\alpha}\right].$$

The flow rate is

$$Q = 2\pi\int_{R_1}^{R_2} rv(r)\,dr = \frac{\pi}{8\eta}\frac{dp}{dx}R_1^4(\alpha^2 - 1)\left(1 + \alpha^2 - \frac{\alpha^2 - 1}{\ln\alpha}\right).$$

8.4.7. Examine the Poiseuille flow in the circular tube whose axis makes an angle θ with the horizontal plane.

Hint: The procedure will be the same as in Sect. 8.2.3. The gravitational force directed vertically must now be taken into account in the Navier-Stokes equation. Nothing will change in the expression for v except the coefficient which will now be $b = -(\partial p/\partial x) - \rho g\sin\theta$.

8.4.8. Determine the velocity and pressure distributions in the Couette flow between two coaxial circular cylinders (radii R_1 and R_2) rotating about its axis at the angular velocities ω_1 and ω_2.

Solution: Use the polar coordinates z, r, φ, the z-axis coinciding with the axis of the cylinders. Obviously, $p = p(r)$, $v_z = v_r = 0$ and $v_\varphi = v(r)$. We have in polar coordinates

$$\nabla = \boldsymbol{e}_r\frac{\partial}{\partial r} + \frac{\boldsymbol{e}_\varphi}{r}\frac{\partial}{\partial\varphi}, \qquad \Delta = r^{-1}\frac{\partial}{\partial r}\left(r\frac{\partial}{\partial r}\right) + r^{-2}\frac{\partial^2}{\partial\varphi^2}, \qquad v_x = -v\sin\varphi,$$

$$v_y = v\cos\varphi, \qquad (\boldsymbol{v}\cdot\nabla) = \frac{v}{r}\frac{\partial}{\partial\varphi}, \qquad \frac{\partial p}{\partial x} = \frac{x}{r}\frac{\partial p}{\partial r} = \frac{\partial p}{\partial r}\cos\varphi, \qquad \frac{\partial p}{\partial y} = \frac{\partial p}{\partial r}\sin\varphi.$$

Now the components of the Navier-Stokes equation are

$$\cos\varphi\left(\frac{1}{\rho}\frac{\partial p}{\partial r} - \frac{v^2}{r}\right) = \nu\sin\varphi\left[\frac{v}{r^2} - \frac{1}{r}\frac{d}{dr}\left(r\frac{dv}{dr}\right)\right],$$

$$\sin\varphi\left(\frac{1}{\rho}\frac{\partial p}{\partial r} - \frac{v^2}{r}\right) = -\nu\cos\varphi\left[\frac{v}{r^2} - \frac{1}{r}\frac{d}{dr}\left(r\frac{dv}{dr}\right)\right].$$

Hence, it follows that

$$\frac{1}{\rho}\frac{dp}{dr} = \frac{v^2}{r}, \quad \frac{1}{r}\frac{d}{dr}\left(r\frac{dv}{dr}\right) - \frac{v}{r^2} = 0 \quad \text{or} \quad \frac{d}{dr}\left[\frac{1}{r}\frac{d(rv)}{dr}\right] = 0.$$

The general solution of the last equation has the form $v(r) = \Omega_0 r + \Omega_p R_1^2/r$. Using the boundary conditions $[v(R_1) = \omega_1 R_1, v(R_2) = \omega_2 R_2]$, we find the integration constants

$$\Omega_0 = (\alpha^2\omega_2 - \omega_1)/(\alpha^2 - 1), \quad \Omega_p = -\alpha^2(\omega_2 - \omega_1)/(\alpha^2 - 1), \quad \alpha = R_2/R_1$$

and

$$p(r) = p(R_1) + \rho \int_{R_1}^{r} \frac{v^2}{r}\, dr$$

$$= p(R_1) + \rho R_1^2\left(2\Omega_0\Omega_p \ln\frac{r}{R_1} + \frac{\Omega_0^2(r^2 - R_1^2)}{2R_1^2} + \frac{\Omega_p^2(r^2 - R_1^2)}{2r^2}\right).$$

The above formula for $v(r)$ implies that fluid particles at a distance r from the axis rotate about it at an angular velocity $\omega(r) = \Omega_0 + \Omega_p R_1^2/r^2$. Hence, it follows that the flow considered is a combination of rotation at a constant angular velocity Ω_0 and potential rotation (Exercise 7.5.1) at an angular velocity $\Omega_p R_1^2/r^2$.

8.4.9. Determine the viscous stresses in the Couette flow between two rotating cylinders (Exercise 8.4.8) and the moment of viscous forces with respect to the axis.

Solution: Viscous stresses are the same at any plane containing the axis. Therefore, we can confine ourselves to, say, $\sigma_{xy}|_{x=0}$. We have, according to (8.6),

$$\sigma_{xy} = \eta\left(\frac{\partial v_x}{\partial y} + \frac{\partial v_y}{\partial x}\right).$$

Substituting v_x and v_y from Exercise 8.4.8 into this formula, we obtain

$$\sigma_{xy} = 2\eta\Omega_p \frac{R_1^2}{r^2}\frac{y^2 - x^2}{r^2}, \quad \sigma_{xy}|_{x=0} = 2\eta\Omega_p \frac{R_1^2}{r^2} = -2\eta\frac{\omega_2 - \omega_1}{\alpha^2 - 1}\frac{R_2^2}{r^2}.$$

The moment of the viscous force is

$$M = 2\pi r\sigma_{xy} r = -4\pi\eta\frac{\omega_2 - \omega_1}{R_2^2 - R_1^2} R_1^2 R_2^2.$$

M is independent of r, as expected. Note that this fact could be used as the starting assumption for obtaining $v(r)$.

8.4.10. Obtain an expression for viscous waves between two walls a distance d apart. One of the walls is at rest, whereas the tangential velocity of the other is $v = v_0 \exp(-i\omega t)$. Find the amplitude of the viscous force on the wall at rest.

Solution: Let the oscillating wall be at $z = 0$, whereas that at rest at $z = d$. The general harmonic solution of the equation (8.49) for viscous waves is ($v = v_x$)

$$v = \left\{ A \sinh\left[(1 + i)\sqrt{\frac{\omega}{2\nu}}(d - z)\right] + B \cosh\left[(1 + i)\sqrt{\frac{\omega}{2\nu}}(d - z)\right]\right\} \exp(-i\omega t).$$

From the condition $v|_{z=d} = 0$, we obtain $B = 0$. The condition at $z = 0$ yields

$$A \sinh[(1 + i)\alpha] = v_0, \qquad \alpha = \sqrt{\frac{\omega}{2\nu}}\, d,$$

$$v = v_0 \frac{\sinh[(1 + i)\alpha(1 - z/d)]}{\sinh[(1 + i)\alpha]} \exp(-i\omega t).$$

The viscous force per unit area on the wall at rest is

$$f_x|_{z=d} = \eta \left.\frac{\partial v}{\partial z}\right|_{z=d} = -\eta v_0 \frac{\alpha}{d} \frac{(1 + i)\exp(-i\omega t)}{\sinh[(1 + i)\alpha]}$$

$$= -\eta v_0 \frac{\alpha}{d} \frac{(1 + i)\exp(-i\omega t)}{\sinh\alpha \cos\alpha + i \cosh\alpha \sin\alpha}.$$

The amplitude of the viscous force $F = |f_x|$ is given by

$$F = \eta v_0 \frac{\alpha\sqrt{2}}{d} (\sinh^2\alpha \cos^2\alpha + \cosh^2\alpha \sin^2\alpha)^{-1/2} = \frac{\sqrt{2}\alpha\eta v_0}{d\sqrt{\cosh^2\alpha - \cos^2\alpha}}.$$

8.4.11. The same as Exercise 7.5.14, assuming: a) That the water is a viscous fluid and the Stokes formula for a drag is true. b) Take into account that for the large velocity V the drag is $\rho V^2 S/2$ (S being the cross section), estimate the real time for the sphere to reach the bottom.

9. Elements of the Theory of Turbulence

All the flows considered above were regular or "*laminar*" ones. The origin of the term is due to the fact that if one marks fluid with dye, then one can see laminar currents in such flows. Flows are always laminar, according to experiment, if their velocities are sufficiently small. On the other hand, a laminar flow always turns into a *turbulent* one if the velocity increases. Turbulent flow is essentially irregular. Its velocity, pressure and other parameters vary in a random way at any point, even when the boundary and all other external conditions are regular. This chapter contains the basic theory of turbulence.

9.1 Qualitative Considerations. Hydrodynamic Similarity

9.1.1 Transition from a Laminar to Turbulent Flow

Such a transition was first observed in 1839 by Hangen in flow through circular pipes. Systematic investigations, however, were carried out by Reynolds in 1843. It was shown by him that it is not the velocity itself which is essential for transition but the nondimensional quantity (*Reynolds number*) $\mathrm{Re} = vD/\nu$, v being the characteristic velocity, D the characteristic space dimension of the flow (for example, the tube's diameter) and ν the kinematic viscosity. We have seen above that the Reynolds number is the ratio of the inertial term to the viscous one in the Navier-Stokes equation. It was established by Reynolds that some *critical* $\mathrm{Re}_{\mathrm{cr}}$ exists when a transition of laminar to turbulent flow occurs.

The laminar-turbulent transition is easily detected by adding a dye into the fluid. Stream-lines in a tube are straight when the flow is laminar. When the transition occurs, however, the stream-lines become distorted and the dye gradually diffuses all over the cross section of the tube. The Reynolds number can be altered in an experiment by velocity as well as by viscosity variations. The latter can be achieved by using another fluid or by heating the same fluid. In the first such experiments, Reynolds obtained $\mathrm{Re}_{\mathrm{cr}} = 12830$ when the entrance into the pipe was smooth. The laminar flow could be maintained up to $\mathrm{Re}_{\mathrm{cr}} = 20000$ under the special precautions taken against distortion of the inward flow (smooth entrance and smooth pipe walls, no vibrations, etc.). It was shown by numerous experiments that flows are always laminar for $\mathrm{Re} \lesssim 2000$. In this case any distortions in the inward flow decay away along the flow.

We have an analogous situation in a boundary layer where $\mathrm{Re}_{\mathrm{h}} = v_0 h(x)/\nu$ (Chap. 8). The critical $\mathrm{Re}_{\mathrm{cr}}$ appears to be of the same order of magnitude as in the case of flow through pipes. It varies from $\mathrm{Re_h} \sim 10^3$ up to $\mathrm{Re_h} \sim 10^4$, depending on the smoothness of the original flow. Since the Reynolds number is now increasing down the stream [$h(x)$ is increasing], some x_{cr} exists where $\mathrm{Re_h}$ achieves its critical value. The transition of a laminar boundary layer into a turbulent one appears to be gradual, however. It begins at some $x < x_{\mathrm{cr}}$ where sporadic turbulent patches appear in the laminar flow and ends at some $x > x_{\mathrm{cr}}$ where flow becomes completely turbulent.

The critical value of $\mathrm{Re_h}$ and x_{cr} depend essentially on the smoothness of the surface. Roughness, comparable in scale with h, promotes the transition from laminar to a turbulent flow.

A theoretical description of the laminar-turbulent transition has not been achieved yet. Evidently, this problem has to be solved on the basis of hydrodynamical equations (about whose correctness we have no doubts). The laminar-turbulent transition must then be due to the loss of stability of the laminar flow, i.e., due to a dramatic increase of the initially small disturbances which are always present in a real laminar flow. Hence, exploring the stability of the laminar solution of hydrodynamical equations, we can find some critical value of the Reynolds number Re_1 when this stability will be lost. According to *Landau's hypothesis*, one will have a new stable flow at $\mathrm{Re} > \mathrm{Re}_1$. This flow will be a combination of the original laminar flow and one harmonic "mode". If the Reynolds number increases further, then this flow will become unstable in its turn at some $\mathrm{Re} = \mathrm{Re}_2$ and the second harmonic mode will appear, etc. According to Landau, this process will continue incessantly with increasing Re. As a result, a great number of modes will exist at a sufficiently large Re. An ensemble of these modes with different frequencies behaves as a very irregular system and very closely resembles so-called *developed turbulence*.

Initially, Landau's hypothesis appeared to be verified by some experiments (see the description below of the flow past a cylinder, for example). The greater the Reynolds number, the more complicated the flow becomes. However, only a few initial steps of the process could be made in the experiments before the flow suddenly became turbulent.

Recently, new ideas have appeared concerning the laminar-turbulent transition. For its understanding we note that the state of a hydrodynamic system can be mathematically described by the expansion of its velocity field in terms of an orthogonal set of functions (modes). Then the coefficients q_i of this expansion can be regarded as generalized coordinates of this system and their derivatives with respect to time $p_i \equiv \dot{q}_i$ as generalized velocities. In space-limited systems, the number of generalized coordinates is countable. Moreoever, only a finite number $\mathcal{N}$ of modes is important, since very high modes (i very large) corresponding to small-scale variations of velocity disappear quickly due to the viscosity. The $2\mathcal{N}$-dimensional coordinate space (q_i, p_i) is called the *phase space of the flow*. A line in this space describing the change of the state of the system is called the *phase trajectory*.

In the case of steady flow, the latter degenerates into a fixed point describing the equilibrium state of a "focus" type. For periodical motion it is a closed line also called the limit cycle. Only these two types of features or so-called *simple attractors* in phase space have been considered in the hypotheses by Landau and others until recently. The turbulent (random) character of the flow was related to the existence of a large number of modes (degrees of freedom). Recently, however, a new hypothesis was suggested which states that the so-called *strange attractors*, which are different from the simple ones, could exist in phase space of hydrodynamical systems. A characteristic example of such an attractors is the Lorenz attractor appearing in the solution of a model system which includes 3 equations and has two fixed points in phase space. The point of the phase space which represents the state of the system ("representative point") moves for some time in the vicinity of one of the fixed points while the system changes. Then it follows untwisted cycles and eventually goes over into the vicinity of the other fixed point at a finite distance away from the first one. Then the process is repeated over and over in a nonregular (pseudorandom) way. Such (strange) attractors were also obtained for other model systems, but their existence in systems described by hydrodynamical equations has not been proved rigorously yet.

The hypothesis about strange attractors fundamentally changes the notion about the appearance and development of turbulence. We observe pseudorandom behaviour of the system even with a small number of degrees of freedom in this case. Of course, describing the flows with a large number of degrees of freedom is required for the turbulence problem. However, such a description is rather far from being achieved. Meanwhile, the practice needs some guidance in treating turbulent flow. In these circumstances, an experiment together with the so-called similarity theory appears to be very useful. It is possible due to the latter to get results for real cases using those obtained by model experiments.

9.1.2 Similar Flows

Let us consider homogeneous flow at infinity of a viscous incompressible homogeneous fluid past some body. We have the set of hydrodynamical equations

$$\rho \frac{\partial \boldsymbol{v}}{\partial t} + \rho(\boldsymbol{v} \cdot \nabla)\boldsymbol{v} = -\nabla p + \eta \Delta \boldsymbol{v}, \qquad \operatorname{div} \boldsymbol{v} = 0 \tag{9.1}$$

with the conditions on the body's surface S and at infinity:

$$\boldsymbol{v}|_S = 0, \qquad \boldsymbol{v}|_{r\to\infty} = v_0 \nabla x, \qquad r^2 = x^2 + y^2 + z^2. \tag{9.2}$$

The surface S can always be specified by $f(x/D, y/D, z/D) = 0$, where D is a characteristic dimension of the body. In the case of a circular cylinder, for example, $f = (x^2 + y^2)/D^2 - 1/4$. Of course, the surface's equation can also include some dimensionless parameters specifying ratios of dimensions of the body in different directions. For an elliptical cylinder, for example, $f = (x^2 + \alpha^2 y^2)D^{-2} - 1/4$, α being the ratio of the principal axes and D one of these axes.

A solution of the problem (9.1, 2) depends, besides coordinates and time, on four dimensional parameters η, ρ, v_0 and D. It would be impossible to perform an experiment for each of the possible combinations of these quantities because the number of these combinations is enormously large. However, for a homogeneous fluid ($\rho = \text{const}$), the number of the dimensional parameters can be reduced to three if we use the equation for vorticity $\boldsymbol{\omega} = \text{curl}\, \boldsymbol{v}$,

$$\partial \boldsymbol{\omega}/\partial t + \text{curl}(\boldsymbol{\omega} \times \boldsymbol{v}) = \nu \Delta \boldsymbol{\omega} \tag{9.3}$$

instead of the Navier-Stokes equation. This equation can be obtained by applying operation curl to (9.1) and by taking into account the vector identity (6.31) (Exercise 8.4.2). Now, our problem includes only three-dimensional parameters ν, D and v_0.

To further reduce the number of parameters, we introduce the dimensionless velocity $\boldsymbol{v}' = \boldsymbol{v}/v_0$, the dimensionless coordinates $x' = x/D$, $y' = y/D$, $z' = z/D$ and the corresponding dimensionless time $t' = v_0 t/D$. In other words, we measure the length, velocity and time hereafter in terms of D, v_0 and D/v_0, respectively. The boundary conditions do not contain any dimensional parameter

$$\boldsymbol{v}'\big|_{f(x',y',z')=0} = 0, \qquad \boldsymbol{v}'\big|_{z'=\infty} = \nabla' x', \qquad r' = r/D, \qquad \nabla' = D\nabla . \tag{9.4}$$

We now rewrite the equation (9.3) for vorticity. Since $\partial/\partial x = D^{-1}\partial/\partial x'$ and analogously for $\partial/\partial y$ and $\partial/\partial z$, we have

$$\boldsymbol{\omega} = \text{curl}\, \boldsymbol{v} = v_0 D^{-1}\, \text{curl}'\, \boldsymbol{v}' = v_0 D^{-1} \boldsymbol{\omega}' ,$$

where curl′ means the corresponding operation in primed coordinates. Further, we have $\partial/\partial t = v_0 D^{-1} \partial/\partial t'$ and (9.3) becomes

$$\frac{v_0^2}{D^2}\frac{\partial \boldsymbol{\omega}'}{\partial t'} + D^{-1}\, \text{curl}'\left(\frac{v_0}{D}\, \boldsymbol{\omega}' \times v_0 \boldsymbol{v}'\right) = \frac{\nu}{D^2}\, \Delta'\left(\frac{v_0}{D}\, \boldsymbol{\omega}'\right).$$

Dividing this by $v_0^2 D^{-2}$ and introducing the Reynolds number $\text{Re} = v_0 D/\nu$, we finally obtain

$$\frac{\partial \boldsymbol{\omega}'}{\partial t'} + \text{curl}'(\boldsymbol{\omega}' \times \boldsymbol{v}') = \text{Re}^{-1}\, \Delta' \boldsymbol{\omega}' . \tag{9.5}$$

This equation, together with the incompressibility condition $\nabla' \boldsymbol{v}' = 0$ the equation $\boldsymbol{\omega}' = \text{curl}'\, \boldsymbol{v}'$ and the boundary condition (9.4), completely defines the problem of flow past the body and includes only *one* parameter, the *Reynolds number* Re. It means that *for a fixed* Re, *all flows past geometrically similar bodies are similar*, i.e., described by the same function of dimensionless variables. In other words, the fluid velocity measured in terms of v_0 depends only on space coordinates and time measured in terms of D and D/v_0, respectively, and on the Reynolds number Re:

$$\boldsymbol{v} = v_0 \boldsymbol{G}(x/D, y/D, z/D, v_0 t/D, \mathrm{Re}), \tag{9.6}$$

where the function $\boldsymbol{G}$ is the same for all geometrically similar bodies. This important result is called the *similarity principle of Reynolds.*

In Sect. 7.3.1 we had a manifestation of this principle for the particular case of an ideal fluid around a cylinder where the corresponding function $\boldsymbol{G}$ did not depend on Re. It is not difficult also to find such functions for Couette's and Pouiseuille's flows considered in the previous chapter. The mean velocity can be taken as v_0 in these cases. These functions also turn out to be independent of the Reynolds number.

It is customary to specify the pressure in fluids in terms of $\rho v_0^2/2$. Dimensionless pressure also turns out to be the same function of dimensionless variables for all geometrically similar bodies (see, for example, Exercise 8.4.4). The same is true for force exerted by the fluid on the body. The unit for this force is $\rho v_0^2 D^2/2$. For this reason the drag coefficient calculated for some flows in the previous chapter turns out to be dependent also only on the Reynolds number.

Model experiments in wind tunnels are practised widely. Using the results of these experiments and the similarity principle we can obtain parameters of flows past real objects.

9.1.3 Dimensional Analysis and Similarity Principle

We now show that the similarity principle can be deduced from a consideration of physical dimensions the so-called dimensional analysis. It was shown in the previous subsection that four independent dimensional parameters exist in the problem of the incompressible fluid flow past a body of arbitrary shape:

$$[\eta] = \mathrm{ML^{-1}T^{-1}}, \quad [\rho] = \mathrm{ML^{-3}}, \quad [D] = \mathrm{L}, \quad [v_0] = \mathrm{LT^{-1}},$$

where $[v_0]$ is the dimension of v_0, etc.; M, L and T are dimensions of mass, length and time, respectively. The surface of the body S at which the boundary conditions must be satisfied can be completely determined by D and some dimensionless function which specifies the shape of the body.

Suppose now that we are looking for the expression of a dimensionless flow velocity in terms of dimensionless coordinates and time. It is clear that this velocity must be independent of the units which we use for measuring the length (centimeters, or meters, for example), density (gramms or tons) or viscosity, etc. This implies that only dimensionless combinations of those four dimensional parameters may arise in the required expression. This seemingly trivial principle is the basis of the application of dimensional analysis in hydrodynamics.

In many cases it permits us to obtain very general results, which are far from trivial. In the case we are considering now, we have to find the dimensionless combination of the dimensional quantities η, D, ρ and v_0. We will try this combination in the form

$$\mathrm{Re} = D^m v_0^n \eta^k \rho^l . \tag{9.7}$$

One can set $m = 1$ without loss of generality. Now substituting the dimensions of all the parameters in (9.7) and taking into account that the summed powers of M, L and T must be zero, we obtain three equations

$$k + l = 0 , \quad n + k = 0 , \quad 1 + n - k - 3l = 0 ,$$

which yield $l = 1$, $n = 1$, $k = -1$. Now (9.7) becomes $\mathrm{Re} = Dv_0\rho/\eta = Dv_0/\nu$, i.e., the Reynolds number found earlier.

Naturally enough, in other more complicated cases the flow would be characterized by more than one dimensionless parameter. Suppose, for example, that the gravitational force must be taken into account. Then the gravitational acceleration g (of dimension $[g] = \mathrm{LT}^{-2}$) must be added to the four abovementioned dimensional parameters. As a result, one more dimensionless combination can be introduced besides the Reynolds number, for example, the so-called *Froude number* $\mathrm{Fr} = v_0^2/gD$. Both numbers Re and Fr must be the same in the case of similar flows.

When the flow is unsteady, a new dimensional parameter which is the characteristic time of the flow τ (oscillation period, for example) must be added. Then the new dimensionless combination of the *Strouhal number* $\mathrm{Sh} = D/v_0\tau$ can be introduced besides Re.

Equations for a compressible fluid flow include one more dimensional parameter, the sound velocity c. The new dimensionless combination widely used is the Mach number $\mathrm{M} = v_0/c$.

If the heat transfer in the fluid must be taken into account, then the heat-transfer equation with thermal conductivity $\varkappa$ must be incorporated into the set of equations. The dimension of $\varkappa$ is the same as that of ν. Hence, the *Prandtl number* $\mathrm{Pr} = \nu/\varkappa$ can be chosen as a new dimensionless combination.

9.1.4 Flow Around a Cylinder at Different Re

We have seen above that if the gravitational force can be neglected, the incompressible ideal fluid flow is completely specified by the Reynolds number Re. The character of the flow, however, is quite different at different Re. We consider, as an example, the flow around a circular cylinder with its axis perpendicular to the flow:

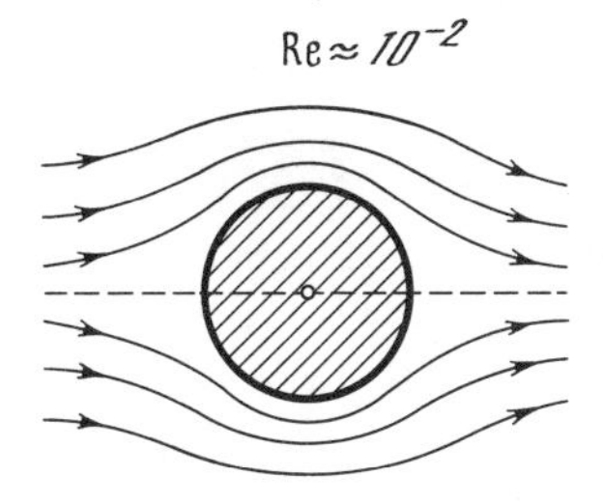

Fig. 9.1. Laminar flow around a cylinder for small Re

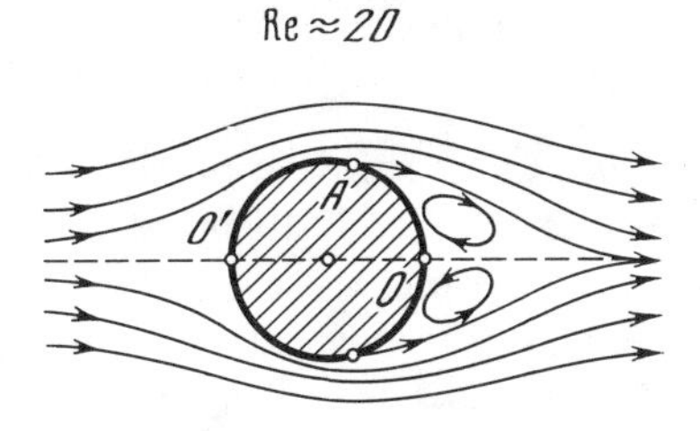

Fig. 9.2. Steady flow past a cylinder with two vortices

i) At small Reynolds number ($\mathrm{Re} \ll 1$), laminar flow takes place around the cylinder, which is very close to that of an ideal fluid (Fig. 9.1).

ii) $1 < \mathrm{Re} < 40$. The first critical Reynolds number Re_1 occurs in the vicinity of $\mathrm{Re} = 1$. At $\mathrm{Re} > \mathrm{Re}_1$, the flow becomes unstable but the new type of flow is formed only at $\mathrm{Re} > 10$ when two eddies are formed behind the cylinder (Fig. 9.2). The flow is still steady and laminar so that at $\mathrm{Re} = \mathrm{Re}_1$, we have only a change of one stable flow for another. The reason for the formation of vortexes becomes clear if we observe the change of pressure and velocity along the surface of the cylinder outside the boundary layer. These quantities can be found from the solution to the ideal fluid equations. According to (7.24, 27), the velocity is zero at the critical points ($\theta = 0, \pi$) and reaches a maximum at the midplane ($\theta = \pm\pi/2$). In accordance with Bernoulli's theorem, the pressure has maxima at critical points and reaches a minimum at the midplane. For this reason the fluid particles downstream of the midplane move against the increasing pressure which leads to their retardation. This retardation will have the strongest effect on the fluid particles close to the surface possessing the smallest velocity. At a point A (Fig. 9.2) downstream from the midplane, these particles come to a standstill; beyond A they will be moving backwards with respect to the fluid particles further from the cylinder which have not yet been retarded. The existence of counterflows near the surface of the body induces the formation of vortexes and also forces the outer flow away from the surface. The latter will result in the separation of the boundary layer from the surface at point A. The higher Re, the closer point A is to the midplane and the greater the drag (the area of high pressure behind the midplane is less).

iii) When Re increases further ($\mathrm{Re} > 40$), the next critical values of this number will be attained, and the flow becomes unsteady. One of the vortexes elongates, then separates and goes down the stream. After that the other vortexes do the same. New vortexes appear at their place and the process is repeated. As a result, the so-called "*Karman path*" is formed in the wake behind the cylinder (Fig. 9.3). The flow is unsteady but periodical since the process of a separation of vortexes is periodical.

iv) At $\mathrm{Re} > 1000$, there will be no time for full formation of the vortexes. Instead one can observe highly turbulent areas rapidly appearing, separating and going down the stream. At $\mathrm{Re} \sim 10^3 – 10^4$, turbulence in the wake becomes greater, the flow is now highly irregular and three dimensional. At $\mathrm{Re} \sim 10^5$, the wake is fully turbulent, beginning just behind the cylinder (Fig. 9.4).

The drag coefficient per unit length of cylinder

$$C_\mathrm{D} = 2F/\rho v_0^2 D$$

(F being the force exerted by the flow on the cylinder) is of considerable interest. Dependence of C_D on Re obtained experimentally is shown schematically in Fig. 9.5. At small Re ($\mathrm{Re} < 1$ is not shown in the figure), the drag coefficient decreases with increasing Re in the same manner as in the case of laminar flow past a sphere, see (8.45). This decrease becomes weaker at higher Re when the point of separation of the boundary layer approaches the midplane but continues approximately up to $\mathrm{Re} \sim 10^3$. At greater Re, some increase in C_D takes place because of the generation of turbulence in the wake and then C_D becomes almost constant. After that (Re is slightly more than 10^5), the drag coefficient drops off drastically. This phenomenon, called *drag crisis*, is due to the generation of turbulence in the boundary layer. Let us consider it closer. At small Re, the boundary layer remains laminar between the front critical point and the

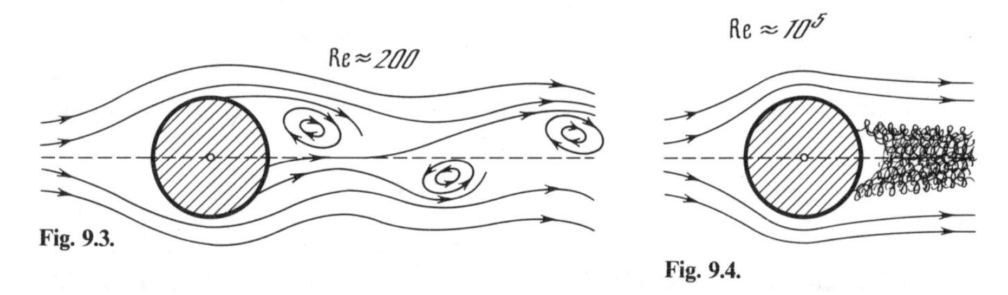

Fig. 9.3.

Fig. 9.4.

Fig. 9.3. Illustrating a Karman street
Fig. 9.4. The flow with a fully developed turbulent wake

Fig. 9.5. The dependence of the Drag coefficient on Re (experiment)

separation point A (Fig. 9.2) where it separates from the cylinder. At $\mathrm{Re} \sim 10^5$, however, the Reynolds number for the boundary layer $\mathrm{Re}_h = v_0 h/\nu$ reaches a critical value somewhere in front of A, and the boundary layer becomes turbulent. It appears that a turbulent boundary layer follows the surface much longer than a laminar one, i.e., the separation point will now be somewhere to the right of A. This is due to the fact that momentum transfer across the boundary layer increases greatly when it becomes turbulent. Therefore, the inflow of fluid from the outer flow to the boundary layer is considerably increased, so that the fluid particles in the boundary layer move in the direction of increasing pressure much longer than in the case of laminar flow. We have seen above, however, that moving the separation point downstream leads to an increase in the area of high pressure to the right of the midplane and consequently to a decrease in drag. When Re increases further, the drag coefficient increases again because the separation point of the boundary layer (now turbulent) again moves to the left.

It is evident enough that the greater the initial distortion of the flow before the body, the earlier the laminar-turbulent transition in the boundary layer occurs and at smaller Re will the drag crisis take place. Prandtl demonstrated this fact by introducing additional distortion to the flow in fitting a thin wire ring around the sphere in the wind tunnel. According to more recent experiments, the drag crisis occurs at $\mathrm{Re}_{\mathrm{cr}} = 2.7 \times 10^5$ if the initial turbulence of the flow is $(\overline{v_{\mathrm{turb}}^2})^{1/2}/v_0 = 5 \times 10^{-3}$ and at $\mathrm{Re}_{\mathrm{cr}} = 1.25 \times 10^5$ for the initial turbulence of 2.5×10^{-2} ($\mathrm{Re}_{\mathrm{cr}}$ was chosen corresponding to $C_{\mathrm{D}} = 0.3$ in this case).

9.2 Statistical Description of Turbulent Flows

9.2.1 Reynolds' Equation for Mean Flow

At first sight it is impossible to reveal any regularity in turbulent flow. The flow velocity and other quantities vary randomly at each point. If the flow is set up repeatedly under the same conditions, exact values of these quantities will be different each time. Reynolds was the first to show that it is possible to obtain reliable relations for average values in turbulent flow considering the latter as a statistical process and using the formalism of probability theory.

Suppose we repeat the same experiment under the same external conditions many times. Because of small uncontrollable disturbances in the flow and in the initial as well as the boundary conditions, we will obtain different results, say, for the fluid velocity at a fixed point during each experiment. We assume further that a large number of results from such experiments constitute a statistical ensemble. The results of each experiment represent a separate "sample value" or "realization" in this ensemble.

We denote by $\langle v_i \rangle$ the *ensemble average* velocity at a fixed point, or the velocity at this point averaged over all the experiments. The actual velocity for a given realization can be expressed as the sum

$$v_i = \langle v_i \rangle + \delta v_i, \qquad \langle \delta v_i \rangle = 0, \tag{9.8}$$

where δv_i is a random deviation of the velocity from the ensemble average value.

Below we will derive Reynolds' equations for average flow using the procedure of averaging over an ensemble. Before that, however, we have to note two important items:

i) We will use some rules for averaging, known in probability theory. Most of them are rather obvious, for example, $\langle \boldsymbol{v}_1 + \boldsymbol{v}_2 \rangle = \langle \boldsymbol{v}_1 \rangle + \langle \boldsymbol{v}_2 \rangle$ where the average of the sum is equal to the sum of to averages. Less trivial is the following rule:

$$\left\langle \frac{\partial v}{\partial s} \right\rangle = \frac{\partial \langle v \rangle}{\partial s}, \tag{9.9}$$

where v is one of the components of velocity or pressure or any other quantity (density, temperature, etc.), and s denotes x_1, x_2, x_3 or t. Equation (9.9) expresses the commutation of the operations of averaging and differentiating.

ii) The question of how to compare the theory developed below with experiment is nontrivial since in the latter, as a rule, we cannot carry out a large number of similar experiments under strictly the same conditions and calculate ensemble average values. Instead, one usually obtains in an experiment only one rather long sample and calculates the *time average* value, for example,

$$\overline{v(t)} = T^{-1} \int_{-T/2}^{T/2} v(t + \tau)\, d\tau \,. \tag{9.10}$$

To distinguish the two kinds of average values, we denote the time average by a bar over the quantity involved.

An important question is what connection is there between these two methods of averaging. Do time averages converge to the ensemble averages when the averaging interval T becomes infinitely long $\lim_{T \to \infty} \overline{v(t)} = \langle v \rangle$? If this requirement is met, such a statistical process is called an *ergodic* one. It is clear that one of the necessary conditions of ergodicity of the process must be the independence of $\overline{v(t)}$ of time t, otherwise $\lim_{T \to \infty} \overline{v(t)}$ does not exist. This requirement must also be met by the statistical moments of different orders formed using the function $v(t)$. For example, the *autocorrelation function* at the points $t_1 = t$ and $t_2 = t + \tau$,

$$b(\tau) = \langle [v(t) - \langle v \rangle][v(t + \tau) - \langle v \rangle] \rangle = \langle \delta v(t)\, \delta v(t + \tau) \rangle \,, \tag{9.11}$$

may depend only on τ but not on t. Such random processes satisfying the condition that all their average characteristics do not change when the time changes, are called *stationary*. For a turbulent flow, the condition of stationarity is satisfied if all average characteristics of the flow (the average velocity distribution, the average temperature and so on) remain unchanged when the time changes. This also implies that all external conditions (external forces, position of bodies, etc.) will remain unchanged.

When a random function $v(\boldsymbol{r}, t)$ depends on time as well as on the space variables, it is possible to obtain *space-average* values by averaging with respect to the coordinates. Analogous to stationarity is *homogeneity* of the function under consideration in this case, when all moments of this function depend only on the vector distance between points but not on $\boldsymbol{r}$. For example, the correlation function is

$$b(\boldsymbol{r}_2 - \boldsymbol{r}_1) = \langle \delta v(t, \boldsymbol{r}_1)\, \delta v(t, \boldsymbol{r}_2) \rangle \,. \tag{9.12}$$

We will proceed further on the assumption that the turbulent flows under consideration are stationary with respect to time and homogeneous with respect to spacial coordinates.

Stationarity (or homogeneity) of the random function is a necessary but not sufficient condition, however, for ergodicity of the process it describes. It can be shown that ergodicity will be secured if the correlation function (9.11) obeys the condition

$$\lim_{\tau \to \infty} b(\tau) = 0 \,. \tag{9.13}$$

In other words, $v(t)$ and $v(t + \tau)$ must be *statistically independent* at large τ.

Now we proceed to the averaging of the hydrodynamic equations for an incompressible fluid. The ensemble average of (8.3) for momentum balance is

$$\rho \frac{\partial \langle v_i \rangle}{\partial t} = -\frac{\partial \langle \Pi_{ik} \rangle}{\partial x_k} \,, \tag{9.14}$$

where ρ is constant and

$$\Pi_{ik} = p\, \delta_{ik} + \rho v_i v_k - \sigma_{ik} \,, \qquad \sigma_{ik} = \eta \left(\frac{\partial v_i}{\partial x_k} + \frac{\partial v_k}{\partial x_i} \right) . \tag{9.15}$$

Taking (9.8) into account we have

$$\langle v_i v_k \rangle = (\langle v_i \rangle + \delta v_i)(\langle v_k \rangle + \delta v_k) = \langle v_i \rangle \langle v_k \rangle + \langle \delta v_i\, \delta v_k \rangle$$

since

$$\langle \langle v_i \rangle \langle v_k \rangle \rangle = \langle v_i \rangle \langle v_k \rangle \,, \qquad \langle \langle v_i \rangle\, \delta v_k \rangle = \langle v_i \rangle \langle \delta v_k \rangle = 0 \,.$$

Hence,

$$\langle \Pi_{ik} \rangle = \langle p \rangle \delta_{ik} + \rho \langle v_i \rangle \langle v_k \rangle - \langle \sigma_{ik} \rangle + \rho \langle \delta v_i \, \delta v_k \rangle \, ,$$
$$\langle \sigma_{ik} \rangle = \eta [(\partial \langle v_i \rangle / \partial x_k) + (\partial \langle v_k \rangle / \partial x_i)] \, . \tag{9.16}$$

As a result, we have the same equation (9.14) for the average quantities $\langle v_i \rangle$ and $\langle p \rangle$, as in the absence of turbulence, with the only difference that an additional term $\rho \langle \delta v_i \, \delta v_k \rangle$ occurs in the momentum flux tensor. Equation (9.14) can also be written in the same form as the Navier-Stokes equation. Indeed, averaging the continuity equation (incompressibility condition) $\operatorname{div} \boldsymbol{v} = 0$, we find $\partial \langle v_k \rangle / \partial x_k = 0$; hence, $\partial (\langle v_i \rangle \langle v_k \rangle) / \partial x_k = \langle v_k \rangle \, \partial \langle v_i \rangle / \partial x_k$. Using (9.14, 16), we now obtain

$$\rho \frac{\partial \langle v_i \rangle}{\partial t} + \rho \langle v_k \rangle \frac{\partial \langle v_i \rangle}{\partial x_k} = -\frac{\partial \langle p \rangle}{\partial x_i} + \frac{\partial}{\partial x_k} \left(\eta \frac{\partial \langle v_i \rangle}{\partial x_k} - \rho \langle \delta v_i \, \delta v_k \rangle \right) . \tag{9.17}$$

9.2.2 Turbulent Viscosity

If turbulence is absent, the only term in the parentheses on the right-hand side in (9.17) is $\eta \, \partial v_i / \partial x_k$ which represents viscous stresses. In the presence of turbulence, the new term

$$\tau_{ik} = -\rho \langle \delta v_i \, \delta v_k \rangle \, , \tag{9.18}$$

called *Reynolds stresses*, must be added to it. Hence, the turbulence can transfer momentum from one part of the fluid to the other in the same manner as viscous forces. The reason for this will be clear if we look at Fig. 9.6. Parallel arrows here represent a value of an average flow velocity in the (x, y) plane for different y, so that the dashed line is the average velocity "profile". Random changes of velocity due to turbulence exist as well as the average current. Then fluid particles cross streamlines of average current and carry momentum, for example, from point O, where momentum of an average flow is larger, to point A where it is less. This process is quite analogous to the transfer of momentum by molecules which causes molecular viscosity.

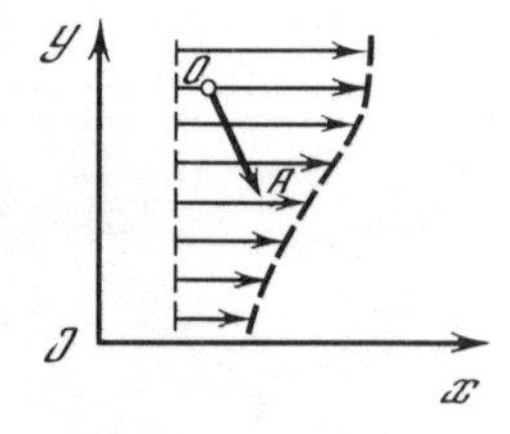

Fig. 9.6. Illustrating the momentum transfer by turbulence

Let us consider a flow past a rigid wall $x_3 = 0$ with the average velocity along x_1. Suppose that the flow is homogeneous in the (x_1, x_2) plane, i.e., all its average characteristics do not depend on x_1 or x_2. Then in (9.17), we have to take into account only one component of the complete stress tensor

$$\sigma'_{13} = \rho(\nu\, \partial\langle v_1\rangle/\partial x_3 - \langle \delta v_1\, \delta v_3\rangle).$$

Formally it can also be written as

$$\sigma'_{13} = \rho(\nu + K)\, \partial\langle v_1\rangle/\partial x_3\,, \qquad K = -\langle \delta v_1\, \delta v_3\rangle(\partial\langle v_1\rangle/\partial x_3)^{-1}. \tag{9.19}$$

K is called the *coefficient of kinematic turbulent viscosity*. It can be different in different parts of the flow and sometimes assumes a negative value. At the wall, in particular, $K = 0$ because of $\delta v_1 = \delta v_2 = 0$. Usually K is larger by many orders of magnitude than the molecular viscosity ν. In an analogous way, turbulence may also transfer heat or some admixture (salt, for example), i.e., *turbulent heat conductivity* or *turbulent diffusion* takes place. The coefficients of turbulent heat conductivity and of diffusion can be defined in the same way as K.

9.2.3 Turbulent Boundary Layer

We consider flow past an infinite plane wall. Let the average velocity be along the x-axis. Average characteristics of the flow are assumed to be independent of x (homogeneous flow). In the case of similar two-dimensional laminar flow in the x-direction, we would have from the Navier-Stokes equation (y-axis is normal to the wall)

$$(\partial^2 v/\partial y^2) = 0\,, \qquad (\partial p/\partial y) = 0\,, \tag{9.20}$$

the same as in the case of Couette flow. From (9.20) we find $v = Ay + B$, $p =$ const. Obviously, $B = 0$ since $v = 0$ at $y = 0$ (at the wall). We denote by τ_0 the viscous (tangential) stress at the wall, then

$$\tau_0 = \eta(\partial v/\partial y)_{y=0} = A\eta\,, \qquad A = \tau_0/\eta\,.$$

Hence, the velocity is proportional to the distance from the wall

$$v = (\tau_0/\eta)y\,. \tag{9.21}$$

Let us now consider turbulent flow. We denote the only nonzero component of the average velocity by $\langle v\rangle = \langle v_x(y)\rangle$. Then we obtain from (9.17)

$$\eta\,\frac{d^2\langle v\rangle}{dy^2} + \frac{d\tau'}{dy} = 0\,, \tag{9.22}$$

where $\tau' = -\rho\langle\delta v_x \delta v_y\rangle$ is the Reynolds stress. Integrating (9.22) and taking into account that the averaged product $\langle\delta v_x \delta v_y\rangle$ may depend only on y, we obtain

$$\eta(d\langle v\rangle/dy) + \tau' = C.$$

At the wall $y = 0$ we have $\delta v_x = \delta v_y = 0$, $\tau' = 0$, hence,

$$C = \eta(d\langle v\rangle/dy)_{y=0} \equiv \tau_0 \quad \text{and finally}$$

$$\eta(d\langle v\rangle/dy) + \tau' = \tau_0. \tag{9.23}$$

We see that the sum of viscous and turbulent stresses is constant (independent of the distance from the wall) and equals τ_0, the viscous stresses at the wall. Very close to the wall (very small y), $\tau' \simeq 0$ and the first term on the left-hand side of (9.23) is the most important (viscous stresses). This region is called the *viscous sublayer*. The average velocity is linear in y, as in the case of laminar flow, see (9.21). It does not mean, however, that the flow in the viscous sublayer is laminar. Considerable turbulent fluctuation may occur here.

At large distances from the wall the second term in (9.23) prevails and we can assume approximately

$$\tau' = -\rho\langle\delta v_x \delta v_y\rangle \simeq \tau_0. \tag{9.24}$$

Momentum flux along the y-axis and the mean velocity gradient are determined here only by the turbulent viscosity, whereas the molecular viscosity is negligible.

We now turn to the dimensional theory and determine on what quantities the gradient $\partial\langle v\rangle/\partial y$ in this region can depend. Since the molecular viscosity is of no importance, we have only three dimensional quantities ρ, y and τ_0. The latter specifies the momentum flux from the wall into the fluid. Using these three quantities, we can form only one combination with the same dimension as that of the velocity gradient, namely, $(\tau_0/\rho)^{1/2}y^{-1}$. Note that, according to (9.24), τ_0/ρ has the same dimension as the velocity squared. The quantity $v_* \equiv (\tau_0/\rho)^{1/2}$, called the *dynamical* or *frictional velocity*, is widely used. Now outside the viscous sublayer we have

$$\frac{d\langle v\rangle}{dy} = A\frac{v_*}{y}, \tag{9.25}$$

where A is some dimensionless universal constant. It is also convenient to use the dimensionless distance $\zeta = yv_*/\nu$ so that (9.25) becomes

$$(d\langle v\rangle/d\zeta) = Av_*\zeta^{-1}. \tag{9.25a}$$

Its general solution is

$$\langle v\rangle = v_*(A\ln\zeta + B); \tag{9.26}$$

A and another dimensionless universal constant B can be determined experimentally. The interval of ζ where (9.26) holds, is called the *logarithmic boundary layer.*

Note that (9.21) for a viscous sublayer can be rewritten in terms of ζ:

$$\langle v \rangle = v_* \zeta . \tag{9.27}$$

According to experiment, (9.27) holds approximately at $\zeta < 8$, whereas the logarithmic law (9.26) holds at $\zeta > 30$ with $A \simeq 2.5$ and $B \simeq 5.5$. In the intermediate region $8 < \zeta < 30$, neither of the equations is a good approximation.

The theory of a turbulent boundary layer was advanced considerably by Millionshikov who has derived an equation which holds in all the regions of intermediate ζ. Using the turbulent viscosity defined by (9.19), (9.23) can be written as

$$(\eta + \rho K) \frac{d\langle v \rangle}{dy} = \tau_0 . \tag{9.28}$$

The molecular viscosity is $\eta = C\rho l_T v_T$, where l_T is the average free path of the molecules, v_T their average thermal velocity and C is some constant of the order of unity. Suppose that for ρK, the same kind of equation holds: $\rho K = \rho l v_*$, where l is the "*mean free path*" or "*mixing length*" in the turbulent flow. The logarithmic law (9.26) is obtained under the assumption that l is proportional to ζ. Millionshikov suggested a modification of this assumption:

$$(v_* l / \nu) = \varkappa (\zeta - \zeta_0) , \tag{9.29}$$

where $\varkappa$ and ζ_0 are dimensionless universal constants, the latter having the meaning of a dimensionless width of the viscous sublayer. Equation (9.29) corresponds to the very natural assumption that l is proportional to the distance from the outer boundary of the viscous sublayer $\zeta = \zeta_0$ and not to the distance from the wall.

Taking (9.29) into account, one can obtain the following equation instead of (9.26):

$$(\langle v \rangle / v_*) = \varkappa^{-1} \ln[1 + \varkappa(\zeta - \zeta_0)] + \zeta_0 , \quad \zeta > \zeta_0 . \tag{9.30}$$

Equation (9.30) and (9.27) at $\zeta < \zeta_0$ give a continuous (with continuous first derivative at $\zeta = \zeta_0$) dependence $\langle v \rangle$ on ζ. Taking into account that (9.30) must coincide with (9.26) at $\zeta \gg 1$, we obtain a relationships between universal constants:

$$A = \varkappa^{-1} , \quad B = \zeta_0 + \varkappa^{-1} \ln \varkappa .$$

The dependence of the average velocity on ζ obtained in such a way is in good agreement with experimental data if one assumes $\varkappa = 0.4$ and $\zeta_0 = 7.8$ ($A \simeq 2.5$, $B \simeq 5.5$). The parameter $\varkappa$ is usually called the *Karman constant*.

9.3 Locally Isotropic Turbulence

9.3.1 Properties of Developed Turbulence

Up to now we have dealt with the average characteristics of turbulent flows. For many purposes, however, statistical properties of the velocity fluctuations $\delta \boldsymbol{v}$ and, in particular, all their correlation functions (*correlation tensor*)

$$B_{ij}(\boldsymbol{r}_1, \boldsymbol{r}_2, t) = \langle \delta v_i(\boldsymbol{r}_1, t)\, \delta v_j(\boldsymbol{r}_2, t) \rangle \tag{9.31}$$

are of considerable importance. In the case of homogeneous turbulence, this tensor will depend only on the position vector $\boldsymbol{r} = \boldsymbol{r}_1 - \boldsymbol{r}_2$. This is the simplest case but the question may arise as to whether it could occur in reality because the homogeneity can be disturbed by many factors such as boundaries, inhomogeneity of the main flow and so on. More thorough considerations show, however, that if the main flow has a sufficiently large Reynolds number, small-scale fluctuations generated in this flow must be homogeneous and isotropic. We are going to show this using the qualitative considerations suggested in the 1920's by Richardson and later developed quantitatively by Kolmogorov to whom the main results of this section are credited.

Suppose we have a fluid flow past a body with the characteristic dimension L_0. Let the characteristic velocity of the flow be v_0 and the Reynolds number of the average flow be sufficiently large, i.e., $\mathrm{Re} = v_0 L_0/\nu \gg \mathrm{Re}_{\mathrm{cr}}$, $\mathrm{Re}_{\mathrm{cr}}$ corresponding to the laminar-turbulent transition. When this transition occurs, turbulent disturbances appear; their scale is $L_1 \sim L_0$ and their characteristic velocity v_1 or quantity $[\langle(\delta v)^2\rangle]^{1/2}$ is comparable with the variation of the average velocity $\delta v_0 \sim v_0$ at the distance L_0. These disturbances are generated directly by the main flow and hence they will be inhomogeneous across the space in approximately the same manner as the main flow. The Reynolds number for these disturbances $\mathrm{Re}_1 = v_1 L_1/\nu$ still greatly exceeds $\mathrm{Re}_{\mathrm{cr}}$. Hence, they appear to be unstable in their turn and generate disturbances with the smaller scale L_2. The latter are also unstable if $\mathrm{Re}_2 = v_2 L_2/\nu \gg \mathrm{Re}_{\mathrm{cr}}$ and generate disturbances with the scale $L_3 < L_2$ and so on. A cascading process will take place as a result, where energy from the larger scales goes to the smaller ones. If the Reynolds number for all scales is very large, viscosity is still of no importance and the process is governed by nonlinear phenomena. Hence, the power taken away from the main flow goes through all of the cascade without loss, i.e., it remains unchanged through the full spectrum of the scales. This power ε *per unit mass* is an important parameter of the developed turbulence. Its order of magnitude can be determined with the help of dimensional analysis; its dimension is $[\varepsilon] = L^2T^{-3}$. Since it does not depend on the turbulence scale, it can depend only on the parameters of the main flow L_0 and v_0. The only combination of these quantities with the required dimensions is

$$\varepsilon \sim v_0^3/L_0 \,. \tag{9.32}$$

The cascade process of the consecutive breaking up of larger scales of the turbulence with energy transition into smaller ones persists until such small scales L_N ($N \gg 1$) will be reached, for which $\mathrm{Re}_N = v_N L_N/\nu \sim 1$ (more exactly, $\mathrm{Re}_N \sim \mathrm{Re}_{cr}$). The flows of such scales are stable so that breaking does not take place. The energy ε received by them from the larger scales transforms into heat due to viscosity. Hence, ε is also the *energy decay rate per unit mass.* It can be shown (Exercise 9.4.9) that its explicit expression is

$$\varepsilon = \frac{1}{2}\nu\left\langle\left(\frac{\partial\delta v_i}{\partial x_k} + \frac{\partial\delta v_k}{\partial x_i}\right)^2\right\rangle .$$

Returning again to the beginning of the cascade process, we note that the velocity in disturbances of scale L_1 generated by the main flow are generally directed in all directions and not only in that of the main flow. In other words, the flow of scale L_1 is more isotropic than the main flow. Analogously, isotropy of flow with scale L_2 will be greater than that of flow with scale L_1 and so on. As a result, after several acts of cascading, the anisotropic character of the main flow will be erased; the *developed turbulence becomes isotropic and statistically homogeneous in space.* The scale $L_1 \sim L_0$ is usually called the *external scale of turbulence* and L_N the *inner or Kolmogorov's* scale. The larger the Reynolds number Re of the main flow, the larger the number of cascading acts. At sufficiently large Re, a wide diversity of scales L_n, $n = 1, 2, \ldots, N$, $N \gg 1$ will exist. For n large compared with unity, but small when compared with N, we have the inequality $L_1 \gg L_n \gg L_N$. An interval of scales satisfying the latter inequality is called an *inertial interval.* Disturbances in this interval have already "forgotten" about the structure of the main flow and the action of viscous forces is still negligible. Some important results for an inertial interval can be obtained by the dimensional theory, taking into account that only two quantities L_n and ε specify the flow in this interval.

Velocity fluctuations v_n of the scale L_n can be easily determined to within an order of magnitude since the only combination of the quantities ε and L_n with the dimension of velocity is

$$v_n \sim (\varepsilon L_n)^{1/3} , \tag{9.33}$$

or using (9.32),

$$v_n \sim v_0(L_n/L_0)^{1/3} . \tag{9.33a}$$

The Reynolds number for an arbitrary scale L_n is

$$\mathrm{Re}_n = \frac{v_n L_n}{\nu} \sim \frac{v_0 L_n^{4/3}}{\nu L_0^{1/3}} \sim \left(\frac{L_n}{L_0}\right)^{4/3} \mathrm{Re}\,, \tag{9.34}$$

where $\mathrm{Re} = v_0 L_0/\nu$ is the Reynolds number of the main flow. If we roughly assume that our results also hold for an inner scale L_N where $\mathrm{Re}_N \sim 1$, one can estimate the order of magnitude of L_N and the corresponding fluctuation velocity v_N is

$$L_N \sim L_0 \mathrm{Re}^{-3/4}\,, \qquad v_N \sim v_0 \mathrm{Re}^{-1/4}\,. \tag{9.35}$$

Hence, L_N and v_N decrease with increasing Re of the main flow as $\mathrm{Re}^{-3/4}$ and $\mathrm{Re}^{-1/4}$, respectively.

9.3.2 Statistical Properties of Locally Isotropic Turbulence

We will consider below only isotropic and statistically homogeneous fields of velocity fluctuations with scales $L_n \ll L_0$. To separate them from the larger-scale fluctuations ($L_n \sim L_0$), we will deal with the relative motions of fluid particles inside small volumes. In fact, we can assume the vector random field $\delta v_i(\boldsymbol{r}_1) - \delta v_i(\boldsymbol{r}_2)$ to be homogeneous and isotropic inside any volume with linear dimensions much less than $L_0(|\boldsymbol{r}_1 - \boldsymbol{r}_2| \ll L_0)$. The turbulence satisfying this condition is called *locally isotropic*. Since the velocity is a vector, we have to consider the so-called *structure tensor* of the vector field

$$D_{ij}(\boldsymbol{r}) = \langle [\delta v_i(\boldsymbol{r}_1) - \delta v_i(\boldsymbol{r}_2)][\delta v_j(\boldsymbol{r}_1) - \delta v_j(\boldsymbol{r}_2)] \rangle\,, \qquad \boldsymbol{r} = \boldsymbol{r}_1 - \boldsymbol{r}_2\,. \tag{9.36}$$

If the $\delta \boldsymbol{v}$ field is statistically homogeneous, an obvious relationship exists between the tensors $D_{ij}(\boldsymbol{r})$ and $B_{ij}(\boldsymbol{r})$ which follows from (9.36) if (9.31) is also taken into account:

$$D_{ij}(\boldsymbol{r}) = 2[B_{ij}(0) - B_{ij}(\boldsymbol{r})]\,. \tag{9.37}$$

In the case of a scalar random field, isotropy means that all average quantities (structure or correlation functions, for example) depend only on the distance $r = |\boldsymbol{r}| = |\boldsymbol{r}_1 - \boldsymbol{r}_2|$. It is not so for a vector random field. Indeed, let us consider the component $D_{11}(\boldsymbol{r})$ of a structure tensor. If the position vector $\boldsymbol{r} = \boldsymbol{r}_1 - \boldsymbol{r}_2$ is directed along the axis x_1, then $D_{11} = D_{rr}$ is a structure function of the *longitudinal* (along $\boldsymbol{r}$) component of $\delta \boldsymbol{v}$. If $\boldsymbol{r}$ is perpendicular to x_1, then $D_{11} = D_{tt}$ describes the correlation of the *transverse* components of $\delta \boldsymbol{v}$. There is no reason, in the general case, for equality of D_{rr} and D_{tt}. In other words, in the case of an isotropic vector field, the components of a structure tensor depend not only on the magnitude but also on the direction of $\boldsymbol{r}$. We will determine this dependence using symmetry considerations while simultaneously determining the number of independent components of the tensor D_{ij}. We imagine arbitrary unit vectors $\boldsymbol{\alpha} = \{\alpha_1, \alpha_2, \alpha_3\}$ and $\boldsymbol{\beta} = \{\beta_1, \beta_2, \beta_3\}$ at the points $\boldsymbol{r}_1$ and $\boldsymbol{r}_2$, respectively. The unit vector along $\boldsymbol{r}$ *is* $\boldsymbol{r}/r$ ($r \equiv |\boldsymbol{r}|$). We now find the projections of the $\delta\boldsymbol{v}$ field on the directions of $\boldsymbol{\alpha}$ and $\boldsymbol{\beta}$ and determine the structure function of scalar fields obtained:

$$D = \langle \alpha_k[\delta v_k(\boldsymbol{r}_1) - \delta v_k(\boldsymbol{r}_2)]\beta_m[\delta v_m(\boldsymbol{r}_1) - \delta v_m(\boldsymbol{r}_2)]\rangle = \alpha_k\beta_m D_{km}(\boldsymbol{r}).$$

In the case of isotropy, D must not change if the vectors $\boldsymbol{r}/r$, $\boldsymbol{\alpha}$ and $\boldsymbol{\beta}$ turn simultaneously through an arbitrary but identical angle. Hence, D may depend only on the quantities $r, \boldsymbol{\alpha} \cdot \boldsymbol{r}/r, \boldsymbol{\beta} \cdot \boldsymbol{r}/r$ and $\boldsymbol{\alpha} \cdot \boldsymbol{\beta}$, which are invariants with respect to such a rotation:

$$D(r, \boldsymbol{\alpha} \cdot \boldsymbol{r}/r, \boldsymbol{\beta} \cdot \boldsymbol{r}/r, \boldsymbol{\alpha} \cdot \boldsymbol{\beta}) = \alpha_k\beta_m D_{km}(\boldsymbol{r}). \tag{9.38}$$

The right-hand side of (9.38) is linear with respect to α_k and β_m, which suggests the following general form of the left-hand side:

$$\begin{aligned} D(r, \boldsymbol{\alpha} \cdot \boldsymbol{r}/r, \boldsymbol{\beta} \cdot \boldsymbol{r}/r, \boldsymbol{\alpha} \cdot \boldsymbol{\beta}) &= R(r)\boldsymbol{\alpha} \cdot \boldsymbol{\beta} + S(r)(\boldsymbol{\alpha} \cdot \boldsymbol{r})(\boldsymbol{\beta} \cdot \boldsymbol{r})/r^2 \\ &= \alpha_k\beta_k R(r) + \alpha_k\beta_m x_k x_m r^{-2} S(r) \\ &= \alpha_k\beta_m[R(r)\,\delta_{km} + S(r)\,r^{-2}x_k x_m]. \end{aligned}$$

Taking into account that the vectors $\boldsymbol{\alpha}$ and $\boldsymbol{\beta}$ are arbitrary and comparing the last formula with (9.38), we obtain

$$D_{ij}(\boldsymbol{r}) = R(r)\,\delta_{ij} + S(r)\,r^{-2}x_i x_j. \tag{9.39}$$

Hence, in an isotropic case we have only two independent components of a symmetric second-order tensor D_{ij} instead of six in the general case. It is convenient to express all the components in terms of the longitudinal and transverse ones considered above (see the example with D_{11}). Choosing the vector $\boldsymbol{r}$ along the x_1-axis for this purpose ($\boldsymbol{r}/r = \{1, 0, 0\}$), we obtain for the longitudinal component $D_{rr} = D_{11} = R + S$. Analogously we find for the transverse components $D_{tt} = D_{22} = D_{33} = R$. Hence, $R = D_{tt}$, $S = D_{rr} - D_{tt}$ and

$$D_{ij} = D_{tt}(r)\,\delta_{ij} + [D_{rr}(r) - D_{tt}(r)]x_i x_j/r^2. \tag{9.39a}$$

One more relationship between D_{tt} and D_{rr} exists due to the incompressibility of the fluid. We determine the latter in the case of an isotropic vector field $\delta\boldsymbol{v}$, for simplicity. We first consider the quantity $\partial B_{ij}(r)/\partial x_i$ (summation is implied in the case of a repeated index). Taking into account that $\boldsymbol{r} = \boldsymbol{r}_1 - \boldsymbol{r}_2$ and denoting $\boldsymbol{r}_1 = \{x_1^{(1)}, x_2^{(1)}, x_3^{(1)}\}$, we obtain $\partial/\partial x_i = \partial/\partial x_i^{(1)}$. Now from (9.31) and the incompressibility condition ($\partial[\delta v_k(\boldsymbol{r}_1)]/\partial x_i^{(1)} = 0$), we find

$$\frac{\partial B_{ij}}{\partial x_i} = \frac{\partial B_{ij}}{\partial x_i^{(1)}} = \left\langle \delta v_j(\boldsymbol{r}_2)\frac{\partial \delta v_i(\boldsymbol{r}_1)}{\partial x_i^{(1)}}\right\rangle = 0.$$

Differentiation of (9.37) *for an isotropic field yields*

$$\frac{\partial D_{ij}}{\partial x_i} = 0 . \tag{9.40}$$

This equation turns out to be true in the general case of locally isotropic turbulence when the correlation tensor may be nonisotropic. We now substitute (9.39a) into (9.40) and take into account the relations

$$\partial r/\partial x_i = x_i/r , \quad \partial (x_k/r)/\partial x_i = \delta_{ik} r^{-1} - x_k x_i r^{-3} .$$

We find

$$\frac{\partial D_{ij}}{\partial x_i} = \frac{dD_{tt}}{dr}\frac{x_j}{r} + \left(\frac{dD_{rr}}{dr} - \frac{dD_{tt}}{dr}\right)\frac{x_i^2 x_j^2}{r^3} + (D_{rr} - D_{tt})\frac{2x_j}{r^2} = 0$$

which immediately gives the required relationship between the transverse and the longitudinal components

$$D_{tt} = D_{rr} + \frac{r}{2}\frac{dD_{rr}}{dr} = \frac{1}{2r}\frac{d}{dr}(r^2 D_{rr}) . \tag{9.41}$$

Spectral representations (Fourier transforms) of correlations and structure functions play a very important role in the theory of random fields. For example, for the components of the correlation tensor (9.31) in the case of homogeneous vector fields we can write

$$B_{ij}(\boldsymbol{r}) = \int_{-\infty}^{\infty} F_{ij}(\boldsymbol{k}) \exp(\mathrm{i}\boldsymbol{k}\cdot\boldsymbol{r})\, d\boldsymbol{k} = \int_{-\infty}^{\infty} F_{ij}(\boldsymbol{k}) \cos \boldsymbol{k}\cdot\boldsymbol{r}\, d\boldsymbol{k} . \tag{9.42}$$

Here we have also used the symmetry conditions

$$B_{ij}(\boldsymbol{r}_1 - \boldsymbol{r}_2) = B_{ij}(\boldsymbol{r}_2 - \boldsymbol{r}_1) , \quad F_{ij}(\boldsymbol{k}) = -F_{ij}(-\boldsymbol{k}) .$$

An analogous representation for D_{ij} will, by virtue of (9.37), be

$$D_{ij}(\boldsymbol{r}) = 2 \int_{-\infty}^{\infty} (1 - \cos \boldsymbol{k}\cdot\boldsymbol{r}) F_{ij}(\boldsymbol{k})\, d\boldsymbol{k} . \tag{9.43}$$

Note that the convergence condition of the integral as $\boldsymbol{k} \to 0$ is weaker for the integral in (9.43) than that for that in (9.42). This is because a wider class of processes can be described by isotropic structure functions than by correlation functions. In an isotropic case the *spectral tensor* $F_{ij}(\boldsymbol{k})$ can be expressed in terms of two scalar functions. Obviously, we can write analogously to (9.39, 39a),

$$F_{ij}(\boldsymbol{k}) = F(k)\,\delta_{ij} + G(k)\frac{k_i k_j}{k^2} = F_{tt}(k)\,\delta_{ij} + (F_{rr} - F_{tt})\frac{k_i k_j}{k^2} .$$

For an incompressible fluid we obtain, substituting (9.43) into (9.40),

$$\int_{-\infty}^{\infty} \sin(\boldsymbol{k}\cdot\boldsymbol{r}) k_i F_{ij}(\boldsymbol{k})\, d\boldsymbol{k} = 0\,,$$

whence $k_i F_{ij}(\boldsymbol{k}) = 0$ follows since $\boldsymbol{r}$ is arbitrary, and finally

$$G(k) = -F(k)\,, \qquad F_{rr}(k) = 0\,, \qquad F_{ij}(\boldsymbol{k}) = \left(\delta_{ij} - \frac{k_i k_j}{k^2}\right) F_{tt}\,. \tag{9.44}$$

The *spatial spectral density* of the velocity field is directly related to the *kinetic energy of turbulent fluctuations.* Note first that in the case of homogeneous isotropic turbulence we have

$$B_{ii}(0) = \langle \delta v_i(\boldsymbol{r})\, \delta v_i(\boldsymbol{r})\rangle = \langle \delta v_1^2\rangle + \langle \delta v_2^2\rangle + \langle \delta v_3^2\rangle = \langle \delta \boldsymbol{v}^2\rangle\,,$$

where $B_{ij}(r)$ is a correlation tensor as always.

On the other hand, if (9.44) is taken into account, (9.42) yields

$$B_{ii}(r) = \int_{-\infty}^{\infty} F_{ii}(\boldsymbol{k}) \cos(\boldsymbol{k}\cdot\boldsymbol{r})\, d\boldsymbol{k}\,, \qquad F_{ii}(\boldsymbol{k}) = (\delta_{ii} - k_i^2/k^2) F_{tt}(k) = 2F_{tt}(k)\,.$$

We use spherical coordinates for the integration over $\boldsymbol{k}$ and the polar axis is along $\boldsymbol{r}$: $d\boldsymbol{k} = k^2 \sin\theta\, d\theta\, d\varphi\, dk$, $\mathbf{k}\cdot\boldsymbol{r} = kr\cos\theta$. For an isotropic case we easily obtain

$$\begin{aligned} B_{ii}(r) &= \int_0^{\infty}\int_0^{\pi}\int_0^{2\pi} \cos(kr\cos\theta) F_{ii}(k) k^2 \sin\theta\, d\theta\, d\varphi\, dk \\ &= 4\pi \int_0^{\infty} \frac{\sin kr}{kr} F_{ii}(k) k^2\, dk\,. \end{aligned} \tag{9.45}$$

Setting $r = 0$ and

$$E(k) = 4\pi k^2 F_{ii}(k)/2 = 4\pi k^2 F_{tt}(k)\,, \tag{9.46}$$

we find for the *average energy of the fluctuations per unit mass*

$$T = \frac{B_{ii}(0)}{2} = \int_0^{\infty} E(k)\, dk\,. \tag{9.47}$$

$E(k)$ is called the *spectral energy density.* According to (9.43, 46), the spectral tensor $F_{ij}(\boldsymbol{k})$ can be expressed in terms of $E(k)$ as

$$F_{ij}(\boldsymbol{k}) = \left(\delta_{ij} - \frac{k_i k_j}{k^2}\right) \frac{E(k)}{4\pi k^2}\,. \tag{9.48}$$

For a structure function we have (Exercise 9.4.2) a formula analogous to (9.45):

$$D_{ii}(r) = 2D_{tt} + D_{rr} = 4 \int_0^\infty \left(1 - \frac{\sin kr}{kr}\right) E(k)\, dk\,. \tag{9.49}$$

we write down the spectral representations for D_{rr} and D_{tt} in terms of $E(k)$ (Exercise 9.4.9):

$$\begin{aligned} D_{rr}(r) &= 4 \int_0^\infty \left(\frac{1}{3} + \frac{\cos kr}{k^3 r^3} - \frac{\sin kr}{k^3 r^3}\right) E(k)\, dk\,, \\ D_{tt}(r) &= 2 \int_0^\infty \left(\frac{2}{3} - \frac{\sin kr}{kr} - \frac{\cos kr}{k^2 r^2} + \frac{\sin kr}{k^3 r^3}\right) E(k)\, dk\,. \end{aligned} \tag{9.50}$$

9.3.3 Kolmogorov's Similarity Hypothesis

It was shown above that small-scale fluctuations ($L_n \ll L_0$) are locally isotropic in the case of developed turbulence. According to *Kolmogorov's first hypothesis*, the turbulence properties for such small scales may depend only on ε and ν. Hence, we have, in particular, $D_{rr} = D_{rr}(\varepsilon, \nu, r)$ (and the same for D_{tt}) for $r \ll L_0$. Combinations of ε and ν with dimensions of velocity and length are, respectively, $v_\nu = (\varepsilon\nu)^{1/4}$ and $L_\nu = (\nu^3/\varepsilon)^{1/4}$. Now, introducing the dimensionless distance r/L_ν, we write the structure function (its dimensions are $[v^2]$) as

$$D_{rr} = v_\nu^2 f_{rr}(r/L_\nu)\,, \qquad D_{tt} = v_\nu^2 f_{tt}(r/L_\nu)\,, \qquad r \ll L_0\,, \tag{9.51}$$

where $f_{rr}(\xi)$ and $f_{tt}(\xi)$ ($\xi = r/L_\nu = r\varepsilon^{1/4}\nu^{-3/4}$) are some *universal functions*. They are related by

$$f_{tt}(\xi) = f_{rr}(\xi) + \xi f'_{rr}(\xi)/2\,,$$

which follows from (9.41).

Naturally, L_ν and v_ν coincide with the minimum turbulence scale L_N and the corresponding velocity v_N defined by (9.35). One can prove this using (9.32). Hence, Kolmogorov's first hypothesis states that *statistical properties of locally isotropic turbulence can be described in a universal way by measuring the distance and the velocity in terms of* L_N *and* v_N, *respectively.*

For the spectral energy density whose dimensions are $[E] = [Lv^2]$, we have

$$E(k) = L_\nu v_\nu^2 \varphi(L_\nu k)\,, \tag{9.52}$$

where $\varphi(\xi)$ is a new *universal* function.

In a narrower spectral interval $L_0^{-1} \ll k \ll L_N^{-1}$ corresponding to an inertial interval, viscosity is also of no importance and the *mean characteristics of tur-*

bulence can be specified in terms of only one quantity ε (the *Kolmogorov's second hypothesis*). We can obtain explicit expressions for the functions f_{rr}, f_{tt} and φ in this interval. Write D_{rr}, for example, in the form

$$D_{rr} = \varepsilon^{1/2}\nu^{1/2}f_{rr}(r\varepsilon^{1/4}\nu^{-3/4})$$

and assume that $f_{rr}(\xi)$ has the form $f_{rr} = C_r\xi^n$ asymptotically at large ξ. The quantity n can be determined from the condition that D_{rr} must be independent of ν, which yields $n = 2/3$. Hence,

$$D_{rr} = C_r\varepsilon^{2/3}r^{2/3}\,, \quad \text{and analogously} \quad D_{tt} = C_t\varepsilon^{2/3}r^{2/3}\,. \tag{9.53}$$

The *universal constants* C_r and C_t are related to each other by $C_t = 4C_r/3$ which follows from (9.41). Hence, we have the following law for developed turbulence with a large Reynolds number: *the average-squared velocity difference between two points a distance r apart is proportional to this distance with power* 2/3 if this distance is neither too small nor too large (inertial interval $L_N \ll r \ll L_0$). This law found first by Kolmogorov in 1941 is known as "*Kolmogorov's 2/3 law.*"

An analogous law for the spectral density $E(k)$ was found by Obukhov, also in 1941, by similarity arguments. We have for the dimensions of $E(k)$, ε and k: $[E(k)] = [Lv^2]$, $[\varepsilon] = \mathrm{L}^2/\mathrm{T}^3$, $[k] = \mathrm{L}^{-1}$. The only combination of ε and k having the same dimension as $E(k)$ is

$$E(k) = C_1\varepsilon^{2/3}k^{-5/3}\,, \qquad L_0^{-1} \ll k \ll L_N^{-1}\,; \tag{9.54}$$

this is the so-called "*5/3*" *law of the turbulence spectrum in an inertial interval.* Equivalence of these two laws can be proved easily using (9.49) (Exercise 9.4.4). The relationship between the constants C_1 and C_r can be found in this way, too, i.e.

$$C_1 = (55/27)C_r/\Gamma(1/3)\,.$$

The works by Kolmogorov and Obukhov concerning the structure of developed turbulence have had a great influence on the further progress in the theory of turbulence. Many experiments were carried out to prove their results and it was shown that experimental data are in good agreement with theory whenever the conditions of isotropy of the turbulence are satisfied.

9.4 Exercises

9.4.1. Integrate (9.28) for a turbulent boundary layer with the coefficient of turbulent viscosity K specified by (9.30) in accordance with Millionshikov's hypothesis.

Solution: Using the dimensionless variables $v' = \langle v \rangle / v_*$, $\zeta = v_* y / \nu$ where $v_* = (\tau_0/\rho)^{1/2}$, we obtain from (9.28), taking (9.30) into account,

$$[1 + \varkappa(\zeta - \zeta_0)]\, dv'/d\zeta = 1 \,.$$

Integration of this equation yields

$$v' = \varkappa^{-1} \ln[1 + \varkappa(\zeta - \zeta_0)] + B \,, \qquad \zeta \geq \zeta_0 \,.$$

This quantity must coincide with that for a viscous sublayer (9.27) at $\zeta = \zeta_0$ which gives $B = \zeta_0$ and finally

$$v' = \zeta_0 + \varkappa^{-1} \ln[1 + \varkappa(\zeta - \zeta_0)] \,.$$

9.4.2. Obtain (9.49) for the convolution $D_{ii}(r)$.

Solution: From (9.43), taking into account that $F_{ii}(k) = E(k)/2\pi k$, we obtain

$$D_{ii} = \int\limits_{-\infty}^{\infty} (1 - \cos \boldsymbol{k} \cdot \boldsymbol{r}) \frac{E(k)}{\pi k^2}\, d\boldsymbol{k} \,.$$

Use spherical coordinates ($d\boldsymbol{k} = k^2 \sin\theta\, d\theta\, d\varphi\, dk$, $\boldsymbol{k} \cdot \boldsymbol{r} = kr \cos\theta$) and integrate over the angles:

$$D_{ii} = \int\limits_0^{\infty} \int\limits_0^{\pi} \int\limits_0^{2\pi} [1 - \cos(kr \cos\theta)] \frac{E(k)}{\pi} \sin\theta\, d\theta\, d\varphi\, dk$$

$$= 2 \int\limits_0^{\infty} E(k) \left\{ \int\limits_0^{\pi} [1 - \cos(kr\cos\theta)] \sin\theta\, d\theta \right\} dk = 4 \int\limits_0^{\infty} E(k) \left(1 - \frac{\sin kr}{kr}\right) dk \,.$$

9.4.3. Obtain the inverse to (9.49): an expression for the spectral density $E(k)$ in terms of $D_{ii}(r)$.

Solution: Apply the operation ∇ to the initial formula of the previous exercise:

$$\nabla D_{ii}(r) = \int\limits_{-\infty}^{\infty} \boldsymbol{k} \sin(\boldsymbol{k} \cdot \boldsymbol{r}) \frac{E(k)}{\pi k^2}\, d\boldsymbol{k} \,.$$

The inverse Fourier transform yields

$$\frac{\boldsymbol{k} E(k)}{\pi k^2} = (2\pi)^{-3} \int\limits_{-\infty}^{\infty} \sin(\boldsymbol{k} \cdot \boldsymbol{r}) \nabla D_{ii}(\boldsymbol{r})\, d\boldsymbol{r}$$

or multiplying by $\boldsymbol{k}$,

$$E(k) = (\boldsymbol{k}/8\pi^2) \int_{-\infty}^{\infty} \sin(\boldsymbol{k} \cdot \boldsymbol{r}) \nabla D_{ii}(\boldsymbol{r}) \, d\boldsymbol{r} \, .$$

In the case of a locally isotropic field $\boldsymbol{k} \cdot \nabla D_{ii}(r) = \boldsymbol{k} \cdot \boldsymbol{r} D'_{ii}$ integration over angular variables in spherical coordinates yields

$$E(k) = (2\pi k)^{-1} \int_0^{\infty} (\sin kr - kr \cos kr) D'_{ii} \, dr = -\frac{k}{2\pi} \frac{d}{dk} \int_0^{\infty} \frac{\sin kr}{k} \frac{dD_{ii}}{dr} \, dr \, .$$

9.4.4. Find the structure function $D_{ii}(r)$ under the assumption that the "5/3 law" for the spectral density $E(k)$ holds for all wave numbers $0 < k < \infty$.

Solution: Substituting $E(k) = C_1 \varepsilon^{2/3} k^{-5/3}$ into (9.49), we obtain

$$D_{ii}(r) = 4C_1 \varepsilon^{2/3} \int_0^{\infty} \left(1 - \frac{\sin kr}{kr}\right) k^{-5/3} \, dk \, .$$

Introducing the dimensionless variables $\xi = kr$, $dk = d\xi/r$ in the integrand, we obtain the "2/3 law" for $D_{ii}(r)$:

$$D_{ii}(r) = 4C_1 A \varepsilon^{2/3} r^{2/3} \, , \qquad A = \int_0^{\infty} (1 - \xi^{-1} \sin \xi) \xi^{-5/3} \, d\xi \, .$$

9.4.5. Using the solution of the previous exercise, find the relationship between the universal constants C_1 in the spectral density (9.54) and C_r in the structure function (9.53) for an inertial interval.

Solution: Using (9.49, 53) and the relation $C_t = 4C_r/3$, find for the convolution $D_{ii}(r)$ in the inertial interval

$$D_{ii}(r) = (11/3) C_r \varepsilon^{2/3} r^{2/3} \, .$$

Comparison of this equation with that obtained in the previous exercise gives $C_r = (12/11) A C_1$. On the other hand,

$$A = \int_0^{\infty} (1 - \xi^{-1} \sin \xi) \xi^{-5/3} \, d\xi = -(3/5) \int_0^{\infty} (\xi - \sin \xi) \, d\xi^{-5/3}$$

$$= -(3/5) \xi^{-5/3} (\xi - \sin \xi) \big|_0^{\infty} + (3/5) \int_0^{\infty} (1 - \cos \xi) \xi^{-5/3} \, d\xi \, .$$

The first term here is equal to zero and the second can be expressed in terms of a Γ-function:

$$A = (3/5) \int_0^\infty (1 - \cos \xi)\xi^{-5/3}\, d\xi = (6/5) \int_0^\infty \xi^{-5/3} \sin^2(\xi/2)\, d\xi$$
$$= -(6/20)\Gamma(-2/3) = (9/20)\Gamma(1/3).$$

Finally, we obtain

$$C_r = (12/11) \cdot (9/20)\Gamma(1/3)C_1 = (27/55)\Gamma(1/3)C_1.$$

9.4.6. It is sometimes instructive to consider locally isotropic turbulence along a fixed line (x_1-axis, for example). In particular, the one-dimensional Fourier transform of the convolution $D_{ii}(r)$ is

$$D_{ii}(r) = 2 \int_{-\infty}^{\infty} (1 - \cos kr) V(k)\, dk.$$

$V(k)$ is called the one-dimensional spectral density. Find $V(k)$ for an inertial interval and relate it to $E(k)$.

Solution: Differentiation of $D_{ii}(r)$ with respect to r gives $D'_{ii}(r) = 2\int_{-\infty}^{\infty} k \sin kr V(k)\, dk$, or using the inverse Fourier transform, $V(k) = (4\pi k)^{-1} \int_{-\infty}^{\infty} \sin kr D'(r)\, dr$. Comparing this formula with the relationship between $E(k)$ and $D_{ii}(r)$ (Exercise 9.4.3), we find

$$E(k) = -k\, dV(k)/dk.$$

Using (9.54) for $E(k)$, we now find for the inertial interval

$$V(k) = C_V \varepsilon^{2/3} k^{-5/3}, \qquad C_V = (3/5)C_1.$$

9.4.7. It is customary in experiments to measure the velocity fluctuations $\delta\boldsymbol{v}(\boldsymbol{r}_0, t)$ only at one point $\boldsymbol{r}_0$ but over a long time; and thereafter to consider the statistical properties of the random field $\boldsymbol{f}(\boldsymbol{r}_0, t, \tau) = \delta\boldsymbol{v}(\boldsymbol{r}_0, t + \tau) - \delta\boldsymbol{v}(\boldsymbol{r}_0, t)$. Relate these properties to spatially statistical properties of $\delta\boldsymbol{v}(\boldsymbol{r})$ using Taylor's hypothesis of "frozen turbulence" [the field $\delta\boldsymbol{v}(\boldsymbol{r})$ is transported by the average flow without change].

Solution: Choose the x_1-axis along the average velocity $\langle \boldsymbol{v} \rangle$ at a point $\boldsymbol{r}_0$. Due to Taylor's hypothesis we have $\delta\boldsymbol{v}(\boldsymbol{r}_0, t + \tau) = \delta\boldsymbol{v}(\boldsymbol{r}_0 - \langle v \rangle \tau, t)$. Introduce a longitudinal structure function with respect to time

$$D^t_{rr}(\tau) = \langle [\delta v_1(\boldsymbol{r}_0, t + \tau) - \delta v_1(\boldsymbol{r}_0, t)]^2 \rangle$$
$$= \langle [\delta v_1(\boldsymbol{r}_0 - \langle \boldsymbol{v} \rangle \tau, t) - \delta v_1(\boldsymbol{r}_0, t)]^2 \rangle = D_{rr}(\langle v \rangle \tau)$$

and analogously for the transverse function $D^{t}_{tt}(\tau) = D_{tt}(\langle v \rangle \tau)$. The convolution of the time structure tensor $D^{t}_{ii}(\tau) = 2D^{t}_{tt} + D^{t}_{rr} = D_{ii}(\langle v \rangle \tau)$ can be represented as a one-dimensional Fourier transform

$$D^{t}_{ii}(\tau) = 2 \int_{-\infty}^{\infty} (1 - \cos \omega\tau) W(\omega)\, d\omega\,.$$

On the other hand, $D_{ii}(r) = 2 \int_{-\infty}^{\infty} (1 - \cos kr) V(k)\, dk$; hence,

$$D^{t}_{ii}(\tau) = D_{ii}(\langle v \rangle \tau) = 2 \int_{-\infty}^{\infty} [1 - \cos(k\langle v \rangle \tau)] V(k)\, dk\,.$$

Introducing the new variable of integration $\omega = \langle v \rangle k$ we find

$$D^{t}_{ii}(\tau) = 2 \int_{-\infty}^{\infty} (1 - \cos \omega\tau) W(\omega)\, d\omega = 2 \int_{-\infty}^{\infty} (1 - \cos \omega\tau) \frac{V(\omega/\langle v \rangle)}{\langle v \rangle}\, d\omega$$

whence follows the relation

$$W(\omega) = \frac{V(\omega/\langle v \rangle)}{\langle v \rangle} \quad \text{or} \quad V(k) = \langle v \rangle W(k\langle v \rangle)\,.$$

In particular, for an inertial interval we have

$$W(\omega) = \frac{C_V}{\langle v \rangle}\, \varepsilon^{2/3} \left(\frac{\omega}{\langle v \rangle} \right)^{-5/3} = C_V \varepsilon^{2/3} \langle v \rangle^{2/3} \omega^{-5/3}\,.$$

9.4.8. Obtain the time rate of kinetic energy $E = \rho \boldsymbol{v}^2/2$ for a viscous incompressible fluid.

Solution: By differentiation we obtain $\partial E/\partial t = \rho \boldsymbol{v}\, \partial \boldsymbol{v}/\partial t = \rho v_i\, \partial v_i/\partial t$. According to the momentum conservation law (8.3),

$$\frac{\partial}{\partial t} (\rho v_i) = \rho \frac{\partial v_i}{\partial t} = - \frac{\partial \Pi_{ik}}{\partial x_k},$$

where $\Pi_{ik} = p\, \delta_{ik} + \rho v_i v_k - \sigma_{ik}$, $\sigma_{ik} = \eta(\partial v_i/\partial x_k + \partial v_k/\partial x_i)$ in the case of an incompressible fluid.

Now we have

$$\frac{\partial E}{\partial t} = - v_i \frac{\partial \Pi_{ik}}{\partial x_k} = - \frac{\partial}{\partial x_k} (v_i \Pi_{ik}) + \Pi_{ik} \frac{\partial v_i}{\partial x_k}$$

$$= - \frac{\partial}{\partial x_k} (p v_k + \rho v_i^2 v_k - v_i \sigma_{ik}) + p \frac{\partial v_k}{\partial x_k} + \rho v_i v_k \frac{\partial v_i}{\partial x_k} - \sigma_{ik} \frac{\partial v_i}{\partial x_k}\,.$$

Here, $\partial v_k/\partial x_k = 0$ due to incompressibility and

$$(\rho/2)v_i^2 v_k = E v_k\,,$$

$$-\frac{\partial}{\partial x_k}\frac{\rho v_i^2 v_k}{2} = -\rho v_i v_k \frac{\partial v_i}{\partial x_k} - E\frac{\partial v_k}{\partial x_k} = -\rho v_i v_k \frac{\partial v_i}{\partial x_k}\,.$$

Hence,

$$\frac{\partial E}{\partial t} = -\frac{\partial}{\partial x_k}[(E+p)v_k - v_i\sigma_{ik}] - \sigma_{ik}\frac{\partial v_i}{\partial x_k}\,.$$

Equating the last term to $\rho\varepsilon$, we obtain

$$\varepsilon = \rho^{-1}\sigma_{ik}\frac{\partial v_i}{\partial x_k} = \nu\left(\frac{\partial v_i}{\partial x_k} + \frac{\partial v_k}{\partial x_i}\right)\frac{\partial v_i}{\partial x_k} = \frac{\nu}{2}\left(\frac{\partial v_i}{\partial x_k} + \frac{\partial v_k}{\partial x_i}\right)^2$$

which is the rate of energy decay per unit mass due to viscosity.

9.4.9. Obtain (9.50) for D_{rr} and D_{tt}.

Solution: Using (9.41, 49), we can easily obtain the relationship between $D_{rr}(r)$ and the convolution $D_{ii}(r)$:

$$D_{ii}(r) = r^{-1}[r^2 D_{rr}(r)]' + D_{rr}(r) = r^{-2}[r^3 D_{rr}(r)]'\,.$$

Integrating this under the condition $D_{rr}(0) = 0$, we obtain $D_{rr} = r^{-3}\int_0^r \xi^2 D_{ii}(\xi)\,d\xi$ or, if (9.49) is taken into account,

$$D_{rr}(r) = 4r^{-3}\int_0^r \xi^2 \int_0^\infty \left(1 - \frac{\sin k\xi}{k\xi}\right)E(k)\,dk\,d\xi\,.$$

Integrating over the variable ξ we find

$$D_{rr} = 4\int_0^\infty \left(\frac{1}{3} + \frac{\cos kr}{k^2r^2} - \frac{\sin kr}{k^3r^3}\right)E(k)\,dk\,.$$

Using the relation $D_{tt} = (D_{ii} - D_{rr})/2$, we find an integral representation for $D_{tt}(r)$:

$$D_{tt} = 2\int_0^\infty \left(\frac{2}{3} - \frac{\sin kr}{kr} - \frac{\cos kr}{k^2r^2} + \frac{\sin kr}{k^3r^3}\right)E(k)\,dk\,.$$

10. Surface and Internal Waves in Fluids

This chapter begins the study of waves in fluids, a very important kind of motion with numerous applications. The main characteristic feature of waves is the possibility of energy transport over considerable distances without mass transport. The diversity of forces acting on the particle in a fluid causes a variety of different types of waves. In the following four chapters we will consider these waves using the linear approximation so that interaction between waves will be disregarded.

In this chapter the basic equations governing wave propagation in fluids will be derived and three sorts of waves related to the gravitational force will be considered, i.e., surface gravity, surface gravity-capillary and internal gravity waves. The first two kinds of waves occur at the fluid surface (the ocean's surface, for example) or near it, whereas the internal waves occur in inner layers with pronounced vertical gradients of the fluid's density.

10.1 Linear Equations for Waves in Stratified Fluids

10.1.1 Linearization of the Hydrodynamic Equations

To describe wave phenomena in fluids we should start from the basic hydrodynamic equations obtained in Chap. 6, namely (6.9, 14, 18):

$$\begin{gathered}\frac{\partial \boldsymbol{v}}{\partial t} + (\boldsymbol{v} \cdot \nabla)\boldsymbol{v} = -\frac{\nabla p}{\rho} - g\nabla z - 2\boldsymbol{\Omega} \times \boldsymbol{v}\,, \\ \frac{\partial \rho}{\partial t} + \nabla(\rho \boldsymbol{v}) = 0\,, \quad \frac{dp}{dt} = c^2 \frac{d\rho}{dt}\,, \quad c^2 = \left(\frac{\partial p}{\partial \rho}\right)_s.\end{gathered} \tag{10.1}$$

We include the Euler equation the gravitational force directed along the vertical and a Coriolis force $-2\boldsymbol{\Omega} \times \boldsymbol{v}$ arising in a coordinate system rotating at the angular frequency $\boldsymbol{\Omega}$ (for example, on the rotating Earth). A centrifugal force which is potential, as has been noted in Sect. 6.2.3, can be included formally into the potential of gravitation and this only changes the direction of the local vertical. Ideal fluids will be considered in what follows.

The nonlinearity of the hydrodynamic equations (10.1) is significant and the *exact theory* of waves in fluids is essentially *nonlinear*. The nonlinearity causes a considerable complication of the theory of wave propagation. However, if fluid-particle displacements are "small" in some sense, the hydrodynamic equations can be linearized with respect to these displacements (or to the corresponding velocities). Wave theory becomes *linear* and the *principle of superposition* will hold (waves propagate independently of each other). An exact definition of how small wave disturbances must be, will be discussed in Chap. 14. Here we just assume that the velocity $\boldsymbol{v}$ in the hydrodynamic equations is small whereas the pressure p and density ρ differ slightly from their equilibrium values p_0 and ρ_0, respectively. Assuming $p = p_0 + p'$, $\rho = \rho_0 + \rho'$, we will keep only those terms which are linear with respect to $\boldsymbol{v} = \{u, v, w\}$, p' and ρ' in (10.1). For an equilibrium state we obtain the hydrostatic equations:

$$\boldsymbol{v}_0 = 0\,, \quad dp_0/dz = -g\rho_0(z)\,. \tag{10.2}$$

The term $\nabla p/\rho$ in the Euler equation transforms as follows:

$$\frac{\nabla p}{\rho} = \frac{\nabla p_0 + \nabla p'}{\rho_0(1 + \rho'/\rho_0)} \simeq \frac{\nabla p_0}{\rho_0} + \frac{\nabla p'}{\rho_0} - \frac{\nabla p_0}{\rho_0}\frac{\rho'}{\rho_0}\,.$$

Analogously the nonlinear term in the equation of continuity becomes

$$\nabla(\rho\boldsymbol{v}) = \nabla[(\rho_0 + \rho')\boldsymbol{v}] \simeq (d\rho_0/dz)w + \rho_0\nabla\boldsymbol{v}$$

and the equation of state [with (6.4)]

$$\frac{dp}{dt} \simeq \frac{\partial p}{\partial t} + \frac{dp_0}{dz}\,w \simeq c^2\left(\frac{\partial\rho}{\partial t} + \frac{d\rho_0}{dz}\,w\right).$$

Now bearing in mind that $\nabla p_0 = (dp_0/dz)\nabla z$ and using the hydrostatic equation (10.2), we can easily write a *linearized set of hydrodynamic equations* (primes for p' and ρ' are omitted):

$$\begin{aligned}
&\frac{\partial\boldsymbol{v}}{\partial t} + \frac{\nabla p}{\rho_0} + g\frac{\rho}{\rho_0}\nabla z + 2\boldsymbol{\Omega}\times\boldsymbol{v} = 0\,,\\
&\frac{\partial\rho}{\partial t} + \frac{d\rho_0}{dz}\,w + \rho_0\nabla\boldsymbol{v} = 0\,,\\
&\frac{\partial\rho}{\partial t} = \frac{1}{c^2}\frac{\partial p}{\partial t} + \rho_0\frac{N^2(z)}{g}\,w\,,
\end{aligned} \tag{10.3}$$

where $N^2(z) = -g(\rho_0^{-1}\,d\rho_0/dz + g/c^2)$ is the *Brunt-Väisälä frequency* introduced above, (6.28).

10.1.2 Linear Boundary Conditions

Assume that the fluid's surface is a plane in the equilibrium state. When this state is violated, two kinds of forces arise which try to return the surface to its initial position: the *gravitational and the surface tension forces.* As a result, a disturbance arises at the fluid surface which propagates in all directions as a wave.

The forces at the fluid surface must be included in the boundary conditions for (10.3). Let $z = 0$ be the free surface in an equilibrium state. Its vertical displacement we denote by $\zeta = \zeta(x, y, t)$. The vertical velocity of the surface particles is $d\zeta/dt$. At the same time these particles can be considered as belonging to the fluid, its vertical velocity being $w(x, y, z, t)$. Hence, we obviously have

$$w|_{z=\zeta} = d\zeta/dt\,. \tag{10.4}$$

This is the so-called *kinematic boundary condition* at a free liquid surface.

In the second (*dynamical*) condition at the free surface, we take into account the existence of a discontinuity of the pressure on both sides of the boundary due to the surface tension. Let $p_0(z) + p$ be the pressure in the fluid and p_a that above the surface. We have by virtue of the well-known Laplace formula

$$p_0(\zeta) + p|_{z=\zeta} - p_a = -\sigma/R\,, \tag{10.5}$$

where σ is the *surface-tension coefficient*, $1/R$ is the sum of principal curvatures of the surface, i.e.,

$$\frac{1}{R} = \nabla[\nabla\zeta/\sqrt{1 + (\nabla\zeta)^2}]\,. \tag{10.6}$$

The boundary conditions (10.4, 5) appear to be nonlinear. We will linearize them assuming ζ to be small. Using (6.4) for the particle time derivative, we obtain from (10.4) by expanding w in a series,

$$w|_{z=0} + \left.\frac{\partial w}{\partial z}\right|_{z=0} \zeta + \cdots - \frac{\partial \zeta}{\partial t} - \boldsymbol{v}\cdot\nabla\zeta|_{z=0} - \cdots \simeq w|_{z=0} - \frac{\partial \zeta}{\partial t} = 0\,.$$

Analogously, from (10.5) with (10.6) and $p_0(0) = p_a$, we obtain

$$\left(\frac{dp_0}{dz}\zeta + p\right)_{z=0} + \cdots = -\sigma\Delta_{\perp}\zeta + \cdots,$$

where $\Delta_{\perp} = \partial^2/\partial x^2 + \partial^2/\partial y^2$ is the Laplacian in a horizontal plane. As a result, confining ourselves only to linear terms and taking into account the hydrostatic equation (10.2), we obtain the *linear boundary conditions* at an undisturbed free surface $z = 0$:

$$w|_{z=0} = \partial\zeta/\partial t\,, \quad p|_{z=0} = g\rho_0(0)\zeta - \sigma\Delta_{\perp}\zeta\,. \tag{10.7}$$

10.1.3 Equations for an Incompressible Fluid

Among the "restoring" forces in a fluid is an elasticity force resulting from the compression of fluid elements. The relationship between the compression and pressure is described by the first two terms in the linearized equation (10.3):

$$\frac{\partial \rho}{\partial t} = \frac{1}{c^2} \frac{\partial p}{\partial t}. \tag{10.8}$$

The elasticity forces are responsible for sound waves propagating with the velocity $c = (\partial p/\partial \rho)_s^{1/2}$. A more detailed treatment of the latter will be given in Chap. 12. In this section we shall examine the waves in an *incompressible fluid.*

The density of a given fluid element must be constant in this case:

$$d\rho/dt = 0. \tag{10.9}$$

Then it follows from an exact equation of continuity (6.9) that

$$\nabla \boldsymbol{v} = 0. \tag{10.10}$$

On the other hand, from the exact state equation (6.18) and taking (10.9) into account, we obtain $c^{-2}\, dp/dt = 0$. In the general case, however, $dp/dt \neq 0$, say, due to hydrostatic pressure change. Therefore, the sound velocity in an incompressible fluid should be assumed infinitely large ($c = \infty$). Now we obtain from (10.3) the *linearized equations of incompressible fluid hydrodynamics:*

$$\begin{gathered} \frac{\partial \boldsymbol{v}}{\partial t} + \frac{\nabla p}{\rho_0} + g \frac{\rho}{\rho_0} \nabla z + 2\boldsymbol{\Omega} \times \boldsymbol{v} = 0, \\ \nabla \boldsymbol{v} = 0, \quad \frac{\partial \rho}{\partial t} - \rho_0 \frac{N^2}{g} w = 0, \end{gathered} \tag{10.11}$$

where the Väisälä frequency is $N^2(z) = -g\rho_0^{-1}\, d\rho_0/dz$. The boundary conditions do not depend on the sound velocity and remain unchanged. The approximation of an incompressible fluid ($c = \infty$) is applicable to those waves whose velocity is much less than the velocity of sound.

10.2 Surface Gravity Waves

10.2.1 Basic Equations

In the case of gravity waves it is the gravitational force that tends to restore the equilibrium if the free surface of the fluid is distorted. The effect of gravity in the boundary conditions (10.7) is described by the term $g\rho_0(0)\zeta$ which is equal to the pressure of a fluid column of height ζ. For the time being we will exclude the effects of surface tension, the Coriolis force and the Archimedes force, assuming in our equations $\sigma = 0$, $\boldsymbol{\Omega} = 0$, $\rho_0 = \text{const}$ ($N^2 = 0$, $\rho = 0$), respectively. Denoting by $\boldsymbol{u} = \{u, v\}$ the horizontal component of particle velocity

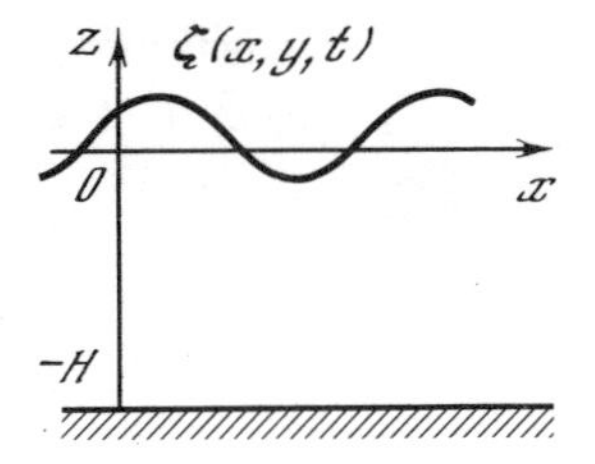

Fig. 10.1. For theory of a surface wave in the presence of a bottom

and $\nabla_- = \boldsymbol{e}_x \partial/\partial x + \boldsymbol{e}_y \partial/\partial y$, from (10.11) we obtain equations for the surface gravity waves:

$$\frac{\partial \boldsymbol{u}}{\partial t} = -\frac{\nabla_- p}{\rho_0}, \quad \frac{\partial w}{\partial t} + \frac{1}{\rho_0}\frac{\partial p}{\partial z} = 0, \quad \nabla_- \boldsymbol{u} + \frac{\partial w}{\partial z} = 0. \tag{10.12}$$

Expressing the horizontal velocity $\boldsymbol{u}$ and the pressure p in terms of the vertical velocity component w using the first and third equations (10.12),

$$\frac{\partial \Delta_- \boldsymbol{u}}{\partial t} = -\nabla_- \frac{\partial^2 w}{\partial t\, \partial z}, \quad \Delta_- p = \rho_0 \frac{\partial^2 w}{\partial t\, \partial z}, \tag{10.13}$$

we obtain for w

$$\frac{\partial}{\partial t} \Delta w = 0. \tag{10.14}$$

Assuming $\sigma = 0$ in (10.7) and taking p from (10.13), we obtain the boundary condition at a free surface $z = 0$ for w:

$$\left(\frac{\partial^3 w}{\partial t^2\, \partial z} - g\Delta_- w\right)_{z=0} = 0. \tag{10.15}$$

If we have a liquid layer of depth H (Fig. 10.1) and the bottom is perfectly rigid, another boundary condition must be imposed on the solution of (10.14)

$$w|_{z=-H} = 0 \tag{10.16}$$

expressing the fact that the vertical component of the particle velocity is zero at the bottom.

10.2.2 Harmonic Waves

We shall look for a solution of (10.14) in the form of a harmonic wave propagating in the horizontal direction:

$$w(x, y, z, t) = \phi(z) \exp[\mathrm{i}(\boldsymbol{k} \cdot \boldsymbol{r} - \omega t)], \tag{10.17}$$

where $\boldsymbol{k} = \{k_x, k_y\}$ is the *horizontal wave vector* and $\boldsymbol{r} = \{x, y\}$. Substitution of this expression into (10.14–16) leads to the boundary value problem for the function $\phi(z)$:

$$\frac{d^2\phi}{dz^2} - k^2\phi = 0\,, \qquad \left(\frac{d\phi}{dz} - \frac{gk^2}{\omega^2}\phi\right)_{z=0} = \phi(-H) = 0\,.$$

The boundary condition for $z = -H$ is satisfied if the solution of the equation for ϕ is chosen to be

$$\phi(z) = b\,\frac{\sinh[k(z+H)]}{\sinh kH}\,. \tag{10.18}$$

Substituting (10.18) into the boundary condition for $z = 0$ yields the relationship between k and ω or the so-called *dispersion law:*

$$\omega^2 = gk \tanh kH\,. \tag{10.19}$$

Hence, the *phase* velocity of a surface gravitational wave depends on the wave number (frequency)

$$c_{\mathrm{Ph}} = \frac{\omega}{k} = \sqrt{gH\,\frac{\tanh kH}{kH}}\,, \tag{10.20}$$

i.e., dispersion takes place (a nonharmonic surface wave changes its shape while propagating). One can easily find the *group* velocity of a surface gravity wave, i.e., the propagation velocity of the envelope of a spectrally narrow wave train (Sect. 2.6.2):

$$c_{\mathrm{g}} = \frac{d\omega}{dk} = \frac{c_{\mathrm{Ph}}}{2}\left(1 + \frac{kH}{\tanh kH} - kH \tanh kH\right), \tag{10.21}$$

which, as we see, depends on the wave number k.

The displacement of the surface $\zeta(x, y, t)$ is determined from the boundary condition (10.7):

$$\frac{\partial\zeta}{\partial t} = -\mathrm{i}\omega\zeta = w|_{z=0} = b\exp[\mathrm{i}(\boldsymbol{k}\cdot\boldsymbol{r} - \omega t)]\,,$$

whence

$$\zeta = a\exp[\mathrm{i}(\boldsymbol{k}\cdot\boldsymbol{r} - \omega t)]\,, \qquad a = \mathrm{i}b/\omega\,. \tag{10.22}$$

Pressure variations at an arbitrary depth z caused by wave motion will be found from (10.13), whence with $\Delta_- = -k^2$ we obtain

$$p = \rho_0 \frac{i\omega}{k \tanh kH} b \frac{\cosh[k(z+H)]}{\cosh kH} \exp[i(\boldsymbol{k} \cdot \boldsymbol{r} - \omega t)] . \tag{10.23}$$

Using the dispersion law (10.19), one can write the last expression in the form:

$$p = \rho_0 g \frac{ib}{\omega} \frac{\cosh[k(z+H)]}{\cosh kH} \exp[i(\boldsymbol{k} \cdot \boldsymbol{r} - \omega t)]$$

$$= \rho_0 g a \frac{\cosh[k(z+H)]}{\cosh kH} \exp[i(\boldsymbol{k} \cdot \boldsymbol{r} - \omega t)] . \tag{10.23a}$$

The horizontal velocity of fluid particles can be found from (10.13):

$$\boldsymbol{u} = i \frac{\boldsymbol{k}}{k} b \frac{\cosh[k(z+H)]}{\sinh kH} \exp[i(\boldsymbol{k} \cdot \boldsymbol{r} - \omega t)] . \tag{10.24}$$

Its direction is that of the $\boldsymbol{k}$-direction of wave propagation. Denoting by ξ the displacement of liquid particles in a horizontal plane ($d\xi/dt = -i\omega\xi = \boldsymbol{u}$) and by η that in the vertical direction ($d\eta/dt = -i\omega\eta = w$), assuming $b = \omega A \exp[i(\alpha - \pi/2)]$ in (10.18, 24) [$a = A \exp(i\alpha)$ is the displacement amplitude of the free surface] and separating the real part, we obtain

$$\xi = -\frac{\boldsymbol{k}}{k} A \frac{\cosh[k(z+H)]}{\sinh kH} \sin(\boldsymbol{k} \cdot \boldsymbol{r} - \omega t + \alpha) ,$$

$$\eta = A \frac{\sinh[k(z+H)]}{\sinh kH} \cos(\boldsymbol{k} \cdot \boldsymbol{r} - \omega t + \alpha) .$$

Now, the time t can be excluded and we find the following equation for paths of particles motions in a surface gravity wave

$$\xi^2/a_\xi^2 + \eta^2/a_\eta^2 = 1 , \quad a_\xi = A \frac{\cosh[k(z+H)]}{\sinh kH} ,$$

$$a_\eta = A \frac{\sinh[k(z+H)]}{\sinh kH} , \tag{10.25}$$

which are *ellipses* with the semi-axes ratio $a_\eta/a_\xi = \tanh[k(z+H)]$.

A *standing* surface wave can be formed by superposition of two *progressing* waves (10.17) with the same frequencies ω and amplitudes b but with opposite wave numbers $\boldsymbol{k}$:

$$w = 2b \frac{\sinh[k(z+H)]}{\sinh kH} \cos(\boldsymbol{k} \cdot \boldsymbol{r}) \exp(-i\omega t) .$$

In the same manner as above, we obtain expressions for the particle displacements:

$$\xi = 2A \frac{\boldsymbol{k}}{k} \frac{\cosh[k(z+H)]}{\sinh kH} \sin(\boldsymbol{k}\cdot\boldsymbol{r})\cos(\omega t - \alpha),$$

$$\eta = -2A \frac{\sinh[k(z+H)]}{\sinh kH} \cos(\boldsymbol{k}\cdot\boldsymbol{r})\cos(\omega t - \alpha).$$

The paths are *straight lines* in this case,

$$\eta = -\{\tanh[k(z+H)]\cdot\cot(\boldsymbol{k}\cdot\boldsymbol{r})\}\xi, \tag{10.26}$$

whose slope depends both on z and $\boldsymbol{r}$. The particles move along vertical straight lines if $\boldsymbol{k}\cdot\boldsymbol{r} = 0, \pm\pi, \pm 2\pi, \dots$ and along horizontal ones if $\boldsymbol{k}\cdot\boldsymbol{r} = \pm\pi/2, \pm 3\pi/2, \dots$.

Using the expressions for the particle velocities (10.17, 18, 24), we can easily prove that $\operatorname{curl}\boldsymbol{v} = 0$, i.e., motion in a surface wave is *potential*. Hence, the velocity can be represented as $\boldsymbol{v} = \nabla\varphi$ where

$$\varphi = \frac{b}{k} \frac{\cosh[k(z+H)]}{\sinh kH} \exp[\mathrm{i}(\boldsymbol{k}\cdot\boldsymbol{r} - \omega t)]. \tag{10.27}$$

10.2.3 Shallow- and Deep-Water Approximations

Two extreme cases are usually considered in the theory of surface gravity waves which differ in the relation between H and the wavelength $\lambda = 2\pi/k$.

i) *Shallow water:* $kH \ll 1$. The depth is small compared with the wavelength ($H/\lambda \ll 1$); we find $\tanh kH \simeq kH$ in (10.19, 20), and hence,

$$\omega = \sqrt{gH}k, \quad c_{\mathrm{Ph}} = \sqrt{gH}. \tag{10.28}$$

Dispersion is absent ($c_{\mathrm{Ph}} = \mathrm{const}$). For the particle's velocity components we have from (10.18, 24) in the same approximation:

$$w = b(1 + z/H)\exp[\mathrm{i}(\boldsymbol{k}\cdot\boldsymbol{r} - \omega t)], \quad \boldsymbol{u} = \mathrm{i}\frac{\boldsymbol{k}}{k} b \frac{\exp[\mathrm{i}(\boldsymbol{k}\cdot\boldsymbol{r} - \omega t)]}{kH}. \tag{10.29}$$

Obviously, $|\boldsymbol{u}| \gg |w|$, i.e., the particle's velocity is practically constant across the liquid layer and parallel to the bottom. This is also seen from the path equation (10.25). The ratio of the ellipse's semi-axes is $a_\eta/a_\xi = k(z+H) < kH \ll 1$, i.e., fluid particles move along elliptic paths which are greatly elongated in the horizontal direction.

ii) *Deep water:* $kH \gg 1$. The depth is large compared with the wavelength ($H/\lambda \gg 1$). We find $\tanh kH \simeq 1$ in (10.19) which yields the *dispersion law* for

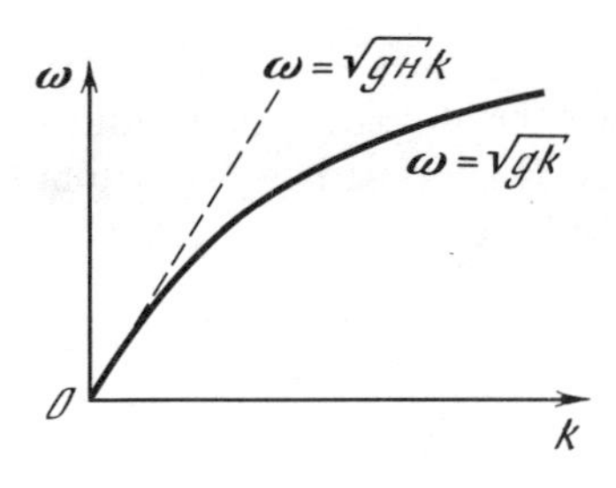

Fig. 10.2. Graphical representation of the dispersion relation for surface gravity waves

waves in deep water:

$$\omega^2 = gk. \tag{10.30}$$

For the phase and group velocities we have

$$c_{\mathrm{Ph}} = \sqrt{\frac{g}{k}} = \frac{g}{\omega}, \qquad c_{\mathrm{g}} = \frac{d\omega}{dk} = \frac{1}{2}\sqrt{\frac{g}{k}} = \frac{c_{\mathrm{Ph}}}{2}. \tag{10.31}$$

Neglecting $\exp(-kH)$ compared with unity, we obtain from (10.17, 18, 23, 24),

$$w = b\exp(kz)\exp[\mathrm{i}(\boldsymbol{k}\cdot\boldsymbol{r} - \omega t)],$$

$$\boldsymbol{u} = \mathrm{i}b(\boldsymbol{k}/k)\exp(kz)\exp[\mathrm{i}(\boldsymbol{k}\cdot\boldsymbol{r} - \omega t)],$$

$$p = \mathrm{i}\rho_0 b(\omega/k)\exp(kz)\exp[\mathrm{i}(\boldsymbol{k}\cdot\boldsymbol{r} - \omega t)]. \tag{10.32}$$

Thus, the particle velocity as well as the pressure variation decrease exponentially with depth. At the depth $|z| = \lambda$ $(z < 0)$, wave motion is practically absent. For the paths of liquid particles, we have from (10.25) $a_\xi = a_\eta = A\exp(kz)$ which yields

$$\xi^2 + \eta^2 = A^2\exp(2kz),$$

circles with radii decrease exponentially with depth.

A liquid layer at a given depth H can be considered either shallow or deep water depending on the length of the propagating waves. For long waves, such as tidal or Tsunami waves, any sea represents shallow water. These waves propagate without dispersion. It is very often convenient to use a graphic representation (Fig. 10.2) of the dispersion law (10.19). Here, the linear nondispersive part of the curve is in the vicinity of the origin. At large frequencies (wave numbers) ($kH \gg 1$), we have the dispersion corresponding to the waves in deep water.

10.2.4 Wave Energy

The energy of any kind of wave is usually defined as an excess of energy of a fixed volume over that at equilibrium. For surface waves it is convenient to take as such a volume a fluid column bounded by the bottom $z = -H$ and the free surface $z = \zeta$ in a vertical direction, and by two planes $x = x_0$ and $x = x_0 + \lambda$ in the direction of wave propagation. In the direction perpendicular to wave propagation (y-direction), the volume is supposed to be of unit width. For the kinetic energy of such a column we have

$$E_k = \frac{\rho_0}{2} \int_{x_0}^{x_0+\lambda} \int_{-H}^{0} v^2 \, dx \, dz \, .$$

v^2 is the sum of the squared real parts of (10.17, 24) and can be written

$$v^2 = B^2 \frac{\sinh^2[k(z+H)]}{\sinh^2 kH} \cos^2(kx - \omega t + \alpha') + B^2 \frac{\cosh^2[k(z+H)]}{\sinh^2 kH} \cos^2(kx - \omega t + \alpha' + \pi/2) \, ,$$

where $B = b\exp(-\mathrm{i}\alpha')$. Note that

$$\int_{x_0}^{x_0+\lambda} \cos^2\left[\frac{2\pi x}{\lambda} - f(t)\right] dx = \frac{\lambda}{2} \, ,$$

where $f(t)$ is an arbitrary function. Hence,

$$E_k = \frac{\rho_0 \lambda B^2}{4 \sinh^2 kH} \int_{-H}^{0} \{\sinh^2[k(z+H)] + \cosh^2[k(z+H)]\} \, dz$$

$$= \frac{\rho_0 \lambda B^2}{4 \sinh^2 kH} \int_{-H}^{0} \cosh[2k(z+H)] \, dz$$

and finally after integration we obtain,

$$E_k = \frac{\rho_0 \lambda B^2}{4k \tanh kH} = \frac{\rho_0 \lambda g}{4} \frac{B^2}{\omega^2} = \frac{\rho_0 \lambda g}{4} A^2 \, . \qquad (10.33)$$

Here, the dispersion law (10.19) was taken into account. We again point out that $A = |a|$ is the amplitude of the free surface displacements.

Analogously we have for the potential-energy increment

$$E_p = \int_0^{\lambda} \left(\int_{-H}^{\zeta} \rho_0 g z \, dz - \int_{-H}^{0} \rho_0 g z \, dz \right) dx = \rho_0 g \int_0^{\lambda} \int_0^{\zeta} z \, dz \, dx = \frac{\rho_0 g}{2} \int_0^{\lambda} \zeta^2 \, dx \, .$$

Substituting the real part of (10.22) $A \cos(kx - \omega t + \alpha)$ into the integral for ζ and integrating, we obtain

$$E_p = \frac{\rho_0 g \lambda}{4} A^2 = E_k \, . \qquad (10.34)$$

Thus, the *kinetic energy* and the *potential energy* of a *progressive surface wave* are *equal* and do not depend on time.

In the case of a standing surface wave

$$\zeta = A \cos kx \cos(\omega t - \alpha),$$

similar calculations lead to

$$\begin{aligned} E_p &= \frac{\rho_0 g A^2 \lambda}{4} \cos^2(\omega t - \alpha) = \frac{\rho_0 g A^2 \lambda}{8} [1 + \cos(2\omega t - 2\alpha)], \\ E_k &= \frac{\rho_0 g A^2 \lambda}{4} \sin^2(\omega t - \alpha) = \frac{\rho_0 g A^2 \lambda}{8} [1 - \cos(2\omega t - 2\alpha)], \\ E_k &+ E_p = \text{const}. \end{aligned} \tag{10.35}$$

The kinetic and potential energies both change in time at twice the frequency, their value oscillating between 0 and $\rho_0 g \lambda A^2/4$. Their sum (total energy of the wave) remains constant. When time-averaged over a period both kinds of energy of the standing wave are equal.

10.3 Capillary Waves

10.3.1 "Pure" Capillary Waves

Now we consider the waves caused by a surface-tension force. The term $-\sigma \Delta_- \zeta$ in the boundary condition (10.7) is responsible for this force. Eliminating any other type of wave motion [assuming $\rho_0 = \text{const}$, $N^2 = 0$, $\boldsymbol{\Omega} = 0$, $g = 0$ in (10.7, 11)], we again obtain (10.4) for the vertical component of the velocity. The boundary condition at the free surface $z = 0$, however, is different now. Assuming $g = 0$ in the second equation (10.7) and applying to this the operator $\Delta_- \partial/\partial t$, we obtain

$$\frac{\partial}{\partial t} \Delta_- p|_{z=0} = -\sigma \Delta_-^2 \frac{\partial \zeta}{\partial t}.$$

Here, $\partial\zeta/\partial t = w|_{z=0}$, according to the first equation in (10.7), and $\Delta_- p$ can be written in terms of w by (10.13). We find as a result

$$\left(\frac{\partial^3 w}{\partial t^2 \partial z} + \gamma \Delta_-^2 w \right)_{z=0} = 0, \tag{10.36}$$

where $\gamma = \sigma/\rho_0$.

We will see below that pure capillary waves (gravitational force is neglected) are *high frequency* ones. Therefore, in all practically interesting cases it can be assumed that $kH \to \infty$, i.e., we can use the deep-water approximation. Instead of a boundary condition at the bottom, we can require that the solution of (10.14) vanishes as $z \to -\infty$. Looking for this solution in the form of a harmonic wave (10.17), we have for $\phi(z)$ and $w(x, y, z, t)$,

$$\phi(z) = b\exp(kz), \qquad w = b\exp(kz)\exp[\mathrm{i}(\boldsymbol{k}\cdot\boldsymbol{r} - \omega t)].$$

Substitution of w into the boundary condition (10.36) yields the *dispersion relation* for *capillary* waves

$$\omega^2 = \gamma k^3. \tag{10.37}$$

The phase and group velocities are

$$c_{\mathrm{Ph}} = \sqrt{\gamma k}, \qquad c_{\mathrm{g}} = \tfrac{3}{2}\sqrt{\gamma k} = \tfrac{3}{2}c_{\mathrm{Ph}}. \tag{10.38}$$

The relations for pressure p and the horizontal component of the velocity $\boldsymbol{u}$ coincide with the corresponding ones for gravity waves in deep water (10.32). Hence, the paths of fluid particles in the case of a capillary wave are circles, too.

10.3.2 Gravity-Capillary Surface Waves

We now consider the joint action of the gravity and surface-tension forces in a homogeneous ($\rho_0 = \mathrm{const}$), incompressible ($c^2 = \infty$) and nonrotating ($\boldsymbol{\Omega} = 0$) fluid. Equation (10.12), the boundary condition (10.16) and the solutions (10.17, 18, 22–24) in the form of harmonic waves still hold in this case. The boundary condition at the free surface, however, will now be a combination of the conditions (10.15, 36):

$$\left(\frac{\partial^3 w}{\partial t^2\,\partial z} - g\,\Delta_{-}w + \gamma\,\Delta_{-}^2 w\right)_{z=0} = 0.$$

This leads to the new dispersion relation

$$\omega^2 = (gk + \gamma k^3)\tanh kH. \tag{10.39}$$

Hence, the squared phase velocity is

$$c_{\mathrm{Ph}}^2 = \left(\frac{g}{k} + \gamma k\right)\tanh kH. \tag{10.40}$$

In particular, for waves in deep water ($kH \gg 1$) we have

$$\omega^2 = gk + \gamma k^3, \qquad c_{\mathrm{Ph}}^2 = \frac{g}{k} + \gamma k. \tag{10.41}$$

For small k (long-wavelength waves), the first term on the right is the principal one. Neglecting the second term, we obtain the case of gravity waves, c.f. (10.9), considered above. The phase velocity decreases with increasing k. On the contrary, for large k (short-wavelength waves), in (10.41) only the second term (pure capillary waves) should be kept. The phase velocity grows with k. The minimum of the phase velocity occurs for a certain $k = k_0 = 2\pi/\lambda_0$. Equating the derivative on the right-hand side of the first equation (10.41) to zero, we find

$$k_0 = \sqrt{g/\gamma}\,, \qquad \lambda_0 = 2\pi\sqrt{\gamma/g}\,, \qquad (c_{\mathrm{Ph}})_{\min} = \sqrt{2}(g\gamma)^{1/4}\,. \tag{10.42}$$

For the water-air interface $\gamma \simeq 73\ \mathrm{cm^3\,s^{-2}}$, $\lambda_0 \simeq 1.714$ cm, the corresponding frequency $f_0 = (c_{\mathrm{Ph}})_{\min}/\lambda_0 \simeq 13.5$ Hz and $(c_{\mathrm{Ph}})_{\min} \simeq 23.1\ \mathrm{cm\cdot s^{-1}}$. The wavelength λ_0 separates the gravity and capillary waves. At $\lambda \gg \lambda_0$, the gravity force prevails whereas at $\lambda \ll \lambda_0$, the capillary force is the principal one.

For the group velocity of gravity-capillary waves in deep water ($kH \gg 1$), we obtain, differentiating (10.41) with respect to k,

$$2\omega c_{\mathrm{g}} = g + 3\gamma k^2\,.$$

Hence,

$$c_{\mathrm{g}} = \frac{g + 3\gamma k^2}{2\omega} = \frac{c_{\mathrm{Ph}}}{2}\,\frac{3k^2 + g/\gamma}{k^2 + g/\gamma} = \frac{c_{\mathrm{Ph}}}{2}\,\frac{k_0^2 + 3k^2}{k_0^2 + k^2}\,. \tag{10.43}$$

In the extreme cases of gravity ($k_0 \gg k$) and capillary waves ($k_0 \ll k$), the last relation coincides with (10.21) considered for $kH \gg 1$ and with (10.38), respectively. Dispersion $\omega(k)$ and phase velocity $c_{\mathrm{Ph}}(k)$ curves for gravity-capillary waves are plotted in Fig. 10.3a, b.

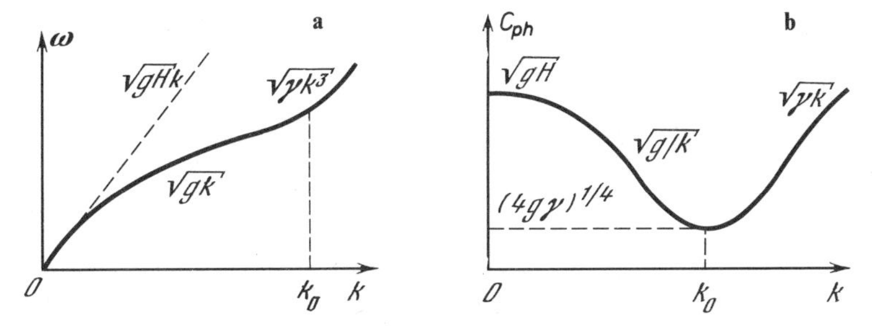

Fig. 10.3. (**a**) Dispersion and (**b**) phase-velocity curves for gravity-capillary waves

10.4 Internal Gravity Waves

10.4.1 Introductory Remarks

A characteristic of internal waves is that vertical particle displacements reach their maximum not at the surface but inside the fluid, say, at the interface of two fluids with different densities. In the sea, an analogous case takes place when low salinity water is above heavier water with greater salinity. Such a case has been called "dead water". It appears that a ship moving in such an area spends a considerable part of its power for internal-wave generation and as a result its speed diminishes.

In such a simple two-fluid model, internal waves are quite analogous to surface ones. They are also concentrated near the interface; the corresponding dispersion law is identical to (10.30) for gravity waves. The only difference is that some "effective" gravity acceleration must be taken instead of g. Indeed, consider the interface $z = 0$ between two liquid half-spaces with densities ρ_1 and ρ_2 ($\rho_2 > \rho_1$). The solution of (10.14) representing a harmonic wave propagating along the x-axis can be written as

$$\begin{aligned} w_1 &= b_1 \exp[-kz + \mathrm{i}(kx - \omega t)], \quad z > 0, \\ w_2 &= b_2 \exp[kz + \mathrm{i}(kx - \omega t)], \quad z < 0. \end{aligned} \tag{10.44}$$

The kinematic and dynamic boundary conditions must be satisfied at the interface. In a linear approximation, the former ($w_j|_{z=\zeta} = d\zeta/dt, j = 1, 2$) gives equality of the vertical velocities:

$$w_1|_{z=0} = w_2|_{z=0} = \partial\zeta/\partial t .$$

The dynamic condition [equality of total pressures at both sides of the boundary ($z = \zeta$)] leads to

$$-g\rho_1\zeta + p_1|_{z=0} = -g\rho_2\zeta + p_2|_{z=0} .$$

The pressure p can be expressed in terms of w by the last equation of (10.13). As a result we obtain two boundary conditions for the vertical components of particle velocity:

$$w_1|_{z=0} = w_2|_{z=0}, \quad gk^2(\rho_2 - \rho_1)w_j|_{z=0} = \omega^2\left(\rho_2 \frac{\partial w_2}{\partial z} - \rho_1 \frac{\partial w_2}{\partial z}\right)_{z=0} . \tag{10.44a}$$

Substituting (10.44), we obtain the relationship between the amplitudes of waves at both sides of the boundary $b_1 = b_2 = b$ and the required dispersion law:

$$\omega^2 = \frac{\Delta\rho}{\rho_1 + \rho_2} gk, \quad \Delta\rho = \rho_2 - \rho_1 . \tag{10.45}$$

It is natural enough that the last relation translates into (10.30) if $\rho_1 = 0$ (the interface between liquid and vacuum).

Internal waves, however, occur not only in the case of a discontinuous variation of density along the vertical, but also when the equilibrium density $\rho_0(z)$ is a continuous function of a vertical coordinate. This is practically always the case in the ocean and the atmosphere. Comparatively large vertical density gradients in the ocean are observed in *season* and *main thermoclines* where internal waves may be produced most often. Calm sea does not imply calmness at depths where internal waves with displacements of many tens of meters can occur.

10.4.2 Basic Equation for Internal Waves. Boussinesq Approximation

We start from the linear hydrodynamic equations (10.11) of an incompressible and nonrotating ($\boldsymbol{\Omega} = 0$) fluid. We should bear in mind that the equilibrium density $\rho_0(z)$ and the Väisälä frequency $N^2(z)$ are functions only of the vertical coordinate in these equations. Assuming $\boldsymbol{v} = \{\boldsymbol{u}, w\}$, we rewrite (10.11) in the form

$$\begin{aligned} &\frac{\partial \boldsymbol{u}}{\partial t} + \frac{\nabla_- p}{\rho_0} = 0\,, \qquad \nabla_- \boldsymbol{u} = -\frac{\partial w}{\partial z}\,, \\ &\frac{\partial w}{\partial t} + \frac{1}{\rho_0}\frac{\partial p}{\partial z} + \frac{g\rho}{\rho_0} = 0\,, \qquad \frac{1}{\rho_0}\frac{\partial \rho}{\partial t} = \frac{N^2}{g}\,w\,. \end{aligned} \tag{10.46}$$

This system is easily reduced to one equation with respect to the vertical particle velocity w. In fact, excluding $\boldsymbol{u}$ from the first two equations and taking into account that $\nabla_-^2 = \Delta_-$, we express the pressure p in terms of w:

$$\Delta_- p = \rho_0\, \partial^2 w/\partial t\, \partial z\,. \tag{10.47}$$

We now apply the operator $\Delta_-\, \partial/\partial t$ to the third equation of (10.46). In the result obtained, we substitute $\partial(\rho/\rho_0)/\partial t$ from the fourth equation and $\Delta_- p$ from (10.47), this finally leads to the required equation:

$$\frac{\partial^2}{\partial t^2}\left(\Delta w + \frac{1}{\rho_0}\frac{d\rho_0}{dz}\frac{\partial w}{\partial z}\right) + N^2\,\Delta_- w = 0\,. \tag{10.48}$$

Here, the Väisälä frequency is $N^2(z) = -g\rho_0^{-1}\, d\rho_0/dz$ [see (6.28) where $c^2 \to \infty$ due to the fluid incompressibility].

Very often, for example, in the theory of ocean waves, a simplified equation is used; the term proportional to $\partial w/\partial z$ in (10.48) is neglected as compared with Δw, i.e.,

$$\frac{\partial^2}{\partial t^2}\,\Delta w + N^2(z)\Delta_- w = 0\,. \tag{10.49}$$

This approximation corresponds to assuming that $\rho_0(z)$ in (10.46) is constant, say, $\rho_{00} = \rho_0(0)$ everywhere except in the expression for $N^2(z)$ and is called the *Boussinesq approximation.* Estimates show that for the linear problem in wave theory considered below, this assumption is completely justified (see also Exercise 10.6.8).

10.4.3 Waves in an Unlimited Medium

Waves specified by (10.49) can also propagate in an unlimited medium. They have the simplest form in the case of a constant Väisälä frequency (N^2 = const). For this case we will consider harmonic plane internal waves of the kind

$$w = b\exp[\mathrm{i}(\boldsymbol{\varkappa}\cdot\boldsymbol{R} - \omega t)] = b\exp[\mathrm{i}(\boldsymbol{k}\cdot\boldsymbol{r} + k_z z - \omega t)], \tag{10.50}$$

where $\boldsymbol{\varkappa} = \{\boldsymbol{k}, k_z\} = \{k_x, k_y, k_z\}$ is the wave vector, ω the wave frequency, b the amplitude and $\boldsymbol{R} = \{\boldsymbol{r}, z\} = \{x, y, z\}$. Substituting (10.50) into (10.49) yields a relationship between $\boldsymbol{\varkappa}$ and ω, the *dispersion relation for internal waves:*

$$\omega^2 = N^2\frac{k^2}{k^2 + k_z^2} = N^2\frac{k^2}{\varkappa^2} = N^2\sin^2\theta, \tag{10.51}$$

where θ is the angle which the vector $\boldsymbol{\varkappa}$ makes with the vertical. It follows from (10.51) that

i) only waves with frequency $\omega < N$ can occur;

ii) if θ is fixed, the frequency ω is uniquely determined by (10.51), but its wavelength, and consequently phase velocity, can be arbitrary.

We choose the x-axis along the horizontal wave vector $\boldsymbol{k}$. Then (10.50) can be written as

$$w = b\exp[\mathrm{i}(\varkappa\xi - \omega t)],$$

where $\xi = x\sin\theta + z\cos\theta$ and $\omega = N|\sin\theta|$. Since ω does not depend on $\varkappa$, it is evident that general expression

$$w = F(\xi)\exp(-\mathrm{i}\omega t) \tag{10.52}$$

with $F(\xi)$ being an arbitrary function can be a solution of (10.49) if N^2 = const. This fact can be proved by substitution of (10.52) into (10.49). One can see that a plane harmonic wave of arbitrary shape (10.52) remains at a fixed position all the time. The reason for this will become clear below after we have considered the group velocity in the three-dimensional case. The considerations will be analogous to those in Sect. 2.6.2. Let the wave frequency depend on the wave vector

$$\omega = \omega(\boldsymbol{\varkappa}) = \omega(k_x, k_y, k_z).$$

We represent a wave disturbance at $t = 0$ by the Fourier transform

$$w(\boldsymbol{R}) = \int_{-\infty}^{\infty} b(\boldsymbol{\varkappa}) \exp(\mathrm{i}\boldsymbol{\varkappa} \cdot \boldsymbol{R})\, d\boldsymbol{\varkappa}. \tag{10.53}$$

As in Sect. 2.6.2, we deal with the so-called narrow-band process where the spectral density $b(\boldsymbol{\varkappa})$ is small except in the close vicinity of some $\boldsymbol{\varkappa} = \boldsymbol{\varkappa}_0$. In this case, (10.53) can also be written as

$$w(\boldsymbol{R}) = F(\boldsymbol{R}) \exp(\mathrm{i}\boldsymbol{\varkappa}_0 \cdot \boldsymbol{R}), \tag{10.53a}$$

where $F(\boldsymbol{R})$ is a slowly varying function compared with $\exp(\mathrm{i}\boldsymbol{\varkappa}_0 \cdot \boldsymbol{R})$. At an arbitrary time t, the function $w(\boldsymbol{R}, t)$, being a solution of a certain wave equation with the dispersion relation $\omega = \omega(\boldsymbol{\varkappa})$ and satisfying the initial conditions (10.53), can be written as

$$w(\boldsymbol{R}, t) = \int_{-\infty}^{\infty} b(\boldsymbol{\varkappa}) \exp\{\mathrm{i}[\boldsymbol{\varkappa} \cdot \boldsymbol{R} - \omega(\boldsymbol{\varkappa})t]\}\, d\boldsymbol{\varkappa}$$

$$\simeq \exp[\mathrm{i}(\boldsymbol{\varkappa}_0 \cdot \boldsymbol{R} - \omega_0 t)] \int_{-\infty}^{\infty} b(\boldsymbol{\varkappa}) \exp\{\mathrm{i}(\boldsymbol{\varkappa} - \boldsymbol{\varkappa}_0)[\boldsymbol{R} - \nabla_{\boldsymbol{\varkappa}}\omega(\boldsymbol{\varkappa})|_{\boldsymbol{\varkappa}_0} t]\}\, d\boldsymbol{\varkappa},$$

where $\omega_0 = \omega(\boldsymbol{\varkappa}_0)$. Here we have used the expansion

$$\omega(\boldsymbol{\varkappa}) \simeq \omega_0 + [\nabla_{\boldsymbol{\varkappa}}\omega(\boldsymbol{\varkappa})]_{\boldsymbol{\varkappa}_0}(\boldsymbol{\varkappa} - \boldsymbol{\varkappa}_0).$$

Comparing the last expression with (10.53a) we find

$$w(\boldsymbol{R}, t) = F[\boldsymbol{R} - \boldsymbol{c}_{\mathrm{g}}(\boldsymbol{\varkappa}_0)t] \exp[\mathrm{i}(\boldsymbol{\varkappa}_0 \cdot \boldsymbol{R} - \omega_0 t)],$$

where the vector $\boldsymbol{c}_{\mathrm{g}}$, called the group velocity of the wave, is

$$\boldsymbol{c}_{\mathrm{g}}(\boldsymbol{\varkappa}) = \nabla_{\boldsymbol{\varkappa}}\omega(\boldsymbol{\varkappa}) = \left\{\frac{\partial\omega}{\partial k_x}, \frac{\partial\omega}{\partial k_y}, \frac{\partial\omega}{\partial k_z}\right\}. \tag{10.54}$$

We see that the envelope of a plane harmonic wave $F(\boldsymbol{R} - \boldsymbol{c}_{\mathrm{g}}t)$ is constant at the planes $\boldsymbol{R} - \boldsymbol{c}_{\mathrm{g}}t = \mathrm{const}$ and moves in space with the group velocity.

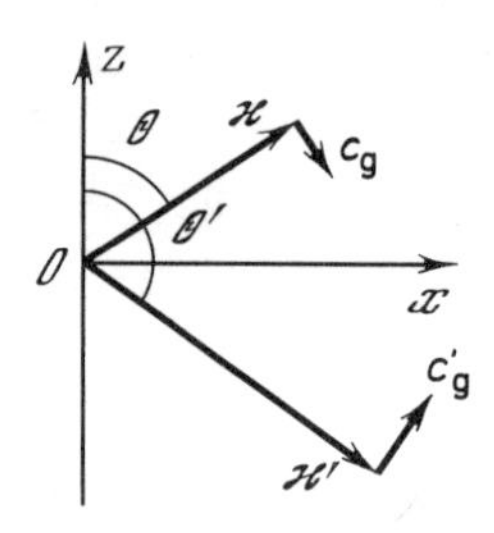

Fig. 10.4. Mutual orientation of the wave vector $\boldsymbol{\varkappa}$ and the group velocity vector $\boldsymbol{c}_g$ for an internal wave. $\boldsymbol{\varkappa}'$ and $\boldsymbol{c}'_g$ refer to downward wave propagation

In the case of plane internal waves considered above, a direct calculation of $\omega(\boldsymbol{\varkappa})$ using (10.51) gives

$$\frac{\partial\omega}{\partial k} = N\frac{k_z^2}{\varkappa^3}, \quad \frac{\partial\omega}{\partial k_z} = -N\frac{kk_z}{\varkappa^3}, \quad \boldsymbol{c}_g = N\frac{kk_z}{\varkappa^3}\left(\frac{k_z\boldsymbol{k}}{k^2} - \nabla z\right). \tag{10.55}$$

We easily prove that $\boldsymbol{\varkappa}\cdot\boldsymbol{c}_g = \omega k_z(k_z - k_z)/\varkappa^2 = 0$, i.e., $\boldsymbol{c}_g$ is perpendicular to the wave vector $\boldsymbol{\varkappa}$; its horizontal projection is in the $\boldsymbol{k}$-direction and the vertical projection is opposite to k_z, see (10.55) and Fig. 10.4.

Now it is apparent that a train of internal waves traveling in space can be obtained only in the case where a plane harmonic wave is modulated in the plane of its wave-front (perpendicular to the direction of propagation). It is only along the wave-front that disturbances and, consequently, energy can move.

For the pressure in a plane harmonic internal wave, we obtain from (10.50) with (10.47),

$$p = -\rho_0 b\frac{\omega k_z}{k^2}\exp[\mathrm{i}(\boldsymbol{k}\cdot\boldsymbol{r} + k_z z - \omega t)]. \tag{10.56}$$

The pressure gradient $\nabla p = \mathrm{i}\boldsymbol{\varkappa}p$ has the same direction as that of wave propagation or $\boldsymbol{\varkappa} = \{\boldsymbol{k}, k_z\}$.

The horizontal component of the fluid-particle velocity can be found from the first equation of (10.46):

$$\boldsymbol{u} = \frac{1}{\mathrm{i}\omega\rho_0}\nabla_{-}p = -b\frac{k_z\boldsymbol{k}}{k^2}\exp[\mathrm{i}(\boldsymbol{k}\cdot\boldsymbol{r} + k_z z - \omega t)]. \tag{10.57}$$

Now, for the total particle-velocity vector we have, taking (10.50) into account, $\boldsymbol{v} = \{\boldsymbol{u}, w\} = (\nabla z - k_z\boldsymbol{k}/k^2)w$. One can easily see that

$$\boldsymbol{v}\cdot\boldsymbol{\varkappa} = \boldsymbol{u}\cdot\boldsymbol{k} + k_z w = w(k_z - k_z) = 0.$$

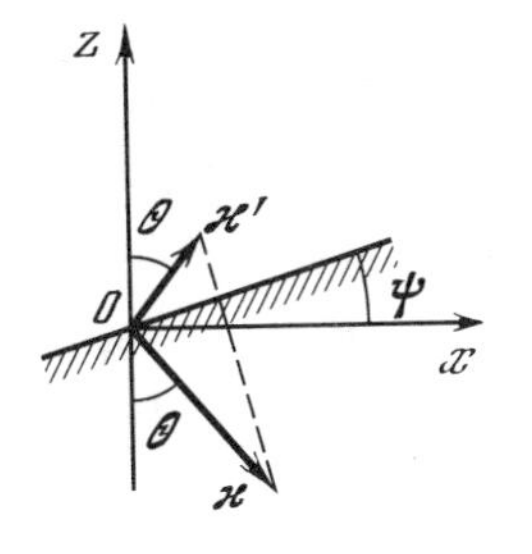

Fig. 10.5. Reflection of an internal wave at an inclined plane boundary

Thus, liquid particles move along straight lines perpendicular to $\boldsymbol{\varkappa}$ in the plane containing the z-axis as well as the vector $\boldsymbol{\varkappa}$. Therefore, internal waves are *transverse*.

Reflection of internal waves at boundaries reveals some interesting features, especially if the boundary is plane but not horizontal (a sloped ocean bottom, for example, Fig. 10.5). The frequency of the wave and the wave vector component along the boundary are conserved, as in the case of any other type of waves. The frequency of an internal wave, however, specifies uniquely the angle between the wave vector and the vertical. This angle must be conserved. As a result, the angle of reflection in the usual sense (the angle between the wave vector and the normal to the boundary) is not equal to the incidence angle. From the condition that projections of wave vectors of the incident $\boldsymbol{\varkappa}$ and reflected $\boldsymbol{\varkappa}'$ waves on the boundary must be equal, one can easily find a relationship between the magnitudes of the vectors. The picture in the plane where both the normal to the boundary and the wave vector of the incident wave lie, is shown in Fig. 10.5. We have, according to this figure,

$$\varkappa' \cos(\pi/2 - \theta - \psi) = \varkappa \cos(\pi/2 - \theta + \psi) \quad \text{or}$$

$$\varkappa' \sin(\theta + \psi) = \varkappa \sin(\theta - \psi) .$$

Thus, the *wavelength* (*wave number*) *changes* in the flection of the wave. This does not contradict the rule of frequency conservation since for a fixed frequency, any wavelength can occur. It is interesting to note also that the energy runs away from the boundary (in the direction of $\boldsymbol{c}_g$) in an "incident" wave (whose wave vector is directed toward the boundary) and toward the boundary in a "reflected" wave. Therefore, it would be more correct to call the reflected wave incident, and vice versa.

The Väisälä frequency is constant, for example, in an isothermal atmosphere. Indeed, we have $\rho_0^{-1}\, d\rho_0/dz$ = const, according to the hydrostatic equation if p_0/ρ_0 = const. Results obtained above are also useful for high-frequency waves when the wavelength is small compared with the vertical scale of the Väisälä frequency changes.

10.5 Guided Propagation of Internal Waves

10.5.1 Qualitative Analysis of Guided Propagation

An example of the typical dependence of the Väisälä frequency on depth in the ocean is given in Fig. 10.6a. A characteristic feature here is the maximum of this parameter at some depth. This depth serves as the waveguide axis for internal waves, as will be shown later.

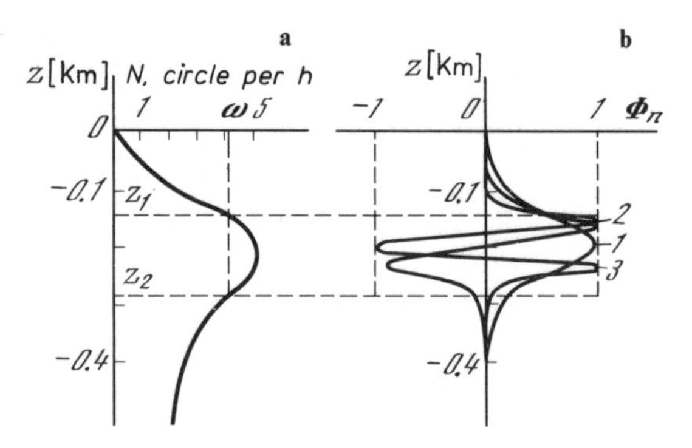

Fig. 10.6. Schematic dependence of the Väisälä frequency on the depth of the ocean (**a**) and the first three internal modes (**b**)

Regarding the Väisälä frequency as a function of only the vertical coordinate z, we look for a solution of (10.49) for a vertical particle velocity in the form of a harmonic wave propagating in a horizontal direction:

$$w = \phi(z)\exp[\mathrm{i}(\boldsymbol{k}\cdot\boldsymbol{r} - \omega t)] . \tag{10.58}$$

Substitution into (10.49) leads to the equation for the function $\phi(z)$:

$$\frac{d^2\phi}{dz^2} + k^2\left(\frac{N^2}{\omega^2} - 1\right)\phi = 0 . \tag{10.59}$$

We assume $\phi(-H) = 0$ to be the boundary condition at the lower boundary $z = -H$. The ocean bottom satisfies this condition fairly well. The boundary condition (10.15) for the surface of the water can be written in terms of ϕ by substituting it into (10.58):

$$\left(\frac{d\phi}{dz} - \frac{gk^2}{\omega^2}\right)_{z=0} = 0 . \tag{10.60}$$

For internal waves in the atmosphere, the condition at the lower boundary (the Earth's surface) is the same as for the ocean. Instead of the condition at the upper boundary, we can require boundedness of the solution as $z \to \infty$.

A qualitative analysis of (10.59) brings out some significant features of guided internal waves. At a given frequency, a solution of (10.59) satisfying the boundary conditions can arise only for some values $k = k_n(\omega)$ called *eigenvalues*. The corresponding solutions $\phi_n(z)$ are termed *eigenfunctions*. Expression (10.58) for the vertical component of the velocity for a given $\phi(z) = \phi_n(z)$ presents a *normal wave* or *mode*. The dependence $k = k_n(\omega)$ or the inverse one $\omega = \omega_n(k)$ is a *dispersion relation*.

In the case of the ocean, a solution $\phi_0(z)$ always exists among the modes with maximum value at $z = 0$, the so-called *surface mode;* $\phi_0(z)$ becomes zero only at the ocean bottom. In the case of deep water ($kH \gg 1$), one can easily prove that (10.48) and condition (10.15) are satisfied by the surface wave (10.32) with a dispersion relation (10.30) for an arbitrary dependence $N^2(z)$ (Exercise 10.6, 10).

Proceeding with the analysis of internal waves, we note that in the depth range $z_2 < z < z_1$ where $\omega < N(z)$ (Fig. 10.6a), ϕ'' and ϕ have different signs, according to (10.59). Hence, the solution of (10.59) has an oscillatory character, and $\phi(z)$ can vanish one or more times in this range. As a result, we have modes of a different order. The latter is the number which exceeds by one the number of zeros of the function $\phi(z)$. We see that this function is analogous to the quantum-mechanical ψ-function for a particle in a potential well.

At $z < z_2$ and $z > z_1$, we have $\omega > N(z)$ (Fig. 10.6a) and the functions ϕ'' and ϕ have the same sign, according to (10.59). Therefore, the function $\phi(z)$ is not an oscillating one and decreases exponentially when the distances $z - z_1$ and $z_2 - z$ increase. In experiments, therefore, internal waves can be observed more easily at depths where the Väisälä frequency is large, say in the ocean's thermoclines.

Figure 10.6b portrays the dependence of the amplitudes of the modes $n = 1, 2, 3$ on z, corresponding to the profile $N(z)$ in Fig. 10.6a. Note that at $z = 0$, all modes have amplitudes very close to zero. This fact allows the use of the simpler condition $w|_{z=0} = 0$ at the surface, i.e., the so-called "*rigid cover*" approximation frequently used in physical oceanography when surface waves have to be eliminated.

Suppose that the eigenfunction $\phi_n(z)$ and the dispersion relation $k = k_n(\omega)$ are to be found for some n. The former is supposed to be normalized in some way, for example, $\max|\phi_n(z)| = 1$. For the vertical velocity in a given mode we have, according to (10.58),

$$w_n(\boldsymbol{r}, z, t) = b_n \phi_n(z) \exp[\mathrm{i}(\boldsymbol{k}_n \cdot \boldsymbol{r} - \omega t)] . \tag{10.61}$$

From the general equations (10.60, 61), we can now easily find the pressure in the wave p_n, the horizontal velocity of fluid particles $\boldsymbol{u}_n$, the density increment

ρ_n and from the relation $\partial\zeta/\partial t = w$, the vertical displacement of fluid particles $\zeta_n(\boldsymbol{r}, z, t)$:

$$\begin{aligned}
p_n &= \mathrm{i}\rho_0 b_n(\omega/k_n^2)\phi_n'(z)\exp[\mathrm{i}(\boldsymbol{k}_n\cdot\boldsymbol{r}-\omega t)]\,,\\
\boldsymbol{u}_n &= \mathrm{i}b_n(\boldsymbol{k}_n/k_n^2)\phi_n'(z)\exp[\mathrm{i}(\boldsymbol{k}_n\cdot\boldsymbol{r}-\omega t)]\,,\\
\rho_n &= \mathrm{i}\rho_0 b_n(N^2/g\omega)\phi_n(z)\exp[\mathrm{i}(\boldsymbol{k}_n\cdot\boldsymbol{r}-\omega t)]\,,\\
\zeta_n &= \mathrm{i}(b_n/\omega)\phi_n(z)\exp[\mathrm{i}(\boldsymbol{k}_n\cdot\boldsymbol{r}-\omega t)]\,.
\end{aligned} \tag{10.62}$$

The density variations at a given point $\rho_n(\boldsymbol{r}, z, t)$ can be presented as $\rho_n = -(d\rho_0/dz)\zeta_n$. Hence, the density changes only due to the advection of particles with different density at this point, as it should be in an incompressible fluid.

The energy of the nth mode in a column of fluid with the height equal to the distance of the surface from the bottom, having a width equal to $\lambda_n = 2\pi/k_n$ in the propagation direction and unity perpendicular to it, can be calculated in the same way as in the case of surface waves (Sect. 10.2.4). Such calculations (Exercise 10.6, 12) show that the kinetic energy equals the potential energy, and the total energy is

$$E_n = \frac{\rho_0}{2}\frac{|b_n|^2}{\omega^2}\lambda_n\Big[g\phi_n^2(0) + \int\limits_{-H}^{0} N^2(z)\phi_n^2(z)\,dz\Big]\,. \tag{10.63}$$

In the case of a surface wave $[n = 0, \phi_0(0) = 1]$ in a homogeneous layer ($\rho_0 =$ const, $N^2 = 0$), (10.63) will just give the sum of (10.33 and 34).

Of all the modes, the simplest is the first one in which, according to Fig. 10.6b, all points of the thermocline at a given $\boldsymbol{r}$ and t move in the same direction and the thermocline as a whole takes the form of a wave. If in this case the wavelength is large compared with the effective thickness of the thermocline $z_1 - z_2$ (Fig. 10.6a) but small compared with the distances from this layer to the boundaries, then the dispersion relation (10.45) holds. The latter was obtained for an infinitely thin layer with a jump-like variation of density ($z_1 = z_2 = z_0, N \to \infty, \rho_2 = \rho_0(z_2) > \rho_0(z_1) = \rho_1$). Wave motions are concentrated near the level $z = z_0$ and decrease exponentially with the distance from this level, see (10.44).

To estimate the order of magnitude of the parameters encountered in reality, we can consider the contact of two homogeneous half spaces with a relative density difference of $\Delta\rho/\rho_0 = (\rho_2 - \rho_1)/\rho_0 \simeq 10^{-3}$ which is typical for the ocean. Then from (10.45) we obtain for the waves with period $T = 2\pi/\omega = 1200$ s, $k = 2\pi/\lambda = (\omega^2/g)2\rho_0/\Delta\rho$, $\lambda = 1.2$ km.

10.5.2 Simple Model of an Oceanic Waveguide

To illustrate the general results we consider the following simple model of a waveguide in the ocean (Fig. 10.7). In the upper layer adjacent to the surface

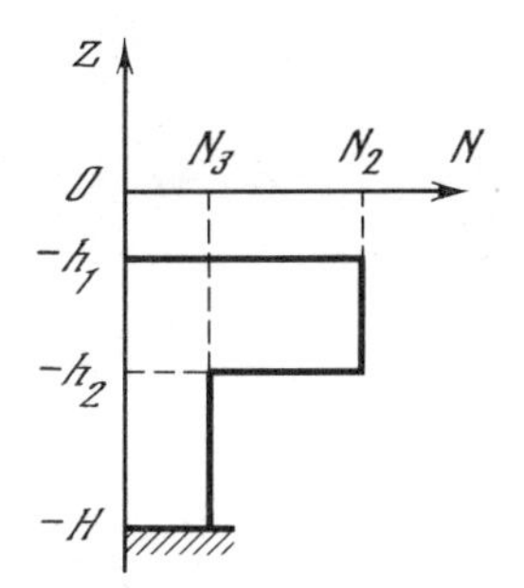

Fig. 10.7. Simple model of a waveguide for internal waves

$(-h_1 < z < 0)$ where the Väisälä frequency is small due to the mixing processes, we assume $N_1 = 0$. The thermocline is visualized by a layer $-h_2 < z < -h_1$ with a constant Väisala frequency N_2. In the bottom layer $(-H < z < -h_2)$, the Väisälä frequency is assumed constant and equal to $N_3 < N_2$. The equilibrium density of the liquid $\rho_0(z)$ is considered continuous at the interfaces of the layers $z = -h_1$ and $z = -h_2$.

The solution of (10.59) in the surface layer $(N_1 = 0)$ is

$$\phi_1(z) = a_1 \sinh kz + b_1 \cosh kz .$$

Using the boundary condition (10.60) at the surface, we find $b_1 = a_1\omega^2/gk$ so that

$$\phi_1(z) = a_1(\sinh kz + \delta \cosh kz), \qquad \delta = \omega^2/gk . \tag{10.64}$$

For the thermocline layer $(-h_2 < z < -h_1)$, the general solution of (10.59) is

$$\phi_2(z) = a_2 \sin[k\alpha_2(z + h_1)] + b_2 \cos[k\alpha_2(z + h_1)], \qquad \alpha_2 = \sqrt{(N_2/\omega)^2 - 1} . \tag{10.65}$$

Writing a solution for the bottom layer in the form

$$\phi_3(z) = a_3 \sinh[k\alpha_3(z + H), \qquad \alpha_3 = \sqrt{1 - (N_3/\omega)^2} , \tag{10.66}$$

it satisfies the condition at the bottom $(z = -H)$.

Boundary conditions at the interfaces $z = -h_1$ and $z = -h_2$ are those of continuity of the vertical velocities $w = \partial\zeta/\partial t$ and of the total pressures $p_0 + p$. Such conditions were considered above, see (10.44a). Taking into account the continuity of $\rho_0(z)$, we obtain from these conditions

$$\begin{aligned} &\phi_1(-h_1) = \phi_2(-h_1), \qquad \phi_1'(-h_1) = \phi_2'(-h_1), \\ &\phi_2(-h_2) = \phi_3(-h_2), \qquad \phi_2'(-h_2) = \phi_3'(-h_2). \end{aligned} \tag{10.67}$$

Here, as usual, the prime refers to the derivative with respect to z. Substitution of $\phi_j(z), j = 1, 2, 3$ into (10.67) leads to a homogeneous system of equations with respect to the amplitudes a_j, b_j:

$$(-\sinh\beta_1 + \delta\cosh\beta_1)a_1 = b_2\,, \qquad (\cosh\beta_1 - \delta\sinh\beta_1)a_1 = \alpha_2 a_2\,,$$
$$-a_2\sin\sigma + b_2\cos\sigma = a_3\sinh\beta_3\,, \qquad a_2\cos\sigma + b_2\sin\sigma = a_3(\alpha_3/\alpha_2)\cosh\beta_3\,, \tag{10.68}$$

where the following abbreviations were introduced:

$$\beta_1 = kh_1\,, \quad \sigma = k\alpha_2 d_2\,, \quad d_2 = h_2 - h_1\,, \quad \beta_3 = k\alpha_3 d_3\,, \quad d_3 = H - h_2\,. \tag{10.69}$$

Equating the determinant of (10.68) to zero or simply eliminating the quantities a_1, b_2 and a_3, we obtain

$$\begin{aligned}&[1 - \delta\tanh\beta_1 + \alpha_2^2(\delta - \tanh\beta_1)\alpha_3^{-1}\tanh\beta_3]\sin\sigma \\ &\quad - \alpha_2[\delta - \tanh\beta_1 - (1 - \delta\tanh\beta_1)\alpha_3^{-1}\tanh\beta_3]\cos\sigma = 0\end{aligned} \tag{10.70}$$

for the determination of the eigenvalues $k_n(\omega)$. From (10.69) we obtain $k = \sigma/\alpha_2 d_2$. Let us now express β_1, β_3 and δ in (10.70) in terms of σ, in such a way obtaining an equation for σ. The roots of these equations $\sigma_n(\omega)(n = 1, 2, \dots)$ determine the eigenvalues $k_n(\omega) = \sigma_n(\omega)/\alpha_2 d_2$ at a fixed frequency ω.

10.5.3 Surface Mode. "Rigid Cover" Condition

The quantities α_2 and σ become imaginary $[\alpha_2 = \mathrm{i}\sqrt{1 - (N_2/\omega)^2}]$ if $\omega > N_2 > N_3$. In this case, $\sin\sigma = \mathrm{i}\sinh|\sigma|$, $\cos\sigma = \cosh|\sigma|$ and no terms within oscillatory dependence on σ appear in (10.70). Therefore, this equation has only one root $k_0(\omega)$ corresponding to a surface wave. As mentioned earlier, in the case of deep water ($k_0 H \gg 1$), a surface wave (10.32) is the solution of our equations for arbitrary dependence $N(z)$.

Now we show that in our model even for $\omega < N_2$, the root σ_0 of (10.70), with the smallest modulus corresponds to a surface wave. Suppose that σ_0 as well as β_1 and β_3 are small, and expand the left-hand side of (10.70) in a Taylor series, keeping only linear terms:

$$\sigma_0 - \alpha_2(\delta - \beta_1 - \beta_3/\alpha_3) = 0\,.$$

Taking (10.69) into account and the expression (10.64) for δ, we obtain the dispersion relation $\omega^2 = gHk^2$ for a surface wave in shallow water. Here we have for σ_0,

$$\sigma_0^2 = k_0^2\alpha_2^2 d_2^2 = \frac{\omega^2}{gH}\left(\frac{N_2^2}{\omega^2} - 1\right)d_2^2 < \frac{N_2^2 d_2}{gH} = \varepsilon\frac{d_2}{H} \ll 1\,,$$

where

$$\varepsilon = \frac{N_2^2 d_2}{g} = -\rho_0^{-1}\frac{d\rho_0}{dz}d_2 \simeq \frac{\Delta\rho}{\rho_0} \ll 1\,. \tag{10.71}$$

Here, $\Delta\rho$ (about 10^{-3} g/cm^3 in reality) is the increment of equilibrium density, $\rho_0(z)$, in the thermocline. Hence, our assumption that σ_0 is small is justified in the case of the ocean. Note that the condition for shallow water was also satisfied

$$k_0^2 H^2 = \omega^2 H/g < \frac{N_2^2 d_2}{g} = \varepsilon \ll 1$$

and the parameter $\delta = \omega^2/gk_0 = k_0 H$ is small, too.

For eigenvalues of higher orders $k_n(\omega)$ (internal modes $n = 1, 2, \ldots$), we have $\sigma_n \gg \sigma_0$. These roots correspond to consecutive branches of $\tan\sigma$ if we write (10.70) in the form $\tan\sigma = F(\sigma)$ and solve it graphically (Fig. 10.8). For the wave numbers we have $k_n = \sigma_n/\alpha_2 d_2 \gg k_0$, and the parameter δ becomes even smaller. Hence, two significant conclusions follow:

i) For the same frequency, internal waves are much shorter (the wavelength is less) than the surface wave. The phase velocity of the latter is much greater than that of internal waves.

ii) For internal waves the "*rigid cover*" condition ($w|_{z=0} = 0$) is well satisfied. Indeed, denoting the amplitude of an internal wave in the thermocline by $A = \sqrt{a_2^2 + b_2^2}$, (10.65), one can easily obtain from (10.68) an estimation for a_1:

$$a_1 \simeq A\alpha_2(\cosh^2\beta_1 + \alpha_2^2\sinh^2\beta_1)^{-1/2}\,.$$

Hence, we have at the free surface $z = 0$, according to (10.64),

$$\phi_1(0) \simeq a_1\sigma \simeq A\,\delta\alpha_2(\cosh^2\beta_1 + \alpha_2\sinh^2\beta_1)^{-1/2} \ll A\,.$$

10.5.4 Internal Modes

After what was said above, we consider internal waves in the "rigid cover" approximation assuming $\delta = 0$. The dispersion relation (10.70) now takes the form

$$\tan\sigma = F(\sigma), \qquad F(\sigma) = \left(\tanh\beta_1 + \frac{\tanh\beta_3}{\alpha_3}\right)\left(\alpha_2^2\tanh\beta_1\frac{\tanh\beta_3}{\beta_3} - 1\right)^{-1}. \tag{10.72}$$

The right-hand side is the function of σ only and vanishes at $\sigma = 0$. Let us consider first the case of rather high frequencies $N_3 < \omega < N_2$ where $\beta_3 = k\sqrt{1 - N_3^2/\omega^2}\,d_3$ is real. In this case, the right-hand side of (10.72) is negative

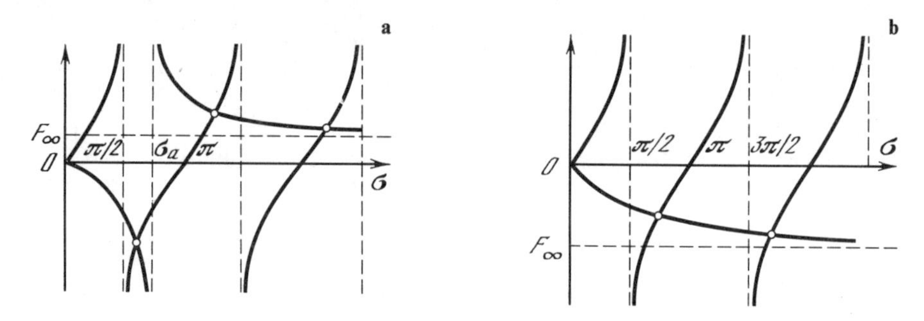

Fig. 10.8a, b. Graphical solution of the dispersion relation for internal waves

for $\sigma \ll 1$ and $F(\sigma) \to F_\infty = \alpha_2(1+\alpha_3)/(\alpha_2^2 - \alpha_3)$ for $\sigma \to \infty$. If $\alpha_2^2 > \alpha_3$ or with (10.65, 66),

$$\omega^2 < \omega_a^2 = N_2^2/(2 - N_3^2/N_2^2), \tag{10.73}$$

the function $F(\sigma)$ asymptotically approaches the value $\sigma = \sigma_a$ which is determined from the relation $\alpha_2^2 \tanh \beta_1 \tanh \beta_3|_{\sigma=\sigma_a} = \alpha_3$.

In Fig. 10.8 the right- and left-hand sides of (10.72) are plotted for $\omega < \omega_a$ and $\omega > \omega_a$. The points of intersection of the curves correspond to the roots of (10.72) $\sigma_n(\omega)$ specifying the eigenvalues $k_n(\omega) = \sigma_n(\omega)/\alpha_2 d_2$. One can see from the figure that the values of $\sigma_n(\omega)$ are confined to the following intervals:

$$\pi(n - 1/2) < \sigma_n(\omega) < n\pi \quad \text{for} \quad \omega > \omega_a \quad \text{or} \quad \omega < \omega_a \quad \text{but} \quad \sigma < \sigma_a$$

$$\pi(n-1) < \sigma_n(\omega) < \pi(n - 1/2) \quad \text{for} \quad \omega < \omega_a \quad \text{and} \quad \sigma > \sigma_a .$$

Using (10.65, 69) we can write the dispersion relation (relationship between ω and k) for internal waves in implicit form:

$$\omega^2 = N_2^2 k^2 d_2^2 [k^2 d_2^2 + \sigma_n^2(\omega)]^{-1} . \tag{10.74}$$

Hence we conclude that for $k \to \infty$ (short-wavelength waves), $\omega \to N_2$, i.e., the frequencies of all internal modes become close to the maximum Väisälä frequency. The right-hand side of (10.72) is then very small $[F(\sigma) \sim \alpha_2 \ll 1]$ and $\sigma_n \to n\pi$.

The eigenfunctions $\phi(z)$ in the frequency range $N_3 < \omega < N_2$ become maximum in the thermocline and decrease exponentially [see (10.64, 66) for $\delta = 0$] while moving away from it. The higher the frequency, the more pronounced is this decrease, i.e., the more pronounced is the concentration of the internal waves in the thermocline.

At low frequencies ($\omega < N_3$), $\alpha_3 = \sqrt{1 - N_3^2/\omega^2}$ as well as β_3 become imaginary. Thus, $\sinh[k\alpha_3(z+H)]$ in (10.66) transforms into $\mathrm{i}\sin[k|\alpha_3|(z+H)]$ and the eigenfunction becomes oscillatory for $z < -h_2$, too. In the dispersion relation (10.72), one should also set $\tanh \beta_3 = \mathrm{i} \tan|\beta_3|$ so that the right-hand side is an oscillatory function of σ. We confine ourselves to an analysis of the case of very low frequencies ($\omega \to 0$, *hydrostatic approximation*) when

$$\alpha_2 \simeq N_2/\omega, \quad \alpha_3 \simeq \mathrm{i}N_3/\omega, \quad \sigma \simeq kN_2 d_2/\omega = N_2 d_2/c_{\mathrm{Ph}},$$
$$\beta_3 \simeq \mathrm{i}N_3 d_3/c_{\mathrm{Ph}} = \mathrm{i}\sigma N_3 d_3/N_2 d_2, \quad k = \sigma\omega/N_2 d_2 \to 0,$$

where $c_{\mathrm{Ph}} = \omega/k$ is the phase velocity of the waves. Rewriting the dispersion relation (10.72) in the form

$$\tan\sigma = (N_3 h_1 + c_{\mathrm{Ph}} \tan|\beta_3|)(N_2 h_1 \tan|\beta_3| - c_{\mathrm{Ph}} N_3/N_2)^{-1}, \tag{10.75}$$

we note that the latter includes only c_{Ph} but not the frequency. This means that at low frequencies, the phase velocities of internal modes [i.e., roots of (10.75)] are independent of frequency, i.e., *low-frequency internal modes propagate along the waveguide without dispersion* [$\omega_n = (c_{\mathrm{Ph}})_n k$].

The eigenfunctions $\phi(z)$ in the upper layer $-h_1 < z < 0$ decrease linearly while moving away from thermocline toward the surface at low frequencies, and become zero at $z = 0$. In the layer below the thermocline, however, the amplitudes of the internal modes can be comparable with those on the thermocline. Moreover, the first internal mode has a maximum amplitude below the thermocline if $N_3 d_3 > N_2 d_2$ ($|\beta_3| > \sigma$), as known for the ocean.

The analysis carried out above allows dispersion curves of the system of surface and internal waves in the ocean to be plotted (Fig. 10.9). It can be shown (Exercise 10.6.17) that in the general case of arbitrary $N(z)$, each dispersion curve is monotonically rising ($c_g = d\omega/dk > 0$). At low frequencies, these curves are linear and their slope decreases as the number of modes increases. Note, however, that in the ocean and in the atmosphere at very low frequencies comparable with the Earth's rotation frequency, the shapes of the dispersion curves will be different (Chap. 11).

In a model for the atmosphere one should assume the layer $z > -h_1$ (Fig. 10.7) to be unbounded in the positive direction with a nonzero Väisälä frequency $N_1 < N_2$. The waveguide's (trapped) modes occur only at frequencies

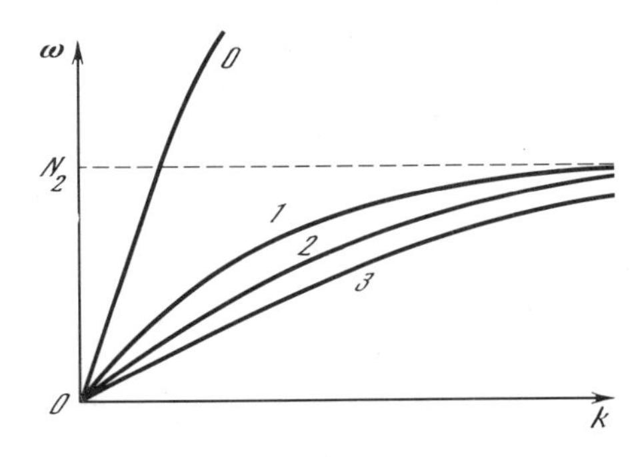

Fig. 10.9. Schematic plot of dispersion curves for surface ($n = 0$) and internal ($n = 1, 2, 3$) modes

$\omega > N_1$. At lower frequencies, waves propagating in the vertical direction (non-trapped by a waveguide) can exist in the upper atmosphere. The energy travels upward from an excitation area (usually low atmospheric layers) in this case.

10.6 Exercises

10.6.1. A train of gravity surface waves is produced at $x = 0$, $w|_{x=0} = F(t)\exp(-\mathrm{i}\omega_0 t)$, containing $\mathcal{N} \gg 1$ periods so that $F(t) = 0$ if $t < 0$ or $t > T = (2\pi/\omega_0)\mathcal{N}$ and $F(t) = 1$ if $0 < t < T$. Assuming the water is deep, determine a) the number of wave crests, $\mathcal{N}_1$, at the surface, which an observer at rest will see; b) how many oscillations an observer in a boat will experience during the wave train passes under him.

Solution: The wave train propagates with the group velocity, hence,

$$w(x, t) = F(t - c_g^{-1}x)\exp[\mathrm{i}(k_0 x - \omega_0 t)], \qquad k_0 = \omega_0^2/g.$$

For an observer at rest, the train occupies the range $L = Tc_g$ at any time t. The number of crests is

$$\mathcal{N}_1 = Lk_0/2\pi = Tk_0 c_g/2\pi = 2\pi(\mathcal{N}/\omega_0)k_0 c_g/2\pi = \mathcal{N} c_g/c_{\mathrm{Ph}}.$$

For waves in deep water, $c_g = c_{\mathrm{Ph}}/2$ hence, $\mathcal{N}_1 = \mathcal{N}/2$. For an observer in a boat at a point x, the wave train passes by during the time T, hence the boat experiences $\mathcal{N}_2 = T\omega_0/2\pi = \mathcal{N}$ oscillations.

10.6.2. An observer is on the bank of a river which is flowing with that speed v_0. Under what conditions does he observe an immobile standing wave ($\omega = 0$) with crests across the river?

Solution: The standing wave for an immobile observer can be written as

$$\zeta = a\cos(kx)\exp(-\mathrm{i}\omega t) = \frac{a}{2}\{\exp[\mathrm{i}(kx - \omega t)] + \exp[-\mathrm{i}(kx + \omega t)]\}.$$

In a coordinate system moving together with the water $x' = x - v_0 t$, we have

$$\zeta = \frac{a}{2}\{\exp[\mathrm{i}(kx' - \omega_1' t)] + \exp[-\mathrm{i}(kx' + \omega_2' t)]\},$$

$$\omega_1' = \omega - kv_0, \qquad \omega_2' = \omega + kv_0.$$

In this coordinate system, however, the usual dispersion relation $(\omega - kv_0)^2 = gk = (\omega + kv_0)^2$ must be satisfied which is possible only for $\omega = 0$ and $k = g/v_0^2$. Hence,

$$\zeta = a\cos(gx/v_0^2) = a\cos[k(x' + v_0 t)].$$

Thus, in the coordinate system moving with the water we have a wave running against the flow of the river with the same velocity as that of the flow. An observer on the shore observes the motionless ($\omega = 0$) disturbed free surface with a spatial period of disturbance $2\pi/k = 2\pi v_0^2/g$.

10.6.3. Show that the centre of gravity of fluid remains at a constant depth in a progressing surface wave whereas it oscillates with twice the frequency in a standing one.

Solution: The centre of gravity of a fluid at rest ($\zeta = 0$) is at the depth $z_0 = -H/2$ (Fig. 10.1). Determine the vertical displacement of the centre of gravity Δz of a fluid column of width λ when $\zeta \neq 0$. According to the definition of the centre of gravity,

$$\int_0^\lambda dx \int_0^\zeta (z - z_0 - \Delta z)\, dz = 0\,.$$

Integrating with accounting for the periodicity of $\zeta(\int_0^\lambda \zeta\, dx = 0)$, we find that $\Delta z = (2H\lambda)^{-1} \int_0^\lambda \zeta^2\, dz$. For a progressive wave $\zeta = a\cos(kx - \omega t)$, we obtain $\Delta z = a^2/4H$. For a standing wave $\zeta = a\cos(kx)\cos\omega t$,

$$\Delta z = (a^2/8H)(1 + \cos 2\omega t)\,.$$

10.6.4. Describe the field of a harmonic gravity-capillary wave reflected at a rigid vertical wall.

Solution: Let the wall at $x = 0$ be perpendicular to the x-axis. Write the incident wave in the form of (10.17) with $\phi(z)$ from (10.18):

$$w_+ = b\phi(z)\exp[\mathrm{i}(k_x x + k_y y - \omega t)]\,,$$

where ω and $k = \sqrt{k_x^2 + k_y^2}$ satisfy the dispersion relation (10.39). The frequency ω (wave number k) and the projection k_y of the wave vector on the boundary are the same for a reflected wave, so the latter can be written in the form

$$w_- = Vb\phi(z)\exp[\mathrm{i}(-k_x x + k_y y - \omega t)]\,,$$

where V is the reflection coefficient. At a rigid wall the total normal velocity u_x is zero. According to (10.24), which also holds for gravity-capillary waves, we have

$$(u_+)_x = \mathrm{i}b(k_x/k^2)\phi'(z)\exp[\mathrm{i}(k_x x + k_y y - \omega t)]\,,$$
$$(u_-)_x = -\mathrm{i}b(k_x/k^2)\phi'(z)\exp[\mathrm{i}(-k_x x + k_y y - \omega t)]\,.$$

From the condition $[(u_+)_x + (u_-)_x]_{x=0} = 0$, we find $V = 1$. Hence, the total wave field for $x < 0$ is

$$w = w_+ + w_- = 2b\phi(z)\cos(k_x x)\exp[i(k_y y - \omega t)].$$

In the case of normal incidence ($k_y = 0$), we obtain a standing surface wave.

10.6.5. Determine the frequencies of the normal modes of vibration of an incompressible fluid in a rectangular basin of depth H and of horizontal dimensions L and D. The basin's walls are rigid. Find the lowest frequency of the oscillations.

Solution: The function $w = b\phi(z)f(x, y)\exp(-i\omega t)$ is the general harmonic solution of (10.14) for waves in the basin, $f(x, y)$ being a combination of the exponential functions $\exp[i(\pm k_x x \pm k_y y)]$, $\omega^2 = gk \tanh kH$. It was shown in the previous exercise that the reflection coefficient of surface waves at a vertical rigid boundary is unity. Hence, $w(x, y, z, t)$ must be a standing wave both in the x and y-directions. If we assume $f(x, y) = \cos(k_x x)\cos k_y y$, then we find the vanishing of the particle velocities $u_x = k^{-2}\,\partial^2 w/\partial x\,\partial z$ and $u_y = k^{-2}\,\partial^2 w/\partial y\,\partial z$ normal to the basin's walls $x = 0$ and $y = 0$, respectively. Requiring these velocities to vanish on the opposite walls $u_x|_{x=L} = 0$ and $u_y|_{y=D} = 0$, we obtain $\sin k_x L = 0$, $(k_x)_n = n\pi/L$, $n = 0, 1, 2, \ldots$ and $\sin k_y D = 0$, $(k_y)_m = m\pi/D$, $m = 0, 1, 2, \ldots$, respectively. Whence for wave numbers and frequencies of eigenoscillations we have

$$k_{nm} = \pi\sqrt{n^2/L^2 + m^2/D^2}\,, \qquad \omega_{nm}^2 = gk_{nm}\tanh k_{nm}H\,.$$

Taking $L > D$, say, we find the lowest frequency $\omega_{\min}^2 = \omega_{10}^2 = (g\pi/L)\tanh(\pi H/L)$, $k_{10} = \pi/L$. For example, for $H = 40$ m, $L = 1$ km, we find $f_{\min} = 10^{-2}$ Hz.

10.6.6. Obtain the dispersion relation for waves in a liquid layer of thickness H, consisting of two liquids with the densities ρ_1 and ρ_2 which have their interface at a depth h. Assume the upper boundary of the layer ($z = 0$) to be free, and the lower one ($z = -H$) to be perfectly rigid.

Solution: Look for a solution to (10.14) in the form $w = \phi(z)\exp[i(\boldsymbol{k}\cdot\boldsymbol{r} - \omega t)]$, where the function $\phi(z)$ satisfies $\phi'' - k^2\phi = 0$. For $-h < z < 0$ and $-H < z < -h$, the general solution of the last equation is

$$\phi_1(z) = b_1 \sinh kz + c_1 \cosh kz\,,$$

$$\phi_2(z) = b_2 \sinh k(z + H)\,,$$

respectively. At the surface $z = 0$, the function $\phi_1(z)$ satisfies the condition $\phi_1' - (gk^2/\omega^2)\phi_1 = 0$. Whence we have the relation $b_1 = gkc_1/\omega^2$. The conditions for $\phi_1(z)$ and $\phi_2(z)$ at the interface result from (10.44a):

$$\phi_1(-h) = \phi_2(-h),$$
$$gk^2 \Delta\rho\phi_j(-h) = \omega^2[\rho_2\phi_2'(-h) - \rho_1\phi_1'(-h)], \quad j = 1, 2, \quad \Delta\rho = \rho_2 - \rho_1,$$

whence we obtain relationships between the constant b_1, c_1, and b_2:

$$-b_1 \sinh kh + c_1 \cosh kh = b_2 \sinh kd, \quad d = H - h,$$
$$gk^2 \Delta\rho b_2 \sinh kd = \omega^2(\rho_2 b_2 \cosh kd - \rho_1 b_1 \cosh kh + \rho_1 c_1 \sinh kh)k.$$

Taking into account the relationship between b_1 and c_1, we find

$$c_1 = \frac{\sinh(kd)}{\cosh(kh) - (gk/\omega^2)\sinh(kh)} b_2 = \frac{\sinh(kd)}{\cosh(kh)} \frac{\omega^2}{\omega^2 - gk\tanh(kh)} b_2.$$

The dispersion relation has the form

$$(\omega^2 - gk\tanh kh)\left(\omega^2 - \frac{\Delta\rho}{\rho_2} gk\tanh kd\right)$$
$$= \frac{\rho_1}{\rho_2}\omega^2 \tanh(kd)[gk - \omega^2\tanh(kh)].$$

10.6.7. Examine the dispersion relation obtained in the previous exercise for the following cases: a) $kh \gg 1$, $kd \gg 1$ and b) $\Delta\rho/\rho_2 \ll 1$.

Solution: The dispersion relation under consideration is quadratic with respect to ω^2. Consequently, it contains two branches of dispersion curves $\omega_0 = \omega_0(k)$ and $\omega_1 = \omega_1(k)$. In case (a), assuming $\tanh kh = \tanh kd = 1$, we easily find $\omega_0^2 = gk$ which is a surface wave in deep water and $\omega_1^2 = gk\,\Delta\rho/(\rho_1 + \rho_2)$ which is a wave at an interface of two media. In case (b) $(\rho_1 \simeq \rho_2 = \rho_0)$ we write the equation for ω^2 in the explicit form:

$$[1 + \tanh(kh)\tanh(kd)]\omega^4 - gk(\tanh kh + \tanh kd)\omega^2$$
$$+ g^2k^2(\Delta\rho/\rho_0)\tanh(kh)\tanh(kd) = 0.$$

Its roots are

$$(\omega^2)_{0,1}$$
$$= \frac{gk}{2[1+\tanh(kh)\tanh(kd)]}\left\{\tanh(kh)+\tanh(kd)\right.$$
$$\left.\pm\sqrt{[\tanh(kh)+\tanh(kd)]^2 - 4\frac{\Delta\rho}{\rho_0}\tanh(kh)\tanh(kd)[1+\tanh(kh)\tanh(kd)]}\right\}$$
$$= gk\frac{\tanh[k(h+d)]}{2}\left\{1 \pm \sqrt{1 - \frac{4\Delta\rho}{\rho_0}\frac{\coth[k(h+d)]}{\coth(kh)+\coth(kd)}}\right\}.$$

Expanding the square root in the Taylor series, we obtain $\omega_0^2 = gk \tanh kH$ which is a surface wave in a layer and

$$\omega_1^2 = \frac{\Delta\rho}{\rho_0} gk(\coth kh + \coth kd)^{-1},$$

an interval wave at the interface influenced by the boundaries $z = 0$ and $z = -H$.

10.6.8. Obtain the dispersion relation for plane internal waves in a liquid with a constant Väisälä frequency without the Boussinesq approximation. Define the criterion of applicability of the latter in this case.

Solution: Look for a solution of (10.48) with $N^2 = -g\rho_0^{-1} \cdot d\rho_0/dz = \text{const}$ in the form of a harmonic wave $w = b\phi(z)\exp[\mathrm{i}(\boldsymbol{k} \cdot \boldsymbol{r} - \omega t)]$, $\boldsymbol{k} = \{k_x, k_y\}$, $\boldsymbol{r} = \{x, y\}$. For $\phi(z)$, we obtain

$$\phi'' - 2\mu\phi' + k^2(N^2/\omega^2 - 1)\phi = 0, \quad \mu = N^2/2g,$$

Substitution of $\phi = \psi \exp(\mu z)$ leads to an equation without a first derivative:

$$\psi'' + [k^2(N^2/\omega^2 - 1) - \mu^2]\psi = 0,$$

whose solution is the function $\psi(z) = b\exp(\mathrm{i}k_z z)$, $k_z^2 = k^2(N^2/\omega^2 - 1) - \mu^2$. As a result, we have for a solution of (10.48),

$$w(\boldsymbol{r}, z, t) = b\exp[\mu z + \mathrm{i}(\boldsymbol{\varkappa} \cdot \boldsymbol{R} - \omega t)], \quad \boldsymbol{\varkappa} = \{\boldsymbol{k}, k_z\}, \quad \boldsymbol{R} = \{\boldsymbol{r}, z\}$$

with the dispersion relation $\omega^2 = N^2k^2/(\varkappa^2 + \mu^2)$. The latter coincides with (10.51) obtained in the Boussinesq approximation if $\mu^2 = N^4/4g^2 = (2\rho_0)^{-2}(d\rho_0/dz)^2$ can be neglected compared with $\varkappa^2$, i.e., when $\varkappa \gg (2\rho_0)^{-1} d\rho_0/dz$. This is the criterion of applicability of the Boussinesq approximation for linear internal waves. In an ocean, this criterion is satisfied for wavelengths up to hundreds of kilometers.

10.6.9. Determine the group-velocity direction for waves with the dispersion relation obtained in Exercise 10.6.8.

Solution: Calculate $\boldsymbol{c}_g = \{(\boldsymbol{k}/k)\,\partial\omega/\partial k, \partial\omega/\partial k_z\}$ using the dispersion relation $\omega = Nk(\varkappa^2 + \mu^2)^{-1/2}$:

$$(\partial\omega/\partial k) = N^2(\varkappa^2 + \mu^2)^{-3/2}(k_z^2 + \mu^2), \quad (\partial\omega/\partial k_z) = -N(\varkappa^2 + \mu^2)^{-3/2}kk_z.$$

Write $\boldsymbol{c}_g$ in the form $\boldsymbol{c}_g = \boldsymbol{A}_\perp + \boldsymbol{A}_\parallel$, where $\boldsymbol{A}_\perp = Nk_z(\varkappa^2 + \mu^2)^{-1/2}\varkappa^{-2}[(k_z/k)\boldsymbol{k} - k\nabla z]$ and $\boldsymbol{A}_\parallel = Nk(\varkappa^2 + \mu^2)^{-3/2}(\mu/\varkappa)^2\boldsymbol{\varkappa}$ are the vectors perpendicular and parallel to the wave-propagation direction $\boldsymbol{\varkappa}$, respectively. Denoting with α the angle between $\boldsymbol{c}_g$ and the direction normal to $\boldsymbol{\varkappa}$ ($\alpha = 0$ in the Boussinesq approximation, Sect. 10.4.3), we obtain $\tan\alpha = |A_\parallel/A_\perp| = k\mu^2/k_z(\varkappa^2 + \mu^2)$. Note that α is large when a wave propagates in a direction close to the horizontal ($k_z \simeq 0$). In this case, the group velocity turns out to be finite (due to $A_\parallel$), whereas when calculated according to (10.55) in the Boussinesq approximation it is very small. Note also that a wave train weakly modulated in the propagation direction will travel slowly in the $\varkappa$-direction at a propagation speed of the envelope equal to

$$|\boldsymbol{A}_{||}| = (Nk/\varkappa)(\varkappa^2 + \mu^2)^{-3/2}\mu^2 = c_{\mathrm{Ph}}\mu^2/(\varkappa^2 + \mu^2), \quad c_{\mathrm{Ph}} = \omega/\varkappa .$$

10.6.10. Show that a solution in the form of a surface wave in a homogeneous ($\rho_0 = \mathrm{const}$) infinitely deep fluid satisfies the equation for internal waves (10.48) and the boundary conditions (10.15) for an arbitrary dependence $N(z)$.

Solution: A direct substitution of the expression for a surface wave

$$w = \exp[kz + \mathrm{i}(\boldsymbol{k} \cdot \boldsymbol{r} - \omega t)], \quad \omega^2 = gk$$

into (10.48) yields the identity

$$\{-\omega^2[-k^2 + k^2 - N^2(z)k/g] - N^2(z)k^2\}w = g^{-1}N^2(z)k(\omega^2 - gk)w = 0 .$$

Thus, the equation is satisfied. The boundary condition is verified in an analogous way:

$$\left(\frac{\partial^3 w}{\partial t^2\,\partial z} - g\Delta_- w\right)_{z=0} = -k^2(\omega^2 - gk)w|_{z=0} = 0 .$$

10.6.11. Find the smallest root σ_1 (for the first internal model) of the dispersion relation (10.72) under the condition that the lower layer in Fig. 10.7 is homogeneous ($N_3 = 0$) and the thickness of the middle layer $d_2 \to 0$ ($N_2^2 d_2$ remains constant). Consider in what cases a layer of finite thickness d_2 can be replaced by an interface of two liquids with different densities.

Solution: For $N_3 = 0$, we have from (10.66, 69) $\alpha_3 = 1$, $\beta_3 = k d_3$. It is more appropriate in this case to consider the dispersion relation (10.72) for a fixed wave number k, therefore we rewrite this relation in the form

$$\tan\sigma = \sigma k d_2 \frac{\tanh\beta_1 + \tanh\beta_3}{\sigma^2\tanh(\beta_1)\tanh(\beta_3) - k^2 d_2^2} .$$

As $d_2 \to 0$, the smallest root of this equation also vanishes $[\sigma_1(k) \to 0]$. Then expanding $\tan\sigma$ in a series and confining ourselves just to the main term, we obtain $\sigma_1^2 \simeq k d_2(\coth\beta_1 + \coth\beta_3)$. Taking into account (10.69) for σ, we find the dispersion relation for the first internal mode

$$\omega_1^2(k) = N_2^2 d_2 k(\coth\beta_1 + \coth\beta_3)^{-1} .$$

Further, using the expression for N_2^2, we obtain at the limit $d_2 \to 0$:

$$N_2^2 d_2 = -g\,d_2\rho_0^{-1}\,d\rho_0/dz \to g\,\Delta\rho/\rho_0 ,$$

where $\Delta\rho = -(d\rho_0/dz)\,d_2 = \rho_3 - \rho_1$ is the difference between the fluids' densities at the lower and upper boundaries of the thermocline. Then the following formula results:

$$\omega_1^2(k) = g(\Delta\rho/\rho_0)k(\coth\beta_1 + \coth\beta_3)^{-1}$$

which is similar to the dispersion relation for waves at the interface (Exercise 10.6.7). If d_2 is small but finite, then the root $\sigma_1(k)$ should be small for the validity of the last formula. As is seen from the relation for σ_1^2, it is the case when a) $k\,d_2 \ll 1$ ($d_2/\lambda \ll 1$), where the thickness of the layer is small compared with the wavelength, or b) $k\,d_2/\beta_1 = d_2/h_1 \ll 1$ and $k\,d_2/\beta_3 = d_2/d_3 \ll 1$, where the thickness of the inhomogeneous layer is small compared with those of the homogeneous layers. Under these conditions, an inhomogeneous layer of finite thickness can be replaced by an interface if only the first internal mode is of interest. Such a replacement is impossible for the higher modes.

10.6.12. Calculate the energy of the nth mode in a vertical column of liquid confined between the surface and the bottom. The width of the column is λ along the propagation direction and one in the perpendicular direction.

Solution: Taking the x-axis along the propagation direction, we have for the kinetic energy

$$E_\mathrm{k} = (\rho_0/2)\int_0^{\lambda}\int_{-H}^{0} v^2\,dz\,dx\,,$$

$v^2 = |u|^2 + |w|^2$. Using the real parts of (10.61, 62), for u and w and integrating over x, we obtain

$$E_\mathrm{k} = \frac{\rho_0\lambda_n}{4}|b_n|^2\int_{-H}^{0}[\phi_n^2 + k^{-2}(\phi_n')^2]\,dz\,, \qquad \lambda_n = 2\pi/k_n\,.$$

Integrating by parts the second term and taking (10.60) into account, we obtain

$$E_\mathrm{k} = \frac{\rho_0|b_n|^2\lambda_n}{4\omega^2}\left[g\phi_n^2(0) + \int_{-H}^{0} N^2(z)\phi_n^2(z)\,dz\right].$$

To calculate the potential energy we note that its density ε_p can be defined as the work done against the Archimedes (buoyancy) force when a fluid particle is transferred from level $z-\zeta$ to level z. By virtue of the incompressibility of the fluid, we have

$$\varepsilon_\mathrm{p} = g\int_{z-\zeta}^{z}[\rho_0(z-\zeta) - \rho_0(\xi)]\,d\xi\,,$$

where $\rho_0(z-\zeta)$ is the density of the transferred particle and $\rho_0(\xi)$ is the density of the surrounding fluid. Expanding the integrand in the Taylor series with respect to $\xi - z + \zeta$ and integrating, we obtain

$$\varepsilon_{\mathrm{p}} = -\frac{g}{2}\frac{d\rho_0}{dz}\zeta^2 = \frac{g}{2}\rho\zeta\,.$$

Here, $\rho = -(d\rho_0/dz)\zeta$ describes the variation of the fluid density at the point z. The potential energy of the nth mode is now

$$E_{\mathrm{p}} = \int_0^{\lambda}\int_{-H}^{\zeta_0} \varepsilon_{\mathrm{p}}\,dz\,dx = \frac{g}{2}\int_{-H}^{\lambda}\left(\int_0^{\zeta_0}\rho\zeta\,dz + \int_{-H}^{0}\rho\zeta\,dz\right)dx\,,$$

where $\zeta_0 = \zeta(x, y, 0, t)$ is the displacement of the surface. To calculate the first integral over z, note that the density change here is $\rho = \Delta\rho = \rho_0(0)$ and that $\zeta = \zeta_0$. Therefore, we have in analogy to the case of surface waves (Sect. 10.2.4),

$$\int_0^{\zeta_0} \rho\zeta\,dz = \rho_0(0)\zeta_0^2\,.$$

Now, substituting the real parts of ζ_n, ρ_n and ζ_{n0} from (10.62) into the relation for E_{p} and integrating over x, we obtain the same expression for the mode's potential energy as for E_{k} above. The total energy of the mode is the sum of $E_{\mathrm{k}} + E_{\mathrm{p}}$:

$$E_n = \frac{\rho_0|b_n|^2\lambda_n}{2\omega^2}\left[g\phi_n^2(0) + \int_{-H}^{0} N^2(z)\phi_n^2(z)\,dz\right].$$

It is natural enough that for a surface wave $[n = 0, \phi_0(0) = 1]$ in the case of a homogeneous layer ($\rho_0 = \text{const}, N^2 = 0$), the expressions for E_{k} and E_{p} coincide with (10.33).

10.6.13. Determine the reflection and transmission coefficients for a plane harmonic internal wave incident on the horizontal interface of two fluids with Väisälä frequencies N_1 and N_2. The density is assumed continuous across the interface.

Solution: The wave frequency ω and the projection of the wave vector on the interface are the same for the incident and the reflected waves. Therefore, the wave field in the medium of incidence (the "first" medium) can be written as

$$w_1 = b\{\exp[\mathrm{i}(kx + k_{1z}z - \omega t)] + V\exp[\mathrm{i}(kx - k_{1z}z - \omega t)]\}\,.$$

Similarly, we have for the wave in the "second" medium $w_2 = bW\exp[\mathrm{i}(kx + k_{2z}z - \omega t)]$. The internal wave's frequency is completely specified by the angle between the wave vector and the vertical: $\omega = N_1\sin\theta_1 = N_2\sin\theta_2$. Hence,

$k_{1z} = k\sqrt{N_1^2/\omega^2 - 1}$, $k_{2z} = k\sqrt{N_2^2/\omega^2 - 1}$. The reflection V and transmission W coefficients can be found by using the boundary conditions which, in the case of a continuous function $\rho_0(z)$, is the continuity of the vertical velocity $(w_1 - w_2)_{z=0} = 0$ and the pressure, see (10.13),

$$\Delta_- p_1|_{z=0} = \rho_0(0)(\partial w_1/\partial t\, \partial z)_{z=0} = \Delta_- p_2|_{z=0} = \rho_0(0)(\partial^2 w_2/\partial t\, \partial z)_{z=0}.$$

Substitution of w_1 and w_2 into these relations yields the equation for V and W: $1 + V = W$, $k_{1z}(1 - V) = k_{2z}W$. Whence we easily find

$$V = \frac{k_{1z} - k_{2z}}{k_{1z} + k_{2z}} = \frac{\sqrt{N_1^2 - \omega^2} - \sqrt{N_2^2 - \omega^2}}{\sqrt{N_1^2 - \omega^2} + \sqrt{N_2^2 - \omega^2}},$$

$$W = \frac{2k_{1z}}{k_{1z} + k_{2z}} = \frac{2\sqrt{N_1^2 - \omega^2}}{\sqrt{N_1^2 - \omega^2} + \sqrt{N_2^2 - \omega^2}}.$$

10.6.14. Find the coefficients V and W analogous to those considered in the previous exercise assuming that the Väisälä frequencies in the media are the same ($N_1 = N_2 = N$) but the densities ρ_1 and ρ_2 are different ($\rho_1 > \rho_2$).

Solution: In the expressions for w_1 and w_2, one should now set $k_{1z} = k_{2z} = k_z = \sqrt{N^2/\omega^2 - 1}$ (the angle of refraction is equal to the angle of incidence). The continuity condition for the vertical velocity at the interface remains as in the previous exercise. The second boundary condition changes, however. In fact, the continuity condition for the pressure at the interface yields $p_{10}(0) - \rho_1 g\zeta + p_1|_0 = p_{20}(0) - \rho_2 g\zeta + p_2|_0$. Hence taking into account that $w_j|_0 = \partial\zeta/\partial t$, $\Delta_- p_j|_0 = \rho_j(\partial^2 w_j/\partial t\, \partial z)_0$ and $p_{10}(0) = p_{20}(0)$, we obtain

$$\rho_1\left(-g\Delta_- w_1 + \frac{\partial^3 w_1}{\partial t^2\, \partial z}\right)_0 = \rho_2\left(-g\Delta_- w_2 + \frac{\partial^3 w_2}{\partial t^2\, \partial z}\right)_0.$$

Substituting the expressions for w_1 and w_2 into the boundary conditions we find

$$1 + V = W, \qquad -\rho_1 g k^2(1 + V) + \rho_1\omega^2 i k_z(1 - V) = -\rho_2 g k^2 W + \rho_2\omega^2 i k_z W.$$

Finally, for V and W we obtain

$$V = \frac{(i\omega\sqrt{N^2 - \omega^2} - gk)\,\Delta\rho}{(\rho_1 + \rho_2)i\omega\sqrt{N^2 - \omega^2} + gk\,\Delta\rho},$$

$$W = \frac{2i\rho_1\omega\sqrt{N^2 - \omega^2}}{(\rho_1 + \rho_2)i\omega\sqrt{N^2 - \omega^2} + gk\,\Delta\rho}.$$

It is natural enough that for a continuous function $\rho_0(z)$ ($\Delta\rho = 0$), we have $V = 0$, $W = 1$.

10.6.15. Using the results of the previous exercise, obtain a "surface" wave at the interface of two fluids with equal Väisälä frequencies but different densities.

Solution: The required wave corresponds to the purely imaginary values $k_z = \pm i|k_z| = i\sqrt{\omega^2 - N^2}/\omega$, i.e., its frequency $\omega > N$. Choose $k_z = i|k_z|$ for both media and refer to the expressions for w_1 and w_2 of the previous exercise. Then,

$$w_1 = b[\exp(-|k_z|z) + V\exp(|k_z|z)]\exp[i(\boldsymbol{k}\cdot\boldsymbol{r} - \omega t)], \quad z < 0,$$

$$w_2 = bW\exp[-|k_z|z + i(\boldsymbol{k}\cdot\boldsymbol{r} - \omega t)], \quad z > 0.$$

This choice of sign for k_z ensures that the terms proportional to bV and bW decrease exponentially at infinity ($z \to \mp\infty$). The first term in w_1, however, tends to infinity as $z \to -\infty$. It can be eliminated by assuming $b = 0$ and that bV and bW are finite. This means that V and W must be infinite. Equating the denominator in the expression for V in Exercise 10.6.14 to zero for $k_z = i|k_z|$, we obtain the dispersion relation $\omega\sqrt{\omega^2 - N^2} = gk\,\Delta\rho/(\rho_1 + \rho_2)$. Solving this equation with respect to ω, we obtain an explicit form of the dispersion relation $\omega^2(k) = \sqrt{[gk\,\Delta\rho/(\rho_1 + \rho_2)]^2 + N^4/4} + N^2/2$ which at $N = 0$, transforms into (10.45).

10.6.16. Prove the orthogonality condition for internal modes with different ω_n.

Solution: The eigenfunction $\phi_n(z)$ satisfies $\phi_n'' + k^2(N^2/\omega_n^2 - 1)\phi_n = 0$ and the boundary conditions $\phi_n(-H) = \phi_n'(0) - gk^2\phi_n(0)/\omega_n^2 = 0$. Multiply the equation for ϕ_n by ϕ_m and integrate the result over z from $-H$ to 0. Integrating the term $\int_{-H}^{0} \phi_n''\phi_m\, dz$ twice by parts, we obtain the relation

$$\phi_m\phi_n'|_{-H}^{0} - \phi_n\phi_m'|_{-H}^{0} + \int_{-H}^{0} \phi_n\phi_m''\, dz + k^2 \int_{-H}^{0} (N^2/\omega_n^2 - 1)\phi_n\phi_m\, dz = 0.$$

We have

$$\phi_n'(0) = (gk^2/\omega_n^2)\phi_n(0), \quad \phi_m'(0) = (gk^2/\omega_m^2)\phi_m(0),$$

$$\phi_n(-H) = \phi_m(-H) = 0, \quad \phi_m'' = -k^2(N^2/\omega_m^2 - 1)\phi_m,$$

since ϕ_m satisfies the same equation and boundary conditions as ϕ_n with ω_m instead of ω_n. Hence, the last relation becomes

$$gk^2(\omega_n^{-2} - \omega_m^{-2})\left[\phi_n(0)\phi_m(0) + g^{-1}\int_{-H}^{0} N^2\phi_n(z)\phi_m(z)\, dz\right] = 0.$$

The expression in brackets must be zero since $\omega_n^2 \neq \omega_m^2$ and we obtain the orthogonality relation for eigenfunctions $\phi_n(z)$ and $\phi_m(z)$. Normalizing these functions in such a way that the expression in brackets is unity for $n = m$, we obtain

$$\phi_n(0)\phi_m(0) + \int_{-H}^{0} \frac{N^2(z)}{g}\phi_n(z)\phi_m(z)\, dz = \delta_{nm} = \begin{cases} 1, & n = m \\ 0, & n \neq m \end{cases}.$$

With such a normalization, an expression for the total energy (Exercise 10.6.12) can be written as $E_n = (\rho_0 g/2)|b_n|^2 \lambda_n/\omega_n^2 = \rho_0 g \lambda_n |a_n|^2/2$ ($a_n = ib_n/\omega_n$ is the amplitude of vertical displacements).

10.6.17. Derive an expression for the group velocity of internal modes in a liquid layer in terms of the integral of the eigenfunctions $\phi_n(z)$ and prove that $\omega_n(k)$ increases monotonically with increasing k.

Solution: Turn again to the boundary problem for the function $\phi_n(z)$ which depends on $\beta = k^2$ (Sect. 10.5.1):

$$\phi_n''(z, \beta) + \beta\left(\frac{N^2(z)}{\omega_n^2(\beta)} - 1\right)\phi_n(z, \beta) = 0,$$

$$\phi_n(-H, \beta) = \phi_n'(0, \beta) - \frac{g\beta}{\omega_n^2(\beta)}\phi_n(0, \beta) = 0.$$

Together with $\phi_n(z, \beta)$, consider the function $F_n(z, \beta) = \partial\phi_n/\partial\beta$ (the derivative of the eigenfunction with respect to the spectrum parameter) which is a solution of the boundary-value problem obtained by differentiation with respect to β of the equation and the boundary conditions for $\phi_n(z, \beta)$:

$$F_n'' + \beta\left(\frac{N^2}{\omega_n^2} - 1\right)F_n = -\left(\frac{N^2}{\omega_n^2} - 1 - \frac{N^2}{\omega_n^4}\beta\frac{d\omega_n^2}{d\beta}\right)\phi_n,$$

$$F_n(-H, \beta) = 0, \quad F_n'(0, \beta) = \frac{g\beta}{\omega_n^2}F_n(0, \beta) + g\left(\omega_n^{-2} - \beta\omega_n^{-4}\frac{d\omega_n^2}{d\beta}\right)\phi_n(0, \beta).$$

Carrying out an operation analogous to that done in Exercise 10.6.16 (multiplying the equation for F_n by ϕ_n, integrating by parts from $-H$ to 0 over z, using the boundary condition and the equation for ϕ_n'' and grouping the terms with $d\omega_n^2/d\beta$), we find

$$\frac{\beta}{\omega_n^2}\left[g\phi_n(0) + \int_{-H}^{0} N^2\phi_n^2\,dz\right]\frac{d\omega_n^2}{d\beta} = g\phi_n^2(0) + \int_{-H}^{0}(N^2 - \omega_n^2)\phi_n^2\,dz.$$

The derivative $d\omega_n^2/d\beta$ can easily be expressed in terms of the group velocity of the nth mode $[(c_g)_n]$. Indeed,

$$\frac{d\omega_n^2}{d\beta} = 2\omega_n\frac{d\omega_n}{dk}\frac{dk}{d\beta} = \frac{\omega_n}{k}(c_g)_n = (c_{\text{Ph}})_n \cdot (c_g)_n,$$

where $(c_{\text{Ph}})_n$ is the phase velocity of the mode. As a result, an expression for the group velocity of an internal wave of the nth mode is

$$(c_g)_n = (c_{\text{Ph}})_n \frac{g\phi_n^2(0) + \int_{-H}^{0}(N^2 - \omega^2)\phi_n^2\,dz}{g\phi_n^2(0) + \int_{-H}^{0} N^2\phi_n^2\,dz}.$$

The latter can be written in another form, if, using the equation for ϕ_n we write $(N^2 - \omega_n^2)\phi_n^2 = -(\omega_n^2/k^2)\phi_n''$ and integrate by parts

$$\int_{-H}^{0} (N^2 - \omega_n^2)\phi_n^2 \, dz = -\omega_n^2 k^{-2} \int_{-H}^{0} \phi_n \phi_n'' \, dz = -\omega_n^2 k^{-2} \phi_n \phi_n' |_{-H}^{0}$$

$$+ \omega_n^2 k^{-2} \int_{-H}^{0} (\phi_n')^2 \, dz = -g\phi_n^2(0) + \omega_n^2 k^{-2} \int_{-H}^{0} (\phi_n')^2 \, dz \, .$$

Doing the same in the integral in the denominator

$$\int_{-H}^{0} N^2 \phi_n^2 \, dz = \int_{-H}^{0} (N^2 - \omega_n^2)\phi_n^2 \, dz + \omega_n^2 \int_{-H}^{0} \phi_n^2 \, dz \, ,$$

we obtain

$$(c_g)_n = (c_{Ph})_n \frac{\int_{-H}^{0} (\phi_n')^2 \, dz}{\int_{-H}^{0} (\phi_n')^2 \, dz + k^2 \int_{-H}^{0} \phi_n^2 \, dz} \, .$$

Hence the inequality $0 < (c_g)_n < (c_{Ph})_n$ follows immediately, i.e., the group velocity is always positive and smaller than the phase velocity of an internal mode. The first inequality $(c_g)_n = d\omega_n/dk > 0$ means that the dispersion curve $\omega_n = \omega_n(k)$ increases with increasing k. Since further,

$$\frac{dc_{Ph}}{dk} = \frac{d}{dk}\left(\frac{\omega}{k}\right) = \frac{(c_g - \omega/k)}{k} = k^{-1}(c_g - c_{Ph}) < 0 \, ,$$

the phase velocity of the mode decreases with an increase in k.

10.6.18. The pressure $P_a \exp[ik(x - Vt)]$ is applied to the surface of infinitely deep water. Consider the process for generating a gravity surface wave, including the case when $V = c_{Ph} = (g/k)^{1/2}$.

10.6.19. Find the criterion of applicability of the Boussinesq approximation, comparing the order of magnitudes of the terms in (10.48).

10.6.20. Find a dispersion relation $\omega = \omega(\boldsymbol{k})$ for the surface and internal modes (Fig. 10.9) when the fluid in a layer flows horizontally with the speed U.

11. Waves in Rotating Fluids

Gravity waves were discussed in the previous chapter. These waves are of great importance for the dynamics of the ocean and the atmosphere. No less important are waves related to the rotation of the Earth. The theory of such waves, as well as the development of the theory of gravity waves which takes the Earth's rotation into account, is the main purpose of this chapter.

We consider a coordinate system rotating together with a fluid at a constant angular velocity $\boldsymbol{\Omega}$. The Coriolis force $-2m\boldsymbol{\Omega} \times \boldsymbol{v}$ acts on the particle (m being the mass of a particle) moving with velocity $\boldsymbol{v}$ relative to this frame of reference. Such a force is orthogonal to $\boldsymbol{v}$ and analogous to the Lorentze force (e/c) $\boldsymbol{H} \times \boldsymbol{v}$ on an electron in a magnetic field. It is well known that the latter gives rise to an additional motion of the electrons around magnetic field lines. In continuous media, particles cannot move independently. Their interaction gives rise to a pressure gradient in the medium. Interplay of the Coriolis force and the pressure force determines the dynamics of the waves under consideration.

We first consider the simplest case of *inertial waves* in a rotating fluid. Then the combined effect of gravity and rotation will be discussed (*gravity-gyroscopic waves*). We will conclude the chapter with the *Rossby waves* which play a critical role in the global dynamics of the ocean and the atmosphere.

11.1 Inertial (Gyroscopic) Waves

11.1.1 The Equation for Waves in a Homogeneous Rotating Fluid

The linearized system of hydrodynamics equations (10.11) for the incompressible homogeneous ($\rho_0 = \text{const}$, $N^2 = 0$) fluid rotating at the constant angular velocity $\boldsymbol{\Omega}$ takes the form

$$\frac{\partial \boldsymbol{v}}{\partial t} + \frac{\nabla p}{\rho_0} + 2\boldsymbol{\Omega} \times \boldsymbol{v} = 0, \quad \frac{\partial \rho}{\partial t} = 0, \quad \nabla \boldsymbol{v} = 0. \tag{11.1}$$

We choose the z-axis along the vector $\boldsymbol{\Omega}$, then $2\boldsymbol{\Omega} \times \boldsymbol{v} = \{-Fv, Fu, 0\}$, where $\boldsymbol{v} = \{u, v, w\}$ and $F = 2\Omega$ is the so-called *Coriolis parameter*. In components, (11.1) can be written as

$$\begin{gathered}\frac{\partial u}{\partial t} - Fv + \frac{1}{\rho_0}\frac{\partial p}{\partial x} = 0, \quad \frac{\partial v}{\partial t} + Fv + \frac{1}{\rho_0}\frac{\partial p}{\partial y} = 0, \\ \frac{\partial w}{\partial t} + \frac{1}{\rho_0}\frac{\partial p}{\partial z} = 0, \quad \frac{\partial u}{\partial x} + \frac{\partial v}{\partial y} + \frac{\partial w}{\partial z} = 0.\end{gathered} \tag{11.2}$$

An equation for the vertical component of the particle velocity w can be obtained from this system. Applying the operator $\partial/\partial x$ to the first equation and $\partial/\partial y$ to the second one, summing up the relations obtained and taking into account the fourth equation in (11.2) yields

$$\frac{1}{\rho_0}\Delta_- p = \frac{\partial^2 w}{\partial t\,\partial z} + F\left(\frac{\partial v}{\partial x} - \frac{\partial u}{\partial y}\right).$$

On the other hand, applying the operator $\partial/\partial y$ to the first equation of (11.2), $\partial/\partial x$ to the second one and subtracting the relations obtained yields

$$\frac{\partial}{\partial t}\left(\frac{\partial v}{\partial x} - \frac{\partial u}{\partial y}\right) = F\frac{\partial w}{\partial z}.$$

Now we easily find

$$\begin{gathered}\frac{1}{\rho_0}\Delta_-\frac{\partial p}{\partial t} = \left(\frac{\partial^2}{\partial t^2} + F^2\right)\frac{\partial w}{\partial z}, \quad \Delta_-\frac{\partial u}{\partial t} = -\left(\frac{\partial^2}{\partial t\,\partial x} + F\frac{\partial}{\partial y}\right)\frac{\partial w}{\partial z}, \\ \Delta_-\frac{\partial v}{\partial t} = \left(-\frac{\partial^2}{\partial t\,\partial y} + F\frac{\partial}{\partial x}\right)\frac{\partial w}{\partial z}\end{gathered} \tag{11.3}$$

and hence the required equation follows

$$\frac{\partial^2}{\partial t^2}\Delta w + F^2\frac{\partial^2 w}{\partial z^2} = 0, \quad \Delta = \Delta_- + \frac{\partial^2}{\partial z^2}. \tag{11.4}$$

This is the basic equation for the discussion of *inertial* or *gyroscopic waves* in a rotating fluid.

11.1.2 Plane Harmonic Inertial Waves

Substituting the harmonic plane-wave solution

$$w = b\exp[\mathrm{i}(\boldsymbol{\varkappa}\cdot\boldsymbol{R} - \omega t)] = b\exp[\mathrm{i}(\boldsymbol{k}\cdot\boldsymbol{r} + k_z z - \omega t)], \tag{11.5}$$

into (11.4), we easily find the dispersion relation for an inertial wave

$$\omega^2 = F^2\frac{k_z^2}{k^2 + k_z^2} = F^2\frac{k_z^2}{\varkappa^2} = F^2\cos^2\theta, \tag{11.6}$$

where θ is the angle which the wave vector $\boldsymbol{\varkappa}$ makes with the vertical (z or Ω-direction). Thus, as in the case of plane internal waves, the angle θ is fixed if ω is fixed. The wavelength, however, may be arbitrary. The frequency of an inertial

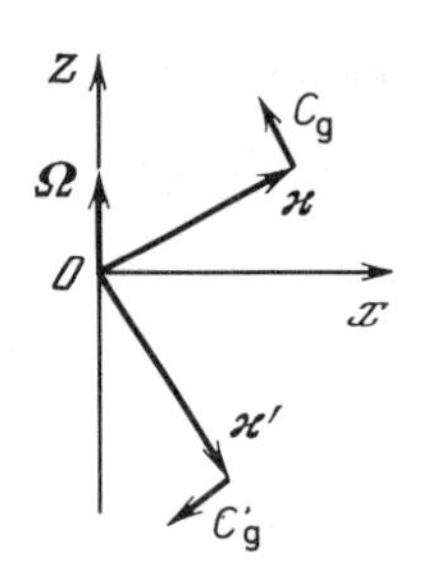

Fig. 11.1. Mutual orientation of the wave vector $\boldsymbol{\varkappa}$ and the group velocity vector $\boldsymbol{c}_g$ for an inertial wave. $\boldsymbol{\varkappa}'$ and c_g' refer to the case of downward wave propagation

wave cannot exceed the quantity $F = 2\Omega$, twice the frequency of the fluid's rotation. For the group velocity $\boldsymbol{c}_g = \nabla_\varkappa \omega = \{(\partial\omega/\omega k)\boldsymbol{k}/k, \partial\omega/\partial k_z\}$, we find $\partial\omega/\partial k_z = Fk^2/\varkappa^3$,

$$\partial\omega/\partial k = -\omega k/\varkappa^2 \quad \text{and}$$

$$\boldsymbol{c}_g = -\frac{\omega}{\varkappa^2}\boldsymbol{k} + \frac{Fk^2}{\varkappa^3}\nabla z, \tag{11.7}$$

or taking into account that $\boldsymbol{k} = \boldsymbol{\varkappa} - k_z\nabla z$, $\boldsymbol{\Omega} = F\nabla z/2$, we obtain

$$\boldsymbol{c}_g = -\frac{\omega}{\varkappa^2}\boldsymbol{\varkappa} + \frac{2\boldsymbol{\Omega}}{\varkappa}. \tag{11.7a}$$

The group velocity is perpendicular to the direction of propagation $\boldsymbol{\varkappa}$ ($\boldsymbol{c}_g \cdot \boldsymbol{\varkappa} = 0$), but unlike internal waves its horizontal projection (not vertical) is opposite to the corresponding projection of the vector $\boldsymbol{\varkappa}$ (Fig. 11.1) There is also an analogy with internal waves when we consider wave reflection at the boundary: the wave vector $\boldsymbol{\varkappa}'$ of a reflected wave is bound to maintain the same angle with the vertical (direction of $\boldsymbol{\Omega}$) as does the wave vector $\boldsymbol{\varkappa}$ of an incident wave. The wavelength of a reflected wave differs from that of an incident one if the reflecting plane is not a horizontal one.

The motion of fluid particles associated with inertial waves differs from that in internal waves. We have seen above that in internal waves, particles move along straight lines perpendicular to the wave vector lying in the same plane as $\boldsymbol{\varkappa}$ and the z-axis. For the wave (11.5), we find the horizontal components of the velocity using (11.3):

$$u = -\frac{k_z}{k^2}\left(k_x + \mathrm{i}\frac{F}{\omega}k_y\right)w, \qquad v = -\frac{k_z}{k^2}\left(k_y \div \mathrm{i}\frac{F}{\omega}k_x\right)w. \tag{11.8}$$

The velocity vector $\boldsymbol{v} = \{u, v, w)$ is again perpendicular to $\boldsymbol{\varkappa}$. Indeed, we have

$$\boldsymbol{\varkappa}\cdot\boldsymbol{v} = k_x u + k_y v + k_z w$$

$$= -k_z\left(\frac{k_x^2}{k^2} + \mathrm{i}\frac{F}{\omega}\frac{k_x k_y}{k^2} + \frac{k_y^2}{k^2} - \mathrm{i}\frac{F}{\omega}\frac{k_x k_y}{k^2}\right)w + k_z w = 0.$$

Without loss of generality, we can set $k_y = 0$ ($k_x = k$). We also assume in (11.5) that $b = B\exp(\mathrm{i}\alpha)$ and $\psi = k_x x + k_y y + k_z z - \omega t + \alpha$. Then for the real parts (11.5, 8), i.e., the only part which have physical meaning, we have

$$u = -\frac{k_z}{k} w = -\frac{k_z}{k} B\cos\psi\,, \qquad v = -\frac{F}{\omega}\frac{k_z}{k} B\sin\psi\,, \qquad w = B\cos\psi$$

and

$$\boldsymbol{v}^2 = u^2 + v^2 + w^2 = \left(\frac{k_z^2}{k^2}\cos^2\psi + \frac{F^2}{\omega^2}\frac{k_z^2}{k^2}\sin^2\psi + \cos^2\psi\right)B^2 = \frac{\varkappa^2 B^2}{k^2} = \text{const}\,.$$

Here we also used the dispersion relation (11.6).

Thus, fluid particles move at a constant velocity in the frontal plane (perpendicular to $\boldsymbol{\varkappa}$). By also taking the periodicity of movement into account, we conclude that particles move along circles of radius $A = \varkappa B/\omega k$ ($A\omega = |\boldsymbol{v}|$) with angular frequency ω. All fluid particles at the given frontal plane move at the same phase, i.e., the plane moves as a whole (without deformation) and in such a way that the trajectory of each of its points is a circle. Different frontal planes move with different phases, therefore a pressure gradient along $\boldsymbol{\varkappa}$ ($\nabla p \sim \boldsymbol{\varkappa}$) rises in the fluid.

Let us consider particle acceleration in an inertial wave and the balance of forces. The acceleration vector of each fluid particle $\boldsymbol{a}_c$ (centripetal acceleration) lies in the frontal plane. According to the Euler equation, $\boldsymbol{a}_c = \partial\boldsymbol{v}/\partial t = -\nabla p/\rho_0 - 2\boldsymbol{\Omega}\times\boldsymbol{v}$, i.e., this acceleration can be caused by the component of the Coriolis force $\boldsymbol{f}_{\parallel}$ which lies in the frontal plane. Its other component balances the pressure gradient along the propagation direction $\boldsymbol{\varkappa}$. Indeed, $\boldsymbol{f}_{\parallel} = -2\boldsymbol{\Omega}_{\perp}\times\boldsymbol{v}$ where $|\boldsymbol{\Omega}_{\perp}| = \Omega\cos\theta$ is the component of vector $\boldsymbol{\Omega}$ along $\boldsymbol{\varkappa}$. Since $\boldsymbol{\Omega}_{\perp}\perp\boldsymbol{v}$, then $f_{\parallel} = 2\Omega v\cos\theta = \omega|\boldsymbol{v}| = \omega^2 A$, the known expression for the centripetal acceleration of particle motion along a circle with the angular velocity ω.

11.1.3 Waves in a Fluid Layer. Application to Geophysics

We consider the simplest case assuming that the angular-velocity vector $\boldsymbol{\Omega}$ is perpendicular to the perfectly rigid boundaries of the layer $z = 0$ and $z = -H$. We look for a solution to (11.4) as a sum of two plane waves of the type (11.5) with opposite signs for k_z:

$$w = b\sin[k_z(z + H)]\exp[\mathrm{i}(\boldsymbol{k}\cdot\boldsymbol{r} - \omega t)]\,, \tag{11.9}$$

where $k = |\boldsymbol{k}|$, and k_z and ω satisfy the dispersion relation (11.6). Obviously, $w|_{z=-H} = 0$. Equating the vertical velocity at the upper boundary to zero

$(w|_{z=0} = 0)$ gives the possible values of k_z:

$$\sin k_z H = 0\,, \quad k_z = n\pi/H\,, \quad n = 1, 2, \ldots .$$

Substituting this into (11.6) we obtain the dispersion relation for the inertial modes in a homogeneous layer

$$\omega^2 = F^2 \frac{n^2\pi^2}{H^2}\left(k^2 + \frac{n^2\pi^2}{H^2}\right)^{-1}. \tag{11.10}$$

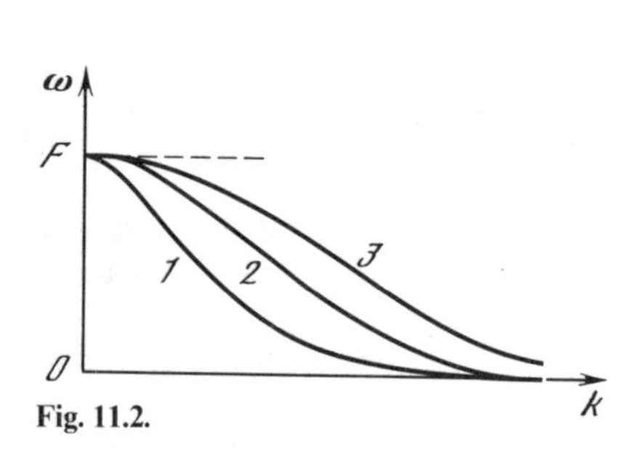

Fig. 11.2.

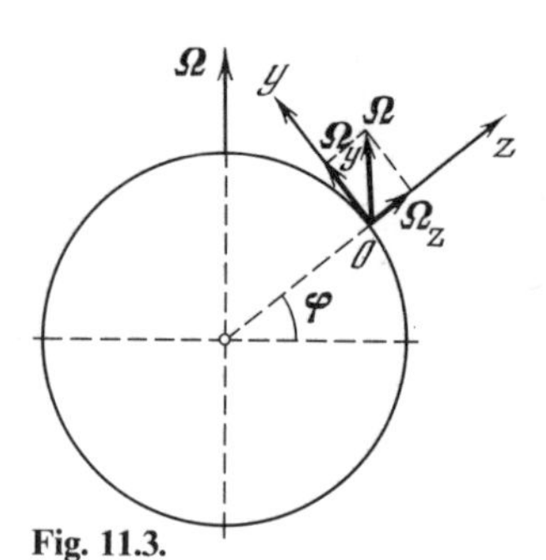

Fig. 11.3.

Fig. 11.2. Schematic behaviour of the dispersion curves for the inertially guided waves in a homogeneous fluid layer obtained in the "traditional approximation"
Fig. 11.3. Geometry for the discussion of inertial waves on the Earth's sphere

It is sketched in Fig. 11.2. The frequency of each mode decreases monotonically with increasing wave number k, i.e., the group velocity $\partial\omega/\partial k$ is negative. It means that a wave train with a narrow spectrum moves in the direction opposite to that of the vector $\boldsymbol{k}$ or the phase velocity.

In nature, inertial waves occur both in the ocean and in the atmosphere. The approximation of a fluid layer with plane-parallel boundaries is sufficiently good for waves of very small wave-length compared to the Earth's radius since the curvature of the Earth can be neglected in this case. It should be noted, however, that the direction of the z-axis (local vertical) is parallel to $\boldsymbol{\Omega}$ only at the poles. This fact brings some additional features into our theory. To elucidate them, we direct the y-axis from south to north and the x-axis from west to east (perpendicular to the plane of Fig. 11.3), as customary done in geophysical problems. In this case, $\boldsymbol{\Omega} = \{0, \Omega_y, \Omega_z\} = \{0, \Omega\cos\varphi, \Omega\sin\varphi\}$, where φ is the latitude. The equation for the vertical velocity component similar to (11.4) is (Exercise 11.4.1)

$$\frac{\partial^2}{\partial t^2}\Delta w + 4(\boldsymbol{\Omega}\cdot\nabla)^2 w = 0\,, \quad \text{where} \tag{11.11}$$

$$(\boldsymbol{\Omega}\cdot\nabla)^2 = \left(\Omega_y\frac{\partial}{\partial y} + \Omega_z\frac{\partial}{\partial z}\right)^2 = \Omega_y^2\frac{\partial^2}{\partial y^2} + 2\Omega_y\Omega_z\frac{\partial^2}{\partial y\,\partial z} + \Omega_z^2\frac{\partial^2}{\partial z^2}.$$

In geophysical problems, the so-called *traditional approximation* is often used when the terms containing Ω_y are neglected. For harmonic waves of the kind (11.5), we have $\partial/\partial y = \mathrm{i}k_y$, $\partial/\partial z = \mathrm{i}k_z$, and this approximation is valid if $|\Omega_y k_y| \ll |\Omega_z k_z|$ or $|k_y| \ll |k_z \tan\varphi|$. In other words, the vertical scale of wave-field variations must be considerably less than the horizontal one and the latitude must not be very low. When this approximation is used, (11.11) transforms into (11.4) with the Coriolis parameter $F = 2\Omega \sin\varphi$.

11.2 Gyroscopic-Gravity Waves

11.2.1 General Equations. The Simplest Model of a Medium

Waves are referred to as *gyroscopic-gravity* waves when besides the Coriolis force, also the gravitational forces play an important role. In the theory which follows, the density stratification of the medium will also be taken into account. The first two and the last equations of (11.2) hold in this case. Changing the third equation and adding the equation of state, we obtain additionally

$$\frac{\partial w}{\partial t} + \frac{1}{\rho}\frac{\partial p}{\partial z} + g\frac{\rho}{\rho_0} = 0\,, \qquad \frac{\partial \rho}{\partial t} - \rho_0\frac{N^2}{g}w = 0\,. \tag{11.12}$$

We will use the Boussinesq approximation assuming the density $\rho_0(z)$ to be constant when it enters the equations in its explicit form. The lower boundary of the layer is assumed rigid ($w|_{z=-H} = 0$) and the upper one free with the boundary conditions (10.7) where $\sigma = 0$:

$$\left(\frac{\partial p}{\partial t} - g\rho_0 w\right)_{z=0} = 0\,. \tag{11.13}$$

These expressions for the pressure and the horizontal velocity still hold since they were obtained without making use of the third equation of (11.2). Utilizing them it is easy to derive from (11.12, 13) an equation and the boundary conditions only in terms of w:

$$\begin{gathered}\frac{\partial^2}{\partial t^2}\Delta w + F^2\frac{\partial^2 w}{\partial z^2} + N^2\Delta_- w = 0\,,\\ w|_{z=-H} = \left[\left(\frac{\partial^2}{\partial t^2} + F^2\right)\frac{\partial w}{\partial z} - g\,\Delta_- w\right]_{z=0} = 0\,.\end{gathered} \tag{11.14}$$

We are looking for a solution of (11.14) in the form of a harmonic wave

$$w = \phi(z)\exp[\mathrm{i}(\boldsymbol{k}\cdot\boldsymbol{r} - \omega t)]\,, \quad \boldsymbol{k} = \{k_x, k_y\}\,. \tag{11.15}$$

On substitution, we obtain the following boundary-value problem for the function $\phi(z)$:

$$\phi'' + k^2 \frac{N^2(z) - \omega^2}{\omega^2 - F^2}\phi = 0, \quad \phi(-\mathrm{H}) = \phi'(0) - \frac{gk^2}{\omega^2 - F^2}\phi(0) = 0. \qquad (11.16)$$

The eigenvalues $k_n(\omega)$ and the corresponding eigenfunctions $\phi_n(z)$ determine the modes of a rotating stratified fluid layer.

First, consider the simplest model assuming the layer to be homogeneous $[N^2(z) = \text{const}]$. This model allows us to get an idea about possible modes in the rotating fluid layer. The solution of (11.16) must vanish at $z = -H$ we have

$$\phi(z) = b\sin[\alpha(z + H)], \qquad \alpha = k\sqrt{\frac{N^2 - \omega^2}{\omega^2 - F^2}}. \qquad (11.17)$$

The parameter α can be real or purely imaginary $\alpha = \mathrm{i}\alpha'$. In the latter case, $\sin[\alpha(z + H)] = \mathrm{i}\sinh[\alpha'(z + H)]$. Substitution of (11.17) into the boundary condition for $z = 0$ leads to the dispersion relation for the modes:

$$\tan\sigma = \frac{\mathrm{H}(\mathrm{N}^2 - \omega^2)}{g\sigma}, \qquad \sigma^2 = \alpha^2 H^2 = k^2 H^2 \frac{N^2 - \omega^2}{\omega^2 - F^2}. \qquad (11.18)$$

We discuss this relation assuming the wave frequency ω to be fixed, and confining ourselves to modes with real k (propagating modes).

11.2.2 Classification of Wave Modes

Note that for any fixed frequency $\omega > F$, (11.18) has at least one root σ_0 (the smallest in magnitude) corresponding to a surface wave. For $kH \ll 1$ (long-wavelength waves), it can be found, assuming $\tan\sigma \simeq \sigma$,

$$\sigma_0^2 \simeq \frac{H(N^2 - \omega^2)}{g}, \qquad \omega^2 = F^2 + gHk^2. \qquad (11.19)$$

Hence, the frequency of the surface wave is always greater than $F = 2\Omega\sin\varphi$. For $gHk^2 \gg F^2$, the usual relation for surface waves in shallow water $(\omega/k)^2 = c_{\mathrm{Ph}}^2 = gH$ holds.

In the case where $\omega^2 > N^2 > F^2$, (11.18) has only one root σ_0. Note further that this equation has no wavelike solutions of the type (11.15) with real ω and k if $N^2 > F^2 > \omega^2$ or $F^2 > N^2 > \omega^2$. In these cases, $\sigma^2 < 0$ and the signs of the right- and left-hand sides of (11.18) are opposite. Hence, two cases remain for discussion:

i) $N^2 > \omega^2 > F^2$, the case of *internal waves* influenced by the Earth's rotation;

ii) $F^2 > \omega^2 > N^2$, the case of *inertial* (*gyroscopic*) *waves* influenced by gravity and stratification.

We will not consider small values of σ, the surface waves occuring in case (i). Note further that for the ocean, $\varepsilon = H(N^2 - \omega^2)/g < HN^2/g \ll 1$, see (10.71), hence the right-hand side of (11.18) is small. As a result, we have the roots $\sigma_n = n\pi$, $n = 1, 2, \ldots$. These roots correspond to the modes for which the rigid cover approximation ($w|_{z=0} \sim \sin\sigma_n \simeq 0$) works well. The respective dispersion relation is obtained from (11.18) by using $\sigma_n = n\pi$:

$$\omega^2 = \frac{n^2\pi^2F^2 + N^2k^2H^2}{n^2\pi^2 + k^2H^2}. \tag{11.20}$$

At high frequencies ($\omega \gg F$ but $\omega < N$), the dispersion relation coincides with (10.74) if $d_2 = H$, $\sigma_n(\omega) = n\pi$. For long-wavelength waves ($kH \ll 1$), rotation of the fluid as a whole leads to dispersion:

$$\omega^2 = F^2 + (N^2 - F^2)(kH/n\pi)^2. \tag{11.21}$$

Differentiating the dispersion relation (11.20) with respect to k we find for the group velocity

$$c_g = \frac{d\omega}{dk} = \frac{n^2\pi^2}{(n^2\pi^2 + k^2H^2)^2}\,\frac{N^2 - F^2}{c_{\mathrm{Ph}}}\,H^2, \tag{11.22}$$

where $c_{\mathrm{Ph}} = \omega/k$ is the mode's phase velocity approaching infinity if $k \to 0$. It follows from the last formula that $c_g > 0$ for internal waves and dispersion curves rise monotonically with increasing k, whereas in the case of inertial waves, $c_g < 0$, i.e., the curves drop monotonically.

11.2.3 Gyroscopic-Gravity Waves in the Ocean

Under natural conditions in the ocean, the pattern is similar to that considered above but both internal and inertial modes can occur for a given $N(z)$. This is due to the fact that not only can the bottom and the surface participate in guiding the waves (as in the case $N = \text{const}$), but water layers can also, because of the specific form of the function of $N(z)$. Consider, for example, the typical

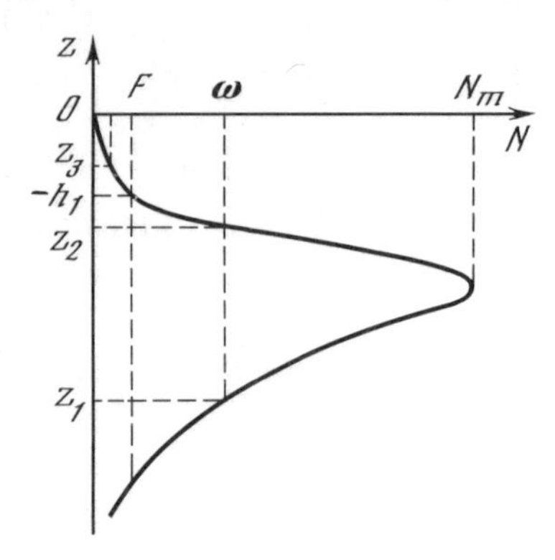

Fig. 11.4. Typical dependence of the Väisälä frequency N on the depth in the ocean

variation of $N(z)$ with depth shown in Fig. 11.4. The Coriolis parameter F corresponds to one of the vertical dashed lines. In the upper mixed ocean layer, the Väisälä frequency is small so that $F > N(z)$ at $z > -h_1$. In the thermocline layer, the Väisälä frequency can attain as high a value as $N_m \sim 6$ cycles per hour ($T \sim 10$ min). Thus, we have the opposite case $N(z) > F$ here. The same happens in the layers below the thermocline and only at very large depths can the Coriolis parameter possibly exceed the Väisälä frequency again. Otherwise, $N(z) > F$ at any depth.

The character of wave motion greatly depends on the frequency. Therefore, we will consider the following cases:

i) ω exceeds the maximum Väisälä frequency ($\omega > N_m$). The coefficient in front of $\phi(z)$ in (11.16)

$$\alpha^2(z) \equiv k^2 \frac{N^2(z) - \omega^2}{\omega^2 - F^2} \tag{11.23}$$

is negative for all z if k is real. Hence, $\phi(z)$ and $\phi''(z)$ have the same sign, i.e., we have a monotonic solution $\phi(z)$ with maximum $|\phi(z)|$ at one of the boundaries. By virtue of $\phi(-H) = 0$, we have $\max|\phi(z)| = |\phi(0)|$. Hence, only a surface mode exists in the fluid in this case.

ii) $F < \omega < N_m$. For $z_1 < z < z_2$ (Fig. 11.4), we have $\alpha^2(z) > 0$. Hence, $\phi(z)$ is an oscillatory function in this interval of depth. Outside this interval ($z < z_1$ and $z > z_2$) we have $\alpha^2(z) < 0$ and the function $\phi(z)$ is monotonic. $|\phi(z)|$ reaches its maximum at the boundary $z = 0$ for small k (surface wave) or in the thermocline (internal modes). A consecutive number is assigned to each of the internal modes which exceeds, by one, the number of zeros of the function $\phi(z)$ in the interval $z_1 < z < z_2$. The higher the number of the mode, the more oscillations has the function $\phi(z)$, the greater the parameter $\alpha^2(z)$ and, consequently, the greater the wave number k, see (11.23). Outside the interval (z_1, z_2), $\phi(z)$

decreases more rapidly in the case of high numbers of modes, i.e., a high wave number k. The eigenvalue can be estimated using the solution of (11.16) in the WKB approximation which yields

$$\int_{z_2}^{z_2} \alpha(z)\,dz \sim n\pi\,, \qquad k_n \sim n\pi\sqrt{\omega^2 - F^2}\left[\int_{z_1}^{z_2} \sqrt{N^2(z) - \omega^2}\,dz\right]^{-1}. \tag{11.24}$$

At high frequencies when $\omega \to N_{\mathrm{m}}$, the wave number $k_n \to \infty$. In the case of low frequencies ($\omega \to F$), one can assume $\omega \ll N(z)$ when estimating the integral in (11.24). As a result, (11.24) transforms into a relation analogous to (11.21):

$$\omega^2 = F^2 + \left[\int_{z_1}^{z_2} N(z)\,dz\right]^2 k^2/n^2\pi^2\,. \tag{11.25}$$

Thus, in the frequency range $F < \omega < N_{\mathrm{m}}$, the system of modes consists of a surface wave concentrated near the surface and a countable set of internal modes trapped by the thermocline.

iii) Consider now the case $\omega < F$. Note first of all that a surface wave does not occur in this case. Indeed, $\phi(0)$ and $\phi'(0)$ have different signs according to (11.16), therefore going away from the surface (negative z), we obtain $|\phi(z)| > |\phi(0)|$ which contradicts the definition of a surface wave. Turning to other modes, we note that for the case under consideration ($\omega < F$) we have $\alpha^2(z) < 0$ in the thermocline [where $\omega < N(z)$] and $\alpha^2(z) > 0$ outside it [where $\omega > N(z)$]. The latter case can occur near the surface or the bottom (Fig. 11.4). Equation (11.16) is solved similarly in both regions. We will confine ourselves to a discussion of the situation in a subsurface mixed layer where $N(z)$ is small and we denote by $N_{\min}$ the minimum value of the Väisälä frequency in this layer. If $\omega < N_{\min}$, then $\alpha^2(z) < 0$ at any z, the function $\phi(z)$ does not oscillate and eigensolutions of (11.16) do not exist. If, however, we have $N_{\min} < \omega < F$, say, in the layer $z_3 < z < 0$ (Fig. 11.4), then the function $\phi(z)$ in such a layer oscillates and a countable set of eigenfunctions for (11.16) exists corresponding to *inertial modes*. For the eigenvalues we obtain an estimate analogous to (11.24):

$$k_n \sim n\pi\sqrt{F^2 - \omega^2}\left[\int_{z_3}^{0} \sqrt{\omega^2 - N^2(z)}\,dz\right]^{-1}.$$

k_n increases when the mode number increases or the frequency ω decreases. For long-wavelength waves [$|kz_3| \ll 1$, $\omega \simeq F$, $\omega \gg N(z)$], we obtain

$$\omega^2 \simeq F^2(1 - z_3^2 k^2/n^2\pi^2)\,.$$

Inertial waves are here trapped in a subsurface layer $z > z_3$. The higher the mode number or the lower the frequency, the more rapidly the mode's amplitude decreases when the reference point moves away this layer.

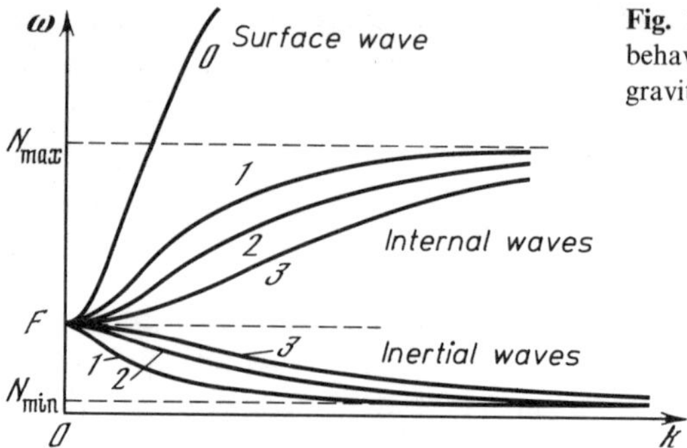

Fig. 11.5. Schematic picture illustrating the behaviour of dispersion curves for gyroscopic-gravity waves

The above discussion allows us to sketch the dispersion curves $\omega(k)$ for gyroscopic-gravity waves, as shown in Fig. 11.5. The monotonic character of these curves follows, for example, from the expression for the group velocity obtained in Exercise 11.4.6.

Note that inertial waves in the ocean (and atmosphere) are rather infrequent phenomena, since the condition $F > N(z)$ is seldom satisfied. However, some sort of degenerate inertial wave motion is often observed in the ocean, the so-called *inertial oscillations*. In this case, the acceleration of particles is due to the Coriolis force ($\nabla p = 0$). The particles move along circular orbits in the horizontal plane with centripetal acceleration equal to the Coriolis force and angular frequency $\omega = F$. Inertial oscillations are the solution of the linear system of equations (10.11a) where $p = w = \rho = 0$, $u = b\exp(-\mathrm{i}Ft)$, $v = -\mathrm{i}b\exp(-\mathrm{i}Ft)$ (Exercise 11.4.7) and are observed at all depths.

11.3 The Rossby Waves

11.3.1 The Tangent or β-Plane Approximation

The Rossby waves, which are also called *planetary* waves, are caused by the latitudinal dependence of the local vertical component of the angular velocity Ω_z. In the simplest version of the theory of these waves, we can refer to (11.2) for a rotating homogeneous fluid written in rectangular coordinates in a local tangent plane with the Coriolis parameter depending linearly on the coordinate y pointing north: $F = F_0 + \beta y$. On the spherical Earth, this is just the first two terms of the expansion of the vertical component of Earth's angular velocity $\Omega_z = \Omega \sin\varphi$ in a series with respect to y (Fig. 11.3). Indeed, we have

$$F = 2\Omega \sin\varphi = 2\Omega(\sin\varphi_0 + \Delta\varphi\cos\varphi_0) = F_0 + \beta y\,,$$
$$F_0 = 2\Omega\sin\varphi_0\,, \quad \beta = \frac{2\Omega\cos\varphi_0}{R_0}\,, \tag{11.26}$$

where φ_0 is the geographical latitude, R_0 the Earth's radius and $y = R_0\,\Delta\varphi$.

The effect caused by the term βy in (11.26) is often called the *β-effect*. The local tangent plane is called the *β-plane* and the Rossby-waves theory for the β-plane is called the *β-plane approximation*. We will confine ourselves to this approximation, too. One has to bear in mind that only the vertical component of the Earth's angular velocity is taken into account, i.e., we remain within the limits of a *traditional approximation*.

11.3.2 The Barotropic Rossby Waves

Let us consider the horizontal flow ($w = 0$) of a homogeneous fluid assuming that all quantities specifying flow do not depend on the vertical coordinate z (*barotropic flow*). Such a flow is two-dimensional which allows us to use a streamline function ψ ($u = \partial\psi/\partial y, v = -\partial\psi/\partial x$). It was shown in Sect. 7.1.2 that

$$\operatorname{curl} \boldsymbol{v} = \left(\frac{\partial v}{\partial x} - \frac{\partial u}{\partial y}\right)\nabla z = -\Delta_{-}\psi\nabla z\,. \tag{11.27}$$

From (11.2) we now easily obtain an equation for ψ:

$$\partial(\Delta_{-}\psi)/\partial t + \beta\,\partial\psi/\partial x = 0\,. \tag{11.28}$$

With (11.27), (11.28) can be written as

$$d(\operatorname{curl} \boldsymbol{v}) = -\beta v\,dt = -\frac{d(2\Omega_z)}{dy}dy\,.$$

This is a linearized form of the angular momentum conservation law in a rotating fluid; the vorticity change along y is related only to the angular velocity gradient in this direction.

We are interested in a plane harmonic wave-type solution of (11.28):

$$\psi = b\exp[i(k_x x + k_y y - \omega t)]\,.$$

Substituting this into (11.28) yields the dispersion relation

$$\omega = -\beta k_x/k^2\,, \quad k^2 = k_x^2 + k_y^2\,, \tag{11.29}$$

which is *anisotropic* in the horizontal plane (k_x, k_y). In particular, a Rossby wave may propagate only in the direction opposite the x-axis (westward) since $k_x < 0$ if $\omega > 0$.

Equation (11.29) can be written as

$$(k_x + \beta/2\omega)^2 + k_y^2 = \beta^2/4\omega^2\,.$$

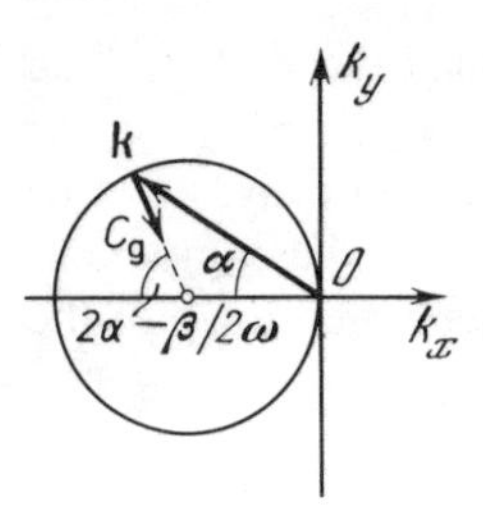

Fig. 11.6. Relationship between the wave vector $\boldsymbol{k}$ for a Rossby wave and the propagation direction α

In the plane (k_x, k_y), it is the equation of a circle of radius $\beta/2\omega$ with the center at the point $(-\beta/2\omega, 0)$ (Fig. 11.6). The vector $\boldsymbol{k}$ describes this circle when the propagation angle α in the (x, y)-plane varies at a fixed ω. Differentiating (11.29) with respect to k_x and k_y gives an expression for the group velocity of the Rossby waves:

$$\boldsymbol{c}_g = \left\{\frac{\partial\omega}{\partial k_x}, \frac{\partial\omega}{\partial k_y}\right\} = \frac{\beta}{k^2}\left\{\frac{k_x^2 - k_y^2}{k^2}, 2\frac{k_x k_y}{k^2}\right\} = \frac{\beta}{k^2}\{\cos 2\alpha, -\sin 2\alpha\}, \tag{11.30}$$

where $k_x = -k\cos\alpha$ and, $k_y = k\sin\alpha$.

It can easily be seen that the group-velocity vector in the plane (k_x, k_y) beginning at point $\boldsymbol{k}$ points at the center of the circle (Fig. 11.6). In particular, if the wave propagates exactly westward ($\alpha = 0$), then the group-velocity vector points eastward. The lines $\psi = \text{const}$ (wave fronts) are perpendicular to $\boldsymbol{k}$ since these lines are also streamlines in the case considered. Fluid particle velocities are tangent to them. Hence, the particle velocity is perpendicular to the wave vector $\boldsymbol{k}$, i.e., the *Rossby wave* is a *transverse* one.

Concerning the reflection of a Rossby wave at a boundary (shore line), we have the same kind of pecularities as in the case of other waves with anisotropic dispersion (internal and inertial waves) (Exercise 11.4.8).

Pressure oscillations caused by the barotropic Rossby wave considered are, according to (11.2),

$$p = -\rho_0(F + i\beta k_y/k^2)\psi .$$

They do not depend on depth and are not zero even at the surface. This is not in accordance with the boundary conditions at the surface where the pressure must be constant and equal to the atmospheric pressure. It means that real barotropic Rossby waves will have some (minor, as will be shown below) deviations from those considered above.

11.3.3 Joint Discussion of Stratification and the β-Effect

We now consider waves in a plane stratified, rotating fluid layer in the β-plane approximation. Assume in (10.11) that $2\Omega = 2\Omega_z\nabla z = F\nabla z$, where $F = F_0 + \beta y$, F_0 and β being specified by (11.26). In components, the system of (10.11) has the form

$$\frac{\partial u}{\partial t} - Fv + \frac{1}{\rho_0}\frac{\partial p}{\partial x} = 0, \quad \frac{\partial v}{\partial t} + Fu + \frac{1}{\rho_0}\frac{\partial p}{\partial y} = 0,$$
$$\frac{\partial w}{\partial t} + \frac{1}{\rho_0}\frac{\partial p}{\partial z} + g\frac{\rho}{\rho_0} = 0, \quad \frac{\partial u}{\partial x} + \frac{\partial v}{\partial y} + \frac{\partial w}{\partial z} = 0, \quad \frac{\partial \rho}{\partial t} = \rho_0\frac{N^2}{g}w. \tag{11.31}$$

Boundary conditions at the bottom $z = -H$ and the surface $z = 0$ are, see (11.13),

$$w|_{z=-H} = \left(\frac{\partial p}{\partial t} - \rho_0 g w\right)_{z=0} = 0. \tag{11.32}$$

To simplify the calculations we use the Boussinesq approximation assuming ρ_0 to be constant.

Keeping in mind that we would like to use the method of separation of variables, we look for a solution of the system (11.31, 32) in the form

$$u = P(z)U(x,y)\exp(-\mathrm{i}\omega t), \quad v = P(z)V(x,y)\exp(-\mathrm{i}\omega t),$$
$$w = \mathrm{i}\omega W(z)G(x,y)\exp(-\mathrm{i}\omega t), \quad p = \rho_0 P(z)G(x,y)\exp(-\mathrm{i}\omega t), \tag{11.33}$$
$$\rho = -\rho_0 g^{-1} N^2(z)W(z)G(x,y)\exp(-\mathrm{i}\omega t).$$

Indeed, the substitution of (11.33) into (11.31) yields the above system in which the first three equations include a function which merely depends on (x, y) or on z. Only in the fourth equation

$$\frac{W'}{P} = \frac{\mathrm{i}}{\omega}G^{-1}(x,y)\left(\frac{\partial U(x,y)}{\partial x} + \frac{\partial V(x,y)}{\partial y}\right) = -\nu$$

do these functions turn out to be interrelated. The constant ν stands for the separation parameter. As a result, we obtain a separate problem in the horizontal plane

$$-\mathrm{i}\omega U - FV + \frac{\partial G}{\partial x} = 0, \quad -\mathrm{i}\omega V + FU + \frac{\partial G}{\partial y} = 0, \quad \frac{\partial U}{\partial x} + \frac{\partial V}{\partial y} = \mathrm{i}\nu\omega G \tag{11.34}$$

and the boundary-value problem for $P(z)$ and $W(z)$:

$$P' = (N^2 - \omega^2)W, \quad W' = -\nu P,$$
$$W(-H) = 0, \quad (P + gW)_{z=o} = 0. \tag{11.35}$$

By simple transformations, the system (11.34) can be reduced to one equation for the function $V(x, y)$, for example,

$$\Delta_{-} V + \mathrm{i}\frac{\beta}{\omega}\frac{\partial V}{\partial x} + \nu(\omega^2 - F^2)V = 0\,. \tag{11.36}$$

The Coriolis parameter $F = F_0 + \beta y$ here depends on y. We assume it to be constant $F = F_0$. However, confining ourselves only to the close vicinity of the straight line $y = 0$ (where $\beta y \ll F_0$, i.e., $y \ll L$, $L \sim F_0/\beta$ being the distance over which), the F changes substantially. On the other hand, L must be large compared with the wavelength or, with (11.26),

$$kL = kR_0 \tan\varphi_0 \gg 1\,. \tag{11.37}$$

Since $kR_0 \gg 1$ (in order for the β-plane approximation to be valid), only very small φ_0 (a very narrow zone along the equator) must be excluded.

Now we look for the solution of (11.36) in the form

$$V = A\exp[\mathrm{i}(k_x x + k_y y)]\,.$$

The substitution here yields the relationship between the separation parameter ν and the components of the horizontal wave vector $\boldsymbol{k} = \{k_x, k_y\}$:

$$\nu = \frac{k^2 + \beta k_x/\omega}{\omega^2 - F_0^2}\,. \tag{11.38}$$

ν depends on k_x, i.e., on the propagation direction in the horizontal plane.

The boundary-value problem (11.35) for the functions $P(z)$ and $W(z)$ can also be reduced to only one equation for $W(z)$:

$$\begin{aligned} &W'' + \nu[N^2(z) - \omega^2)W = 0\,, \\ &W(-H) = 0\,, \quad W'(0) - g\nu W(0) = 0\,, \end{aligned} \tag{11.39}$$

where ν is given by (11.38). In the case of very high frequency, we can neglect the term $k_x\beta/\omega$ in (11.38), thus obtaining the boundary-value problem for gyroscopic-gravity waves already considered in Sect. 11.2. This approximation is possible if $k^2 \gg |\beta k_x/\omega|$ or, by virtue of $k_x \sim k$ and (11.26),

$$\omega \gg \omega_\beta \equiv \frac{2\Omega\cos\varphi_0}{kR_0}\,. \tag{11.40}$$

The frequency ω_β is very small (kR_0 is very large) and $\omega_\beta \ll F_0$ due to (11.37). Hence, condition (11.40) is satisfied if $\omega \gtrsim F_0$ which is the case for high frequency (large wavelength) inertial waves and even more for surface and internal modes with frequencies $\omega > F_0$. If the minimum Väisälä frequency in the ocean is also much higher than ω_β ($N_{\min} \gg \omega_\beta$), which is usually the case, then the variation of the Coriolis parameter can be neglected for inertial waves at any frequency.

11.3.4 The Rossby Waves in the Ocean

We now consider very low frequencies so that the first term in the numerator in (11.38) can be neglected compared with the second one:

$$\omega < \left|\frac{\beta k_x}{k^2}\right| = \frac{\beta|\cos\alpha|}{k} = \frac{2\Omega\cos\varphi_0}{kR_0}|\cos\alpha| = \omega_\beta|\cos\alpha|, \tag{11.41}$$

where $k_x = -k\cos\alpha$ and α is the angle which $\boldsymbol{k}$ makes with the negative (westward) direction of x. Now, the separation parameter v is positive for waves propagating westward ($k_x < 0, \cos\alpha > 0$) and the solution of (11.39) in the region $N^2(z) > \omega_\beta^2 > \omega^2$ is of an oscillatory nature. The corresponding set of eigenfunctions (modes) describe the Rossby waves in the ocean. On the other hand, the parameter v turns out to be negative for eastward propagating waves ($k_x > 0$) when $\omega < F_0$. Hence, no eigensolutions of (11.39) exist corresponding to real ω and $\boldsymbol{k}$ in this case, i.e., the *Rossby waves only propagate westward*, as in the simple case considered in Sect. 11.3.2.

The eigenvalues v_n of (11.39) are functions of ω in general. For very low frequencies $[\omega^2 < \omega_\beta^2 \ll N^2(z)]$ in which we are interested, however, the term ω^2 in (11.39) can be neglected. Then the eigenvalues $0 < v_0 < v_1 < \cdots < v_n < \cdots$ do not depend on frequency and are specified only by stratification $N^2(z)$ (*hydrostatic approximation*). Substitution of v_n into (11.38), where ω can also be neglected as compared with F_0, yields the *dispersion relation for the Rossby waves:*

$$\left(k_x + \frac{\beta}{2\omega}\right)^2 + k_y^2 = \frac{\beta^2}{4\omega^2} - v_n F_0^2, \tag{11.42}$$

which describes again a circle in the (k_x, k_y) plane like in Sect. 11.3.2. Its center is at the point $(-\beta/2\omega, 0)$, too; but its radius is less here. This radius decreases with increasing ω and becomes zero at $(\omega_n)_{\max} = \beta/2F_0\sqrt{v_n}$ which is the *maximum possible frequency* of the nth-order Rossby wave. A wave of this frequency travels westward ($k_y = 0$). Its wave number $(k_n)_{\min} = -k_x = \beta/2\omega = F_0\sqrt{v_n}$ increases with increasing n ($v_n > v_{n-1}$) and its frequency $(\omega_n)_{\max}$ diminishes at the same time.

The dispersion relation (11.42) is solvable for ω:

$$\omega_n = -\frac{\beta k_x}{k^2 + F_0^2 v_n} = \beta\frac{k\cos\alpha}{k^2 + F_0^2 v_n}. \tag{11.43}$$

It follows from (11.43) that at a fixed $\lambda = 2\pi/k$, the maximum frequency occurs for a wave propagating westward ($k_y = 0$). If the angle α is fixed, the maximum frequency and the corresponding wave number are

$$(\omega_n)_{\max}(\alpha) = \frac{\beta\cos\alpha}{2F_0\sqrt{v_n}}, \qquad (k_n)_{\min} = F_0\sqrt{v_n}. \tag{11.44}$$

The Rossby wave for $n = 0$ is called *barotropic.* Its eigenvalue ν_0 can be found under the assumption that W_0 is a linear function of z:

$$W_0(z) = b(1 + z/H), \quad \nu_0 = 1/gH. \tag{11.45}$$

Using a dimensionless coordinate $\xi = z/H$, we can estimate the order of magnitude of the main term in the coefficient of W in (11.39); it turns out to be $N^2H/g \simeq \Delta\rho/\rho$, i.e., small. Hence, the solution (11.45) satisfies the boundary conditions (rigorously) as well as (11.39) (with very high accuracy). Note that a barotropic Rossby wave can also occur in a homogeneous layer. The difference between this wave and that discussed in Sect. 11.3.2 is small when the quantity F_0^2/gHk^2 is small, compare (11.29, 43). This is the case if the wavelength is not too large.

Waves with $n > 0$ are called the *baroclinic Rossby waves.* Stratification of the fluid $[N^2(z) \neq 0]$ is essential for the existence of these waves. Their eigenfunctions have at least one maximum inside the layer where the horizontal component of the particle velocity reverses its direction. The "rigid cover" approximation ($w|_{z=0} = 0$) is satisfied with high accuracy for these waves. By analogy with (11.45), the eigenvalues ν_n for baroclinic waves can be written as $\nu_n = (gH_n)^{-1}$, $H = H_0 > H_1 > \cdots > H_n > \cdots$. The quantities H_n are called *equivalent depths.*

In conclusion, a schematic representation of dispersion curves for all kinds of waves in an incompressible rotating fluid is given in Fig. 11.7. For the Rossby waves we must consider the dispersion surface $\omega = \omega(k_x, k_y)$ since these waves are anisotropic. A cut of the dispersion surface by the plane (ω, k_x) is shown in Fig. 11.7. We obtain the whole surface by rotating this picture about the ω-axis, bearing in mind that the frequency of a Rossby wave decreases with increasing α according to the factor $\cos\alpha$, see (11.43). Dispersion surfaces for other kinds of waves will practically be symmetric about the ω-axis. Typical time scales for different kinds of waves in the ocean are also shown in the figure.

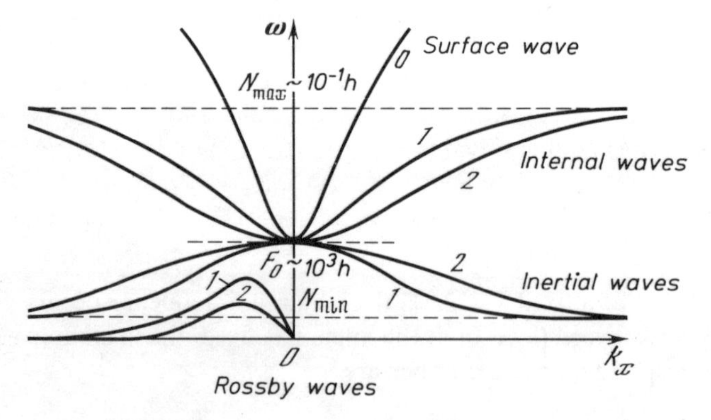

Fig. 11.7. Schematic representation of dispersion curves for all kinds of waves in an incompressible rotating fluid and typical periods for waves in the ocean

11.4 Exercises

11.4.1. Derive an equation for waves in a homogeneous, rotating fluid in the case where the angular velocity $\boldsymbol{\Omega}$ has arbitrary orientation with respect to the coordinate axes.

Solution: Applying the operation curl to (11.1) gives in components with $\operatorname{curl}(\nabla p) = 0$, $\operatorname{curl}(\boldsymbol{\Omega} \times \boldsymbol{v}) = -(\boldsymbol{\Omega} \cdot \nabla)\boldsymbol{v}$,

$$\frac{\partial}{\partial t}\left(\frac{\partial w}{\partial y} - \frac{\partial v}{\partial z}\right) - 2(\boldsymbol{\Omega} \cdot \nabla)u = 0\,, \qquad \frac{\partial}{\partial t}\left(\frac{\partial u}{\partial z} - \frac{\partial w}{\partial x}\right) - 2(\boldsymbol{\Omega} \cdot \nabla)v = 0\,,$$
$$\frac{\partial}{\partial t}\left(\frac{\partial v}{\partial x} - \frac{\partial u}{\partial y}\right) - 2(\boldsymbol{\Omega} \cdot \nabla)w = 0\,, \qquad \frac{\partial u}{\partial x} + \frac{\partial v}{\partial y} + \frac{\partial w}{\partial z} = 0\,. \tag{11.46}$$

Differentiating the first equation with respect to y, the second with respect to x and subtracting one from the other, we obtain

$$\frac{\partial}{\partial t}\Delta_{-}\omega - \frac{\partial^2}{\partial t\,\partial z}\left(\frac{\partial v}{\partial y} + \frac{\partial u}{\partial x}\right) + 2(\boldsymbol{\Omega} \cdot \nabla)\left(\frac{\partial v}{\partial x} - \frac{\partial u}{\partial y}\right) = 0\,.$$

Application of the operator $\partial/\partial t$ and using the third and the fourth equations of (11.46) yield the required equation:

$$\frac{\partial^2}{\partial t^2}\Delta w + 4(\boldsymbol{\Omega} \cdot \nabla)^2 w = 0\,.$$

11.4.2. Find the modes in a homogeneous layer with plane horizontal rigid boundaries at $z = 0$, $-H$ without the use of the common approximation (the vector $\boldsymbol{\Omega}$ is not vertical).

Solution: Look for the solution of the last equation in the previous exercise in the form

$$w = \phi(z)\exp[\mathrm{i}(\boldsymbol{k} \cdot \boldsymbol{r} - \omega t)]\,.$$

Then we obtain for $\phi(z)$:

$$\phi'' + 2\mathrm{i}k\,\frac{F_z F_{\|}}{F_z^2 - \omega^2}\,\phi' + k^2\,\frac{\omega^2 - F_{\|}^2}{F_z^2 - \omega^2}\,\phi = 0\,,$$
$$F_z = 2\Omega_z\,, \qquad F_{\|} = 2\boldsymbol{\Omega} \cdot \boldsymbol{k}/k\,.$$

The general solution is

$$\phi(z) = b_+ \exp(\mathrm{i}a_+ z) + b_- \exp(\mathrm{i}a_- z)\,,$$

where a_+ and a_- are the roots of the characteristic equation

$$a^2 + 2k\frac{F_zF_{||}}{F_z^2 - \omega^2}a - k^2\frac{\omega^2 - F_{||}^2}{F_z^2 - \omega^2} = 0.$$

We find $a_\pm = k_z' \pm k_z''$, for these roots, where

$$k_z' = -k\frac{F_zF_{||}}{F_z^2 - \omega^2}, \qquad k_z'' = k\sqrt{\frac{\omega^2(F_z^2 + F_{||}^2 - \omega^2)}{(F_z^2 - \omega^2)^2}}.$$

The boundary condition at $z = 0$ $[\phi(0) = 0]$ is satisfied if we assume

$$\phi(z) = b\exp(ik_z'z)\sin(k_z''z).$$

Obviously, the condition at $z = -H$ $[\phi(-H) = 0]$ can be satisfied if k_z'' is real, i.e., if

$$\omega^2 < F_z^2 + F_{||}^2 = 4\Omega_z^2 + 4(\boldsymbol{\Omega}\cdot\mathbf{k}/k)^2.$$

In this case we have $k_z''H = n\pi$, $n = 1, 2, 3, \ldots$. Note that the solution obtained differs from (11.9), corresponding to the common approximation ($\boldsymbol{\Omega}\cdot\mathbf{k} = 0$ in the case under considered). In particular, the frequency may be larger than the Coriolis parameter $F_z = 2\Omega_z$.

11.4.3. Derive dispersion relations for modes in the case considered in Exercise 11.4.2.

Solution: These relations can be obtained by solving $k_z''H = n\pi$ of the previous exercise for ω. We have

$$\frac{\omega^2(F_z^2 + F_{||}^2 - \omega^2)}{(F_z^2 - \omega^2)^2} = \left(\frac{n\pi}{kH}\right)^2$$

whence the quadratic equation for ω^2 follows:

$$\left(1 + \frac{n^2\pi^2}{k^2H^2}\right)\omega^4 - \left(2\frac{n^2\pi^2}{k^2H^2}F_z^2 + F_z^2 + F_{||}^2\right)\omega^2 + \frac{n^2\pi^2}{k^2H^2}F_z^4 = 0$$

with the roots

$$\omega_\pm^2 = \frac{2n^2\pi^2F_z^2 + (F_z^2 + F_{||}^2)k^2H^2 \pm kH\sqrt{(F_z^2 + F_{||}^2)^2k^2H^2 + 4n^2\pi^2F_z^2F_{||}^2}}{2(n^2\pi^2 + k^2H^2)}.$$

Hence, there are two branches of the dispersion curves. Note first that for the case $\boldsymbol{\Omega}\cdot\mathbf{k} = 0$ ($\boldsymbol{\Omega}$ is along the z-axis), we have $\omega_+^2 = F_z^2 = 4\Omega_z^2$, $\omega_-^2 = F_z^2n^2\pi^2(n^2\pi^2 + k^2H^2)^{-1}$ and only ω_- corresponds to the inertial waves, see (11.10). We consider, however, the case $\boldsymbol{\Omega} \cdot \boldsymbol{k} \neq 0$ in more detail. As $kH \to 0$ (long-wavelength), we have

$$\omega_{\pm}^2 = F_z^2 \pm \frac{F_z F_{\|}}{n\pi} kH = 4\Omega_z^2 \pm 4 \frac{kH}{n\pi} \Omega_z(\boldsymbol{\Omega} \cdot \mathbf{k}/k) .$$

Both dispersion curves tend to the same frequency $2\Omega_z$ but with different slopes (group velocities) $c_g = \pm F_{\|}H/2n\pi$. For a real case $\Omega = 2\pi/24\ \mathrm{h}^{-1}$, $H = 10^4$ m, we have $C_g \simeq 0.5$ m/s. For short-wavelengths $kH \to \infty$, we obtain

$$\omega_+^2 \to F_z^2 + F_{\|}^2 = 4\Omega_z^2 + 4(\boldsymbol{\Omega} \cdot \mathbf{k}/k)^2 , \qquad \omega_-^2 \to 0 .$$

Note the horizontal anisotropy of the dispersion relations (dependence on the wave propagation direction $\boldsymbol{k}$). Figure 11.8 shows schematically dispersion curves in this case which differ considerably from those obtained in the common approximation (Fig. 11.2).

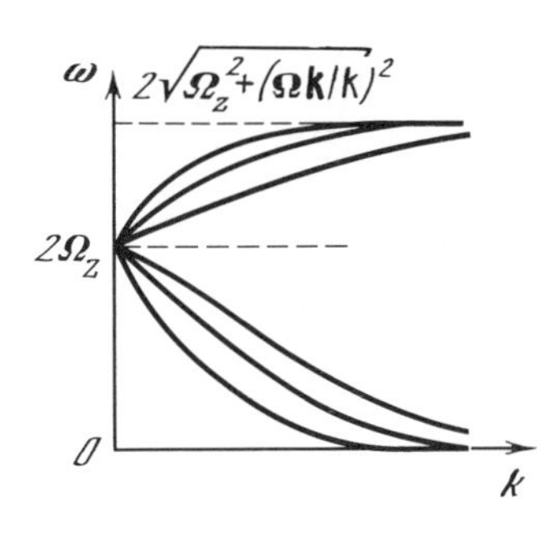

Fig. 11.8. Dispersion curves for inertial waves obtained without using the "traditional approximation" (compare with Fig. 11.2)

11.4.4. Obtain in the Boussinesq approximation an equation for waves in a stratified medium rotating with the angular velocity $\boldsymbol{\Omega}$ and oriented arbitrarily.

Solution: Use the system (10.11) in the same manner as in Exercise 11.4.1. Take into account that

$$\operatorname{curl}\left(g \frac{\rho}{\rho_0} \nabla z\right) = g \frac{\partial}{\partial y}\left(\frac{\rho}{\rho_0}\right)\nabla x - g \frac{\partial}{\partial x}\left(\frac{\rho}{\rho_0}\right)\nabla y , \qquad \operatorname{curl}\left(\frac{\nabla p}{\rho_0}\right) = 0$$

($\rho_0 = $ const in the Boussinesq approximation).

The answer is

$$\frac{\partial^2}{\partial t^2} \Delta w + 4(\boldsymbol{\Omega} \cdot \nabla)^2 w + N^2(z)\, \Delta_- w = 0 .$$

11.4.5. Using the equation obtained in the previous exercise, give a qualitative discussion of gyroscopic-gravity modes in a layer with rigid boundaries.

Solution: Substitution of $w = \phi(z)\exp[i(\boldsymbol{k}\cdot\boldsymbol{r} - \omega t)]$ into the above equation yields

$$\phi'' - 2ik\frac{F_{\parallel}F_z}{\omega^2 - F_z^2}\phi' + k^2\frac{N^2 - \omega^2 + F_{\parallel}^2}{\omega^2 - F_z^2}\phi = 0,$$

where $F_{\parallel} = 2\boldsymbol{\Omega}\cdot\boldsymbol{k}/k$ and $F_z = 2\Omega_z$. Now write $\phi(z)$ in the form

$\phi(z) = \exp[ikF_{\parallel}F_z z/(\omega^2 - F_z^2)]\psi(z)$ and substitute it into the above differential equation. We obtain the equation for $\psi(z)$, which does not include the first derivative

$$\psi'' + k^2\frac{(N^2 - \omega^2)(\omega^2 - F_z^2) + \omega^2 F_{\parallel}^2}{(\omega^2 - F_z)^2}\psi = 0.$$

The eigensolutions (modes) should vanish on the boundaries $\psi(0) = \psi(-H) = 0$. This is possible only when $\psi(z)$ is an oscillatory function. Hence, the coefficient of ψ in the last equation must be positive.

$$[N^2(z) - \omega^2](\omega^2 - F_z^2) + \omega^2 F_{\parallel}^2 > 0.$$

This yields a condition to be imposed on possible frequencies $\omega_-^2 < \omega^2 < \omega_+^2$ where $\omega_\pm^2(z) = [N^2 + F_z^2 + F_{\parallel}^2 \pm \sqrt{(N^2 + F_z^2 + F_{\parallel}^2)^2 - 4N^2F_z^2}]/2 = [N^2 + F_z^2 + F_{\parallel}^2 \pm \sqrt{(N^2 - F_z^2 - F_{\parallel}^2)^2 + 4N^2F_{\parallel}^2}]/2$. Note that $\omega_-^2 \le F_z^2 \le \omega_+^2$ for any $N^2(z) > 0$. The frequency F_z represents a singular point of the equation for ψ separating different branches of the dispersion curves. If

$$F_z^2 < \omega^2 < \max\omega_+^2(z) = \tfrac{1}{2}[N_{\max}^2 + F_z^2 + F_{\parallel}{}^2 + \sqrt{(N_{\max}^2 - F_z^2 - F_{\parallel}^2)^2 + 4N_{\max}^2F_{\parallel}^2}]$$

$[N_{\max} = \max N(z)]$, the modes are located in the layers with large $N^2(z)$ and correspond to internal waves. If

$$\min\omega_-^2(z) = \tfrac{1}{2}[N_{\min}^2 + F_z^2 + F_{\parallel}^2 - \sqrt{(N_{\min}^2 - F_z^2 - F_{\parallel}^2)^2 + 4N_{\min}^2F_{\parallel}^2}] < \omega^2 < F_z^2$$

$[N_{\min} = \min N(z)]$, modes are located in the layers with a small Väisälä frequency and correspond to inertial waves. For long wavelengths ($k \to 0$), we obtain from the equation for ψ

$$(\omega^2 - F_z^2)^2/k^2 \to \alpha = \text{const} \qquad (\text{if } F_{\parallel} = 2\boldsymbol{\Omega}\cdot\boldsymbol{k}/k \neq 0).$$

Hence, as $k \to 0$, we have for $\psi(z)$,

$$\psi'' + \alpha^{-2}F_z^2F_{\|}^2\psi = 0\,, \quad \psi = b\sin(\alpha^{-1}F_zF_{\|}z)\,, \quad F_zF_{\|}\alpha^{-1} = n\pi/H$$

and the dispersion relation becomes

$$\omega^2 = F_z^2 \pm (F_zF_{\|}/n\pi)kH\,, \quad \omega \simeq F_z \pm (F_{\|}/2n\pi)kH\,,$$

which is analogous to the case of a homogeneous fluid (Exercise 11.4.3).

11.4.6. Derive an expression for the group velocity of modes for the boundary-value problem (11.16) in terms of an integral of the eigenfunctions.

Hint: Proceed as in (10.19). The answer is

$$c_g = \frac{d\omega}{dk} = \frac{\omega^2 - F^2}{\omega k} \int_{-H}^{0} \left(\frac{d\phi}{dz}\right)^2 dz \Big/ \int_{-H}^{0} \left[\left(\frac{d\phi}{dz}\right)^2 + k^2\phi\right] dz\,.$$

11.4.7. Prove that the relations for inertial oscillations $u = b\exp(-\mathrm{i}Ft)$, $v = -\mathrm{i}b\exp(-\mathrm{i}Ft)$, $w = 0$ and $F = 2\Omega_z$ satisfy (10.11) for a rotating, stratified fluid without the common approximation.

Solution: First, it is obvious that the continuity equation $\nabla \boldsymbol{v} = 0$ in (10.11) is satisfied. Further, it follows from the equation of state $\partial(\rho/\rho_0)/\partial t = g^{-1}N^2w = 0$ that $\rho = 0$. The Euler equation yields in turn

$$\partial p/\partial x = 0\,, \quad \partial p/\partial y = 0\,, \quad \rho_0^{-1}(\partial p/\partial z) + 2(\Omega_x v - \Omega_y u) = 0\,.$$

Hence,

$$p = p(z,t) = -2b(\mathrm{i}\Omega_x + \Omega_y)\exp(-\mathrm{i}Ft)\int_z^0 \rho_0(\xi)\,d\xi\,.$$

For $b = 1$ m/s, $\Omega = (2\pi/24)\ \mathrm{h}^{-1}$ and $H = 10^4$ m, for example, we have

$$p_{\max} = p(-H) \sim 10^{-2}\ \mathrm{atm}$$

which is very small compared with the atmospheric pressure. Hence, it is reasonable to assume $p \equiv 0$ and the inertial oscillations are the same as in the traditional approximation".

11.4.8. Consider the reflection of a Rossby wave at a rigid vertical wall.

Solution: Write the incident and reflected waves in the form

$$\psi_1 = a\exp[\mathrm{i}(k_{1x}x + k_{1y}y - \omega t)]\,, \quad \psi_2 = aV\exp[\mathrm{i}(k_{2x}x + k_{2y}y - \omega t)]\,.$$

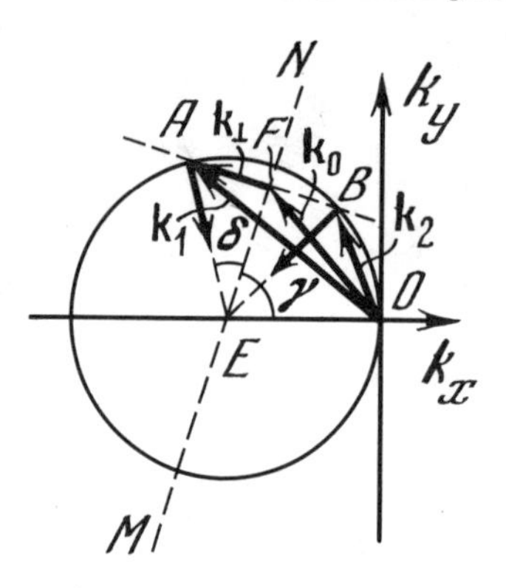

Fig. 11.9. Reflection of a Rossby wave at the boundary

The incident-wave vector $\boldsymbol{k}_1 = \{k_{1x}, k_{1y}\}$ ends at point A on a circle of radius $\beta/2\omega$ with its center at the point $(-\beta/2\omega, 0)$ (Fig. 11.9). Suppose that the straight line MFN, which makes an angle γ with the x-axis, is parallel to the reflecting wall. The frequency ω as well as the projection of the wave vector onto the wall are conserved. Hence, the wave vector of the reflected wave $\boldsymbol{k}_2 = \{k_{2x}, k_{2y}\}$ must end at B on the same circle. This is the point of intersection of the straight line AB perpendicular to the wall with the circle due to the equality of the projection of the wave vectors $\boldsymbol{k}_1$ and $\boldsymbol{k}_2$ on the wall. It is obvious that the law "the angle of reflection is equal to the angle of incidence" does not hold in the general case. This law holds, however, for the group-velocity vectors $(\boldsymbol{c}_g)_j$, $j = 1, 2$. In fact, these vectors point from the ends of the wave vectors to the circle's center and make equal angles ($\angle$AEF $= \angle$FEB) with the wall since EF $\perp$ AB. We introduce the new vectors $\boldsymbol{k}_0 = \mathbf{OF}$ and $\boldsymbol{k}_\perp = \mathbf{FA}$, then $\boldsymbol{k}_1 = \boldsymbol{k}_0 + \boldsymbol{k}_\perp$ and $\boldsymbol{k}_2 = \boldsymbol{k}_0 - \boldsymbol{k}_\perp$. Now the sum of the incident and the reflected waves can be written as

$$\psi = a[\exp(\mathrm{i}\boldsymbol{k}_\perp \cdot \boldsymbol{r}) + V\exp(-\mathrm{i}\boldsymbol{k}_\perp \cdot \boldsymbol{r})]\exp[\mathrm{i}(\boldsymbol{k}_0 \cdot \boldsymbol{r} - \omega t)].$$

The reflection coefficient V can be found from the evident condition that the wall is one of the streamlines, for example, $\psi = 0$ at the line MN. Assuming also that the origin ($x = 0$, $y = 0$) is at the wall, we obtain $\boldsymbol{k}_\perp \cdot \boldsymbol{r} = 0$ at MN and $V = -1$, hence

$$\psi = 2\mathrm{i}a\sin(\boldsymbol{k}_\perp \cdot \boldsymbol{r})\exp[\mathrm{i}(\boldsymbol{k}_0 \cdot \boldsymbol{r} - \omega t)].$$

11.4.9. Consider the barotropic Rossby waves between two parallel rigid walls Γ_1 and Γ_2 a distance L apart.

Solution: Let the origin be at Γ_1. Then the last relation in the previous exercise determines the streamline function ψ which is zero at Γ_1. The other wall (Γ_2), where $\boldsymbol{k}_\perp \cdot \boldsymbol{r} = k_\perp L$, is also a streamline, i.e.,

$$\psi(\Gamma_2) = 0 \quad \text{or} \quad \sin k_\perp L = 0, \quad k_{\perp n} = n\pi/L, \quad n = 1, 2, \ldots.$$

Denote $\delta = \angle AEF$ (Fig. 11.9), then

$$k_\perp = (\beta/2\omega)\sin\delta\,, \qquad 0 \le \delta \le \pi$$

(the case $-\pi < \delta < 0$ can be obtained by interchanging $\boldsymbol{k}_1 \rightleftarrows \boldsymbol{k}_2$). Hence, we have a finite number of eigen-angles δ_n for a fixed frequency so that

$$\sin\delta_n = 2n\pi\omega/\beta L\,, \qquad n \le \beta L/2\pi\omega\,.$$

No propagating modes exist if $\omega > \beta L/2\pi$. The limiting frequency for the nth-mode is $\omega_n = \beta L/2n\pi$.

11.4.10. Determine particle paths in a field composed of two barotropic Rossby waves with equal frequencies and amplitudes propagating at the angles $\pm\alpha$ with the x-axis.

Solution: The field of two such modes is

$$\psi = a\exp[\mathrm{i}(k_x x - \omega t)][\exp(\mathrm{i}k_y y) + \exp(-\mathrm{i}k_y y)]\,, \qquad a = A\exp(\mathrm{i}\theta)\,.$$

The components of the fluid particle displacements ξ, η is (real parts)

$$\xi = \frac{u}{-\mathrm{i}\omega} = \frac{1}{-\mathrm{i}\omega}\frac{\partial\psi}{\partial y} = \frac{2A}{\omega}k_y\sin(k_y y)\sin\varphi\,, \qquad \eta = \frac{2A}{\omega}k_x\cos(k_y y)\cos\varphi\,,$$

where $\varphi = k_x x - \omega t + \theta$. The particle's paths is obtained by eliminating t from the last equations. Taking into account that $k = (\beta/\omega)\cos\alpha$, $k_x = -k\cos\alpha$ and $k_y = k\sin\alpha$, we have

$$(\xi^2/a_x^2) + (\eta^2/a_y^2) = 1\,.$$

This is an ellipse with the semi-axes:

$$a_x = 2\frac{A\beta}{\omega^2}\cos\alpha\left|\sin\alpha\sin\left(\frac{\beta y}{2\omega}\sin 2\alpha\right)\right|, \qquad a_y = 2\frac{A\beta}{\omega^2}\cos^2\alpha\left|\cos\left(\frac{\beta y}{2\omega}\sin 2\alpha\right)\right|.$$

11.4.11. Find relations for the streamlines $\psi = \text{const}$ for the wave field considered in the previous exercise.

Solution: Taking into account the real part of ψ, we have from the relation $\psi = C = \text{const}$,

$$2A\cos(k_y y)\cos(k_x x - \omega t + \theta) = C\,.$$

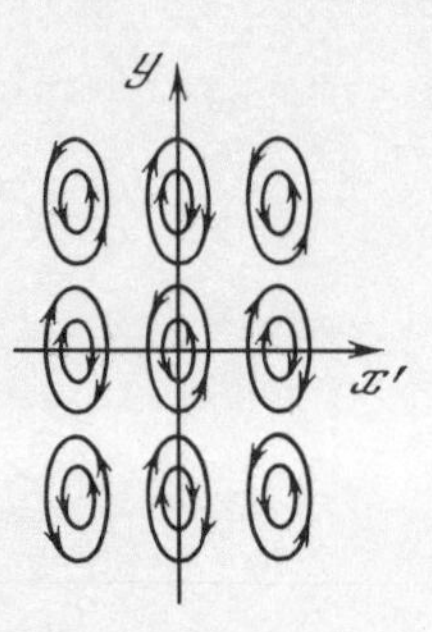

Fig. 11.10. Flow streamlines for two superimposed Rossby waves with different propagation directions

Exclude time t by passing to the coordinate system $x' = x - (\omega t/k_x) + \theta/k_x$ moving with velocity ω/k_x. The streamline relations in this system is $\cos(k_y y)\cos(k_x x') = C/2A$. A streamline pattern is shown schematically in Fig. 11.10. Arrows correspond to the directions of particle motion. As a result, we have obtained a periodic (chessboard-like) pattern with alternating cyclonic (counterclockwise rotation in the case of the north hemisphere) and anticyclonic (clockwise) eddies. In the laboratory coordinate system (x, y) this pattern moves westward with velocity ω/k_x. The path of any given particle is an ellipse (Exercise 11.4.10).

11.4.12. Find the eigenvalue $\nu_n(\omega)$ of the boundary-value problem (11.39) for a constant Väisälä frequency and $\omega < N$.

Solution: At $N = \text{const}$, the solution of (11.39) vanishing at $z = -H$ is

$$W(z) = b\sin[\sqrt{\nu(N^2 - \omega^2)}(z + H)] .$$

Using the condition at $z = 0$ we get

$$\tan\sigma = \frac{(N^2 - \omega^2)H}{g\sigma}, \qquad \sigma = \sqrt{\nu(N^2 - \omega^2)}H .$$

For typical values of N^2 and H, we have $(N^2 - \omega^2)H/g \ll 1$ if $\omega < N$. Hence, the first root can be found by a series expansion of $\tan\sigma$:

$$\sigma_0^2 = \nu_0(N^2 - \omega^2)H^2 = g^{-1}(N^2 - \omega^2)H , \qquad \nu_0 = (gH)^{-1} .$$

For higher n σ_n differs only slightly from the roots of function $\tan\sigma$ or $\sigma_n \simeq n\pi$ $(n = 1, 2, \ldots)$. Hence, $\nu_n \simeq n^2\pi^2/H^2(N^2 - \omega^2)$ and $\nu_n \simeq n^2\pi^2/N^2H^2$ if $N^2 \gg \omega^2$.

11.4.13. Derive a dispersion relation for the baroclinic Rossby waves in a fluid layer with constant N. Discuss under what conditions the inequality $\omega \ll F_0$ holds for the Rossby waves [and hence (11.43) is valid].

Solution: Substituting $v_n = n^2\pi^2/N^2H^2$, derived in the previous exercise, into (11.38) for v, we obtain

$$\frac{n^2\pi^2}{N^2H^2} = \frac{k^2 + \beta k_x/\omega}{\omega^2 - F_0^2}.$$

Denote $f = \omega/N$, $f_0 = F_0/N$ and take into account that $k_x = k\cos\alpha$. Then we have the third degree equation for the determination of the dimensionless frequency f:

$$\frac{n^2\pi^2}{k^2H^2}f^3 - \left(1 + \frac{n^2\pi^2}{k^2H^2}f_0^2\right)f + \delta = 0,$$

where $\delta = (\beta/Nk)\cos\alpha = (2\Omega/NkR_0)\cos\varphi_0\cos\alpha$ according to (11.26). This parameter must be small ($\delta \ll 1$) since $kR_0 \gg 1$ (the latter inequality is necessary for the applicability of the β-plane approximation). The term f^3 in the last equation can be neglected if $\omega \ll F_0$ and we obtain

$$f = \delta[1 + f_0^2 n^2\pi^2/k^2H^2]^{-1},$$
$$\omega = (2\Omega/kR_0z)(1 + F_0^2 n^2\pi^2/N^2k^2H^2)^{-1}\cos\varphi_0\cos\alpha.$$

The applicability condition for this approximation is

$$f/f_0 = (\delta/f_0)(1 + f_0^2 n^2\pi^2/k^2H^2)^{-1} \ll 1,$$

which is satisfied if $\delta/f_0 \ll 1$ or if $kR_0\tan\varphi_0 \gg 1$.

11.4.14. Show that the potential and kinetic energies of the gyroscopic-gravity waves does not equal each other (compare with the waves of Exercise 10.6.12).

11.4.15. Consider the dispersion relation of the Exercise 11.4.12 for the case $\omega > N$.

12. Sound Waves

In Chaps. 10, 11 we have neglected fluid compressibility. This was reasonable because the compressibility affects very little the equations governing surface and other types of hydrodynamic waves considered above.

In this chapter we investigate acoustic or sound waves where fluid compressibility must be taken into account as a main feature. By their nature, sound waves are analogous to longitudinal waves in the solids considered above (Chaps. 2, 4). In both cases elastic forces related to deformations in a medium turn out to be the main factor.

Sound waves have many important applications in the ocean, e.g., underwater telemetry and communications, acoustical exploration of the ocean and so on. Their role in the underwater world is approximately the same as that of electromagnetic waves above the Earth's surface. This is because underwater sound waves can cover distances of many thousands of kilometers whereas electromagnetic waves decay in the sea water almost completely over the first kilometer. In the atmosphere, sound waves of very low frequencies (infrasound) can also propagate over thousands of kilometers.

In this chapter we derive *linear acoustic equations* and a *wave equation* for the sound waves that govern the sound propagation in fluids. The simple solutions of these equation in homogeneous as well as in inhomogeneous fluids will be considered. The main feature of guided propagation (in an underwater sound channel, for example) will be discussed as well as the generation of sound by the most simple sources.

12.1 Plane Waves in Static Fluids

12.1.1 The System of Linear Acoustic Equations

An elasticity force plays the role of a restoring force in the case of acoustic waves. Assuming that all other kinds of forces can be neglected ($g = 0, N^2 = 0, \boldsymbol{\Omega} = 0$), we obtain from the linearized system of hydrodynamic equations (10.3) the following for acoustic (sound) waves:

$$\frac{\partial \boldsymbol{v}}{\partial t} + \frac{\nabla p}{\rho_0} = 0\,, \qquad \frac{\partial \rho}{\partial t} + \rho_0 \nabla \boldsymbol{v} = 0\,, \qquad \frac{\partial \rho}{\partial t} = \frac{1}{c^2}\frac{\partial p}{\partial t}\,. \tag{12.1}$$

The last equation here is the state equation and $c^2 = (\partial p/\partial \rho)_s$ (s stands for entropy) is the *adiabatic sound velocity* squared. We will omit the adjective "adiabatic" in what follows, for brevity.

Excluding the density ρ from (12.1), we obtain two equations of *linear acoustics*

$$\frac{\partial \boldsymbol{v}}{\partial t} + \frac{\nabla p}{\rho_0} = 0\,, \qquad \frac{1}{\rho_0 c^2}\frac{\partial p}{\partial t} + \nabla \boldsymbol{v} = 0\,. \tag{12.2}$$

These equations can be reduced to one for p:

$$\frac{1}{c^2}\frac{\partial^2 p}{\partial t^2} - \Delta p + \frac{1}{\rho_0}\nabla \rho_0 \cdot \nabla p = 0\,. \tag{12.2a}$$

The last term on the left-hand side is zero or small compared with the second one if the density of the medium ρ_0 is constant or the space scale of its variation L is large compared with the wavelength λ. Indeed, this term is of the order of magnitude of $kp/L = 2\pi p/\lambda L$, whereas the second one is $k^2 p = (2\pi/\lambda)^2 p$. It can be neglected as compared with the second term if $L \gg \lambda$. Assuming this to be the case, we obtain our *basic wave equation*

$$\Delta p - \frac{1}{c^2}\frac{\partial^2 p}{\partial t^2} = 0 \tag{12.3}$$

which will be used below. The sound velocity here may be a function of position $c = c(\boldsymbol{r})$. The solution of (12.3) being known, the particle velocity $\boldsymbol{v}(\boldsymbol{r}, t)$ at any t can be found by integrating the first equation of (12.2):

$$\boldsymbol{v}(\boldsymbol{r}, t) = \boldsymbol{v}(\boldsymbol{r}, t_0) - \rho_0^{-1}\nabla \int_{t_0}^{t} p(\boldsymbol{r}, \tau)\, d\tau\,. \tag{12.4}$$

12.1.2 Plane Waves

The wave equation (12.3) has a plane-wave solution if $c =$ const. This type of solution was considered in detail in Sect. 4.1.1. for the case of longitudinal elastic waves. Assuming that the x-axis points in the wave-propagation direction, we have for a solution of (12.3)

$$p(x, t) = f(x - ct) + g(x + ct)\,, \tag{12.5}$$

where f and g are arbitrary functions. The term $f(x - ct)$ represents a wave propagating in the positive direction (with respect to x) and $g(x + ct)$ is a wave propagating in the opposite direction.

A simple relationship exists between the particle velocity and the pressure in a plane wave. Indeed, taking, for example, the first term in (12.5) and substituting it into (12.4), we find

$$\begin{aligned}\boldsymbol{v}(\boldsymbol{r},t) &= \boldsymbol{v}(\boldsymbol{r},t_0) - \rho_0^{-1}\nabla\int_{t_0}^{t} f(x-c\tau)\,d\tau \\ &= \boldsymbol{v}(\boldsymbol{r},t_0) + (\rho_0 c)^{-1}\nabla\int_{x-ct_0}^{x-ct} f(\xi)\,d\xi\,.\end{aligned}$$

Applying the operation ∇ to the integral with variable limits we obtain for the only nonzero velocity component v_x:

$$v_x(x,t) = (\rho_0 c)^{-1} f(x-ct)\,. \tag{12.6}$$

An analogous relation holds for the backward wave:

$$v_x(x,t) = -(\rho_0 c)^{-1} g(x+ct)\,. \tag{12.7}$$

The quantity $\rho_0 c$, which is the ratio between the pressure and the particle velocity in a plane sound wave, is referred to as the *characteristic acoustic impedance* of the medium. The inverse quantity $(\rho_0 c)^{-1}$ is termed the *characteristic acoustic admittance.*

Harmonic waves are of considerable importance in acoustics. The sound pressure can be written as $p = \psi(\boldsymbol{r})\exp(-\mathrm{i}\omega t)$ where ω is the wave frequency, and the wave equation (12.3) is reduced to the so-called *Helmholtz equation:*

$$\Delta\psi + k^2\psi = 0\,, \qquad k^2 = \omega^2/c^2\,. \tag{12.8}$$

The simplest solution of this equation is a harmonic plane wave $\psi = A\exp(\mathrm{i}\boldsymbol{k}\cdot\boldsymbol{r})$, where the relation

$$k_x^2 + k_y^2 + k_z^2 = \omega^2/c^2 \tag{12.9}$$

holds for the components of the *wave vector* $\boldsymbol{k} = \{k_x, k_y, k_z\}$. Hence, we have for the sound pressure

$$p = A\exp[\mathrm{i}(\boldsymbol{k}\cdot\boldsymbol{r} - \omega t)]\,. \tag{12.10}$$

The particle velocity is, according to the first equation in (12.2),

$$\boldsymbol{v} = (\mathrm{i}\omega\rho_0)^{-1}\nabla p\,, \tag{12.11}$$

or in the case of plane wave with (12.10),

$$\boldsymbol{v} = A(\rho_0 c)^{-1}(\boldsymbol{k}/k)\exp[\mathrm{i}(\boldsymbol{k}\cdot\boldsymbol{r} - \omega t)]\,. \tag{12.12}$$

Hence, the velocity $\boldsymbol{v}$ is collinear with the vector $\boldsymbol{k}$. Two harmonic plane waves with equal frequencies and amplitudes but propagating in opposite directions form a *standing wave:*

$$\begin{aligned} p &= 2A\cos(\boldsymbol{k}\cdot\boldsymbol{r})\exp(-\mathrm{i}\omega t)\,, \\ \boldsymbol{v} &= 2\mathrm{i}A(\rho_0 c)^{-1}(\boldsymbol{k}/k)\sin(\boldsymbol{k}\cdot\boldsymbol{r})\exp(-\mathrm{i}\omega t)\,. \end{aligned} \tag{12.13}$$

The *nodal planes* of the pressure ($p = 0$) are determined by $\boldsymbol{k}\cdot\boldsymbol{r} = \pi(n + 1/2)$ and are spaced at half wavelengths. The particle velocity is maximum at these planes. On the contrary, the planes of maximum pressure ($\boldsymbol{k}\cdot\boldsymbol{r} = n\pi$) are the nodal planes of the velocity.

12.1.3 Generation of Plane Waves. Inhomogeneous Waves

Plane waves can be generated by normal velocity or pressure distributions over some plane. In the simplest case the normal velocity is constant over the plane (sound radiation by an oscillating infinite piston). To consider this case we assume that $v_z(t) = f(t)$ is specified at the plane $z = 0$. One can easily prove that the solution of (12.2) is then

$$v_x = v_y = 0\,, \qquad v_z = f(t - z/c)\,, \qquad p = \rho_0 c f(t - z/c)\,.$$

In the case of an arbitrary distribution of normal velocity or pressure over a plane, it is advisable to represent this distribution in the form of a Fourier integral with respect to x, y and t and analyse the sound radiation by a single Fourier component. For example, at $z = 0$ such a component of the pressure is

$$p|_{z=0} = A\exp[\mathrm{i}(k_x x + k_y y - \omega t)]\,. \tag{12.14}$$

The solution of (12.3) turning into (12.14) at $z = 0$ is a harmonic plane wave

$$p(x, y, z, t) = A\exp[\mathrm{i}(k_x x + k_y y + k_z z - \omega t)]\,, \qquad k_z = \sqrt{k^2 - k_\perp^2}\,, \tag{12.15}$$

where $k_\perp = \sqrt{k_x^2 + k_y^2}$ is the projection of the wave vector $\boldsymbol{k}$ onto the plane $z = 0$. It is said frequently that (12.14) is the *trace* of a wave (12.15) at the plane $z = 0$. This trace is a wave of the same frequency and amplitude, its wave vector being the projection of vector $\boldsymbol{k}$ onto the plane. In a similar manner we can construct the wave field in the halfspace $z > 0$ for each Fourier harmonic.

Let us consider an interesting generalization of (12.15) for the case when $k_\perp > k = \omega/c$, i.e., when the wave vector's component along the plane exceeds the vector's magnitude. The quantity $k_z = \sqrt{k^2 - k_\perp^2} = \mathrm{i}\sqrt{k_\perp^2 - k^2}$ is then purely imaginary and the sound pressure in the halfspace $z > 0$ is

$$p(x, y, z, t) = A\exp(-|k_z|z)\exp[\mathrm{i}(k_x x + k_y y - \omega t)]\,. \tag{12.16}$$

Such a wave is termed an *inhomogeneous plane wave*. Its fronts are the planes $k_x x + k_y y = \text{const}$. Its amplitude is not constant in space but decreases exponentially with the distance from the plane $z = 0$. Planes of constant amplitude $z = \text{const}$ are orthogonal to the wave fronts.

We can assume $k_y = 0$ without loss of generality. The fluid-particle velocity, which is in the (x, y) plane ($v_y = 0$), is now not collinear with the propagation direction (x-axis) but has also a component perpendicular to it. One has, according to (12.11),

$$\begin{aligned} v_x &= \frac{1}{\mathrm{i}\omega\rho_0}\frac{\partial p}{\partial x} = \frac{k_x}{\omega\rho_0} A \exp(-|k_z|z)\exp[\mathrm{i}(k_x x - \omega t)], \\ v_z &= \frac{1}{\mathrm{i}\omega\rho_0}\frac{\partial p}{\partial z} = \mathrm{i}\frac{|k_z|}{\omega\rho_0} A \exp(-|k_z|z)\exp[\mathrm{i}(k_x x - \omega t)]. \end{aligned} \tag{12.17}$$

The component v_z has a phase delay of $\pi/2$, as compared with v_x, which means that the particle paths are ellipses.

A plane inhomogeneous wave can be written in the general form (12.10) if we assume the wave vector $\boldsymbol{k} = \boldsymbol{k}' + \mathrm{i}\boldsymbol{k}''$ to be a complex but still obeying (12.9) with real positive $k^2 = \omega^2/c^2$. The expression $p = \exp[-\boldsymbol{k}''\boldsymbol{r} + \mathrm{i}(\boldsymbol{k}' \cdot \boldsymbol{r} - \omega t)]$ obtained in such a way represents the wave propagating along $\boldsymbol{k}'$ with the phase velocity $c_{\mathrm{Ph}} = \omega/k'$, its amplitude varying in the (x, z) plane. Substituting $\boldsymbol{k} = \boldsymbol{k}' + \mathrm{i}\boldsymbol{k}''$ into (12.9) and separating the imaginary and real parts, we obtain

$$\begin{aligned} &\boldsymbol{k}' \cdot \boldsymbol{k}' = k'_x k''_x + k'_y k''_y + k'_z k''_z = 0, \\ &(k')^2 - (k'')^2 = k^2 = \omega^2/c^2. \end{aligned} \tag{12.18}$$

Hence it follows that planes of constant phase $\boldsymbol{k}' \cdot \boldsymbol{r} = \text{const}$ and planes of constant amplitude $\boldsymbol{k}'' \cdot \boldsymbol{r} = \text{const}$ are mutually orthogonal. In the same manner as in the simple case (12.16), the amplitude gradient of the wave is perpendicular to the propagation direction. It also follows from (12.18) that $k^2 < (k')^2$ or $c_{\mathrm{Ph}}^2 = (\omega/k')^2 < (\omega/k)^2 = c^2$, i.e., the phase velocity of an inhomogeneous wave is always less than the sound velocity in free space. Hence, the wavelength $\lambda = (2\pi/k') < 2\pi/k$ is also less than that of the ordinary plane wave with the same frequency.

An inhomogeneous wave cannot exist in infinite free space because its amplitude increases infinitely along one of the axes. The concept of such waves, however, turns out to be very useful in the presence of boundaries or sources. Inhomogeneous waves must be taken into account, for example, when sound generation by an arbitrary distribution of normal velocity or pressure over the plane is considered. We will show below that an ordinary plane wave refracted at an interface between two media can be transformed into an inhomogeneous one and vice versa.

12.1.4 Sound Energy

We now refer to the relation for energy per unit volume given in Sect. 6.4.4 which, in the absence of external forces, is

$$E = \rho v^2/2 + \rho\varepsilon\,,$$

where $\rho = \rho_0 + \rho'$ as well as $p = p_0 + p'$ are the total density and pressure (not their variations). Since we consider hydrodynamical equations in a linear approximation, it is reasonable to confine ourselves to second-order terms in the relation for energy. Hence, we have for the kinetic energy per unit volume $E_k = (\rho_0 + \rho')v^2/2 = \rho_0 v^2/2$. Concerning the internal energy $E_i = (\rho_0 + \rho')\varepsilon$, its increment in transition from an equilibrium state ρ_0, p_0 to a disturbed one $\rho = \rho_0 + \rho'$, $p = p_0 + p'$ is, according to (6.33), $E_i = \rho\int_{\rho_0}^{\rho}(p/\rho^2)\,d\rho$. But $p(\rho) = p_0 + c^2\rho'$, hence,

$$\begin{aligned} E_i &= \rho\int_{\rho_0}^{\rho}(p_0/\rho^2)\,d\rho + c^2\rho\int_0^{\rho'}\rho^{-2}\rho'\,d\rho' \\ &\simeq (p_0/\rho_0)\rho' + (c^2/2\rho_0)(\rho')^2 = (p_0/\rho_0)\rho' + p'\rho'/2\rho_0\,. \end{aligned}$$

Here, the first term which is linear with respect to ρ' represents the internal energy increment corresponding to the work against the equilibrium pressure. The second quadratic term corresponds to the work against the acoustic pressure and is termed the *internal energy of the acoustic wave.* Note that the time as well as the space average of the first term is zero in a harmonic plane wave.

For the sound energy density we obtain, omitting primes at p' and ρ',

$$\begin{aligned} &E = E_k + E_i\,, \qquad E_k = \rho_0 v^2/2\,, \\ &E_i = (c^2/2\rho_0)\rho^2 = p\rho/2\rho_0 = p^2/2\rho_0 c^2\,. \end{aligned} \tag{12.19}$$

For the plane *progressive* wave, where pressure and velocity are related by the simple equation $v = \pm p/\rho_0 c$, we have

$$E_k = E_i = p^2/2\rho_0 c^2\,, \qquad E = p^2/\rho_0 c^2 = \rho_0 v^2\,. \tag{12.20}$$

We now look for a mathematical form of the energy conservation law in acoustics. Differentiating E in (12.29) with respect to time and taking (12.2) into account gives

$$\frac{\partial E}{\partial t} = \rho_0\boldsymbol{v}\cdot\frac{\partial\boldsymbol{v}}{\partial t} + \frac{p}{\rho_0 c^2}\frac{\partial p}{\partial t} = -\boldsymbol{v}\cdot\nabla p - p\nabla\boldsymbol{v} = -\nabla(p\boldsymbol{v})\,. \tag{12.21}$$

Integrating this relation over some fixed volume V and applying the Gauss' divergence theorem we find

$$\frac{\partial}{\partial t}\int_V E\,dV = -\int_S p v_n\,dS\,. \tag{12.21a}$$

Hence, it follows that $\boldsymbol{I} = p\boldsymbol{v}$ is the *acoustic specific energy flux vector*. The same could also be deduced by another kind of reasoning, namely $p\boldsymbol{n}\,dS$ is the sound pressure force on the area dS with the normal $\boldsymbol{n}$ and $p\boldsymbol{v} \cdot \boldsymbol{n}\,dS$ is the work per unit time performed by this force, hence, $\boldsymbol{I} = p\boldsymbol{v}$ is the specific flux of energy per unit time.

In the case of a plane wave propagating in the $\boldsymbol{n}$-direction, we have

$$p = p(\boldsymbol{n} \cdot \boldsymbol{r} - ct), \qquad \boldsymbol{v} = \boldsymbol{n}p(\boldsymbol{n} \cdot \boldsymbol{r} - ct)/\rho_0 c$$

and the specific flux of energy is

$$\boldsymbol{I} = \frac{p^2}{\rho_0 c}\boldsymbol{n} = \rho_0 c v^2 \boldsymbol{n} = Ec\boldsymbol{n}. \tag{12.22}$$

The vector $\boldsymbol{I}$ points along the propagation direction, its magnitude is equal to the energy density times the sound velocity. In other words, the energy in a progressive sound wave is translated at the speed of sound which is only natural since dispersion is absent.

In the case of harmonic waves, we must use the real parts of (12.10) for p and $\boldsymbol{v}$ when calculating the quadratic quantities E and $\boldsymbol{I}$. For an inhomogeneous wave we obtain, for example, from (12.16) assuming $k_y = 0$, taking (12.17) into account, denoting $A = |A|\exp(\mathrm{i}\alpha)$ and $\varphi = k_x x - \omega t + \alpha$:

$$\begin{aligned} I_x &= |A|\exp(-|k_z|z)\cos\varphi \frac{k_x|A|\exp(-|k_z|z)}{\omega\rho_0}\cos\varphi \\ &= \frac{|A|^2 k_x}{2\omega\rho_0}\exp(-2|k_z|z)(1 + \cos 2\varphi), \\ I_z &= -\frac{|A|^2|k_z|}{2\omega\rho_0}\exp(-2|k_z|z)\sin 2\varphi. \end{aligned} \tag{12.23}$$

Note that I_z changes its sign after each half period $T/2$ ($T = 2\pi/\omega$). Hence, the flux along the z-axis averaged over a period is zero. In the x-direction the average flux is

$$\bar{I}_x = \frac{|A|^2 k_x}{2\omega\rho_0}\exp(-2|k_z|z). \tag{12.24}$$

It has been shown above that the inhomogeneous wave (12.16) is related to the pressure distribution (12.14) over the plane $z = 0$, with a speed of propagation less than the sound velocity ($\omega/k_\perp < c$). Hence, such a disturbance does not generate an energy flux in the z-direction. In the case of an ordinary harmonic plane wave (12.10) with real $\boldsymbol{k}$, analogous calculations lead to

$$\bar{I} = \frac{|A|^2}{2\rho_0 c}\frac{\boldsymbol{k}}{k}, \qquad \bar{I}_x = \bar{I}\frac{k_x}{k}, \qquad \bar{I}_y = \bar{I}\frac{k_y}{k}, \qquad \bar{I}_z = \bar{I}\frac{k_z}{k}. \tag{12.25}$$

12.2 Sound Propagation in Inhomogeneous Media

12.2.1 Plane Wave Reflection at the Interface of Two Homogeneous Media

The simplest problem of space inhomogeneity is two homogeneous media in contact along an infinite plane. We discuss sound-wave reflection and refraction at such an interface, the density and the sound velocity being ρ_1 and c_1 in the medium occupying the halfspace $z < 0$ and ρ_2, c_2 for $z > 0$ (Fig. 12.1).

The wave

$$p_1^+ = A\exp[\mathrm{i}(\xi x + k_{1z}z - \omega t)], \quad z < 0 \tag{12.26}$$

is assumed to be incident from the halfspace $z < 0$ upon the interface $z = 0$. The wave vector $\boldsymbol{k}_1$ is chosen in the (x, z)-plane, so that $k_{1x} = \xi$, $k_{1y} = 0$ and $k_{1z} = \sqrt{(\omega/c_1)^2 - \xi^2}$. We have seen earlier (Sect. 4.1.3) that for the case of a static interface, the boundary conditions can be satisfied for arbitrary x and t only if the frequencies and the components of wave vectors along the plane are the same for the incident reflected and the refracted waves. Hence, the reflected wave can be written as

$$p_1^- = VA\exp[\mathrm{i}(\xi x - k_{1z}z - \omega t)], \quad z < 0, \tag{12.27}$$

where V is the *reflection coefficient*. The wave vector of the reflected wave $\boldsymbol{k}_1' = \{\xi, 0, -k_{1z}\}$ is shown in Fig. 12.1.

First let us suppose that the second medium ($z > 0$) is a vacuum, i.e., $z = 0$ is the free boundary of medium 1. Then the sum of pressures in the incident and the reflected waves must vanish at $z = 0$. It immediately gives $V = -1$. The normal component of the total velocity at the boundary is, according to (12.11),

$$v_{1z}|_{z=0} = \frac{2A}{\rho_1 c_1}\frac{k_{1z}}{k_1}\exp[\mathrm{i}(\xi x - \omega t)],$$

i.e., twice the normal velocity of the incident wave.

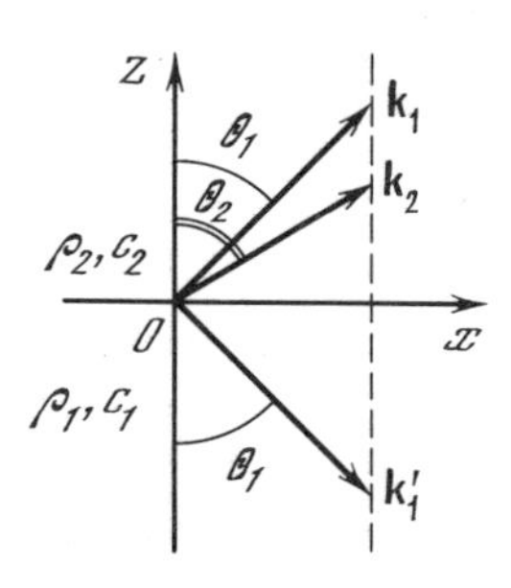

Fig. 12.1. The reflection of a sound wave at an interface between two media. $\boldsymbol{k}_1$, $\boldsymbol{k}_1'$ and $\boldsymbol{k}_2$ are the wave vectors of incident, reflected and refracted waves, respectively

The reflection at a rigid plane boundary is considered quite analogously. The total normal velocity must be zero at $z = 0$ in this case. This yields $V = 1$; for the reflection coefficient the pressure on the boundary is twice as large as that in the incident wave.

In the general case we have to assume the existence of a refracted wave in the second medium

$$p_2 = WA \exp[\mathrm{i}(\xi x + k_{2z}z - \omega t)], \quad z > 0 \tag{12.28}$$

moving away from the boundary. Here, W is the *refraction* (*transmission*) *coefficient* and $\boldsymbol{k}_2 = \{\xi, 0, k_{2z}\}$ ($k_2 = \omega/c_2$, $k_{2z} = \sqrt{k_2^2 - \xi^2}$) is the wave vector in the second medium. The projections $\boldsymbol{k}_1$ and $\boldsymbol{k}_2$ onto the interface are the same (Fig. 12.1) and equal to ξ. The pressure and normal particle velocity must be continuous at the interface. Hence, with (12.11) we have the boundary conditions

$$\begin{aligned} &(p_1^+ + p_1^-)_{z=0} = p_2|_{z=0}, \\ &\rho_1^{-1}[(\partial p_1^+/\partial z) + (\partial p_1^-/\partial z)]_{z=0} = \rho_2^{-1}(\partial p_2/\partial z)_{z=0}. \end{aligned} \tag{12.29}$$

After substitution of (12.26–28), this yields two equations for V and W:

$$\begin{aligned} &1 + V = W, \\ &\rho_1^{-1}k_{1z}(1 - V) = \rho_2^{-1}k_{2z}W. \end{aligned} \tag{12.30}$$

We introduce the angle of incidence, θ_1, and the angle of refraction, θ_2 (Fig. 12.1). Then the equality of the boundary components of the wave vectors $\boldsymbol{k}_1$ and $\boldsymbol{k}_2$ results in the well-known *law of refraction*

$$k_1 \sin\theta_1 = k_2 \sin\theta_2, \tag{12.31}$$

or since $k_1 = \omega/c_1$ and $k_2 = \omega/c_2$,

$$\frac{\sin\theta_1}{\sin\theta_2} = n, \quad n = \frac{k_2}{k_1} = \frac{c_1}{c_2}. \tag{12.31a}$$

Now, taking into account that $k_{1z} = k_1 \cos\theta_1$, $k_{2z} = k_2 \cos\theta_2$ and introducing the new parameter $m = \rho_2/\rho_1$, we obtain from (12.30) the *Fresnel reflection and refraction formulas*

$$\begin{aligned} V &= \frac{m\cos\theta_1 - n\cos\theta_2}{m\cos\theta_1 + n\cos\theta_2} = \frac{m\cos\theta_1 - \sqrt{n^2 - \sin^2\theta_1}}{m\cos\theta_1 + \sqrt{n^2 - \sin^2\theta_1}}, \\ W &= \frac{2m\cos\theta_1}{m\cos\theta_1 + n\cos\theta_2} = \frac{2m\cos\theta_1}{m\cos\theta_1 + \sqrt{n^2 - \sin^2\theta_1}}. \end{aligned} \tag{12.32}$$

In the case of normal incidence ($\theta_1 = \theta_2 = 0$) we have

$$V = \frac{m-n}{m+n} = \frac{\rho_2 c_2 - \rho_1 c_1}{\rho_2 c_2 + \rho_1 c_1}, \qquad W = \frac{2\rho_2 c_2}{\rho_2 c_2 + \rho_1 c_1}, \tag{12.32a}$$

The reflection and refraction coefficients are expressible in terms of the *characteristic impedances* $Z_1 = \rho_1 c_1$ and $Z_2 = \rho_2 c_2$ or,

$$V = \frac{Z_2 - Z_1}{Z_2 + Z_1}, \qquad W = \frac{2Z_2}{Z_2 + Z_1}. \tag{12.33}$$

We can see from (12.32) that the last formulas also hold in the case of oblique incidence if we define Z_1 and Z_2 as

$$Z_1 = \frac{\rho_1 c_1}{\cos\theta_1}, \qquad Z_2 = \frac{\rho_2 c_2}{\cos\theta_2}. \tag{12.34}$$

We can easily prove that Z_1 and Z_2, defined in such a way, are ratios between pressure and normal velocity in the incident and refracted waves, respectively:

$$Z_1 = p_1^+/v_{1z}^+, \qquad Z_2 = p_2/v_{2z}. \tag{12.35}$$

This is why they are termed *normal impedances.*

To determine the reflection coefficient V, it is sufficient to know the so-called *input impedance* Z_{i} of the second medium which is equal to the normal impedance Z_2 at $z = 0$. Since the normal impedance must be continuous at the interface $z = 0$ (because pressure p and normal velocity v_z are continuous), the normal impedance of the total field in the first medium taken at $z = 0$ must be equal to the input impedance Z_{i}. In the case under consideration, however, Z_2 does not depend on z and we have

$$\left.\frac{p_1}{v_{1z}}\right|_{z=0} = \frac{1+V}{1-V}\frac{\rho_1 c_1}{\cos\theta_1} = \frac{1+V}{1-V} Z_1 = \left.\frac{p_2}{v_{2z}}\right|_{z=0} = Z_2 \, .$$

Hence, we obtain (12.33) for V. In the general case, $\rho_2(z)$ and $c_2(z)$ can be arbitrary functions of z. Equation (12.33) for V still holds with the change $Z_2 \to Z_{\mathrm{i}}$ where Z_{i} can be found separately.

12.2.2 Some Special Cases. Complete Transparency and Total Reflection

Let us consider some most interesting special cases.

i) If $\theta_1 \to \pi/2$, we have, according to (12.32), $V \to -1$, $W \to 0$ whence $p_1 = p_1^+ + p_1^- \to 0$, $p_2 \to 0$. This means that a sound wave propagating along the interface of two fluids cannot occur.

ii) The reflection coefficient V is zero if θ_1 satisfies $m\cos\theta_1 = \sqrt{n^2 - \sin^2\theta_1}$ according to (12.32). Solving this equation, we find the angle of *complete transparency:*

$$\theta' = \arcsin\sqrt{(m^2 - n^2)/(m^2 - 1)}\,. \tag{12.36}$$

Evidently, such an angle exists only at those values of m and n when the radicand is positive and less than unity.

iii) Let the sound velocity c_2 in the second medium be greater than c_1 $(n < 1)$. Define the angle of critical incidence, θ_{cr}, by the relation $\sin\theta_{cr} = n$. We can easily see from (12.31) that real θ_2 cannot exist if $\theta_1 > \theta_{cr} = \arcsin n$. The relations (12.32) still hold, however, for harmonic waves and can be written as

$$V = \frac{m\cos\theta_1 - \mathrm{i}\sqrt{\sin^2\theta_1 - n^2}}{m\cos\theta_1 + \mathrm{i}\sqrt{\sin^2\theta_1 - n^2}}\,, \qquad W = \frac{2m\cos\theta_1}{m\cos\theta_1 + \mathrm{i}\sqrt{\sin^2\theta_1 - n^2}} \tag{12.37}$$

or if we assume $V = |V|\exp(\mathrm{i}\varphi)$, then

$$|V| = 1\,, \quad \varphi = -2\arctan(\sqrt{\sin^2\theta_1 - n^2}/m\cos\theta_1)\,. \tag{12.38}$$

Here we have the so-called *total internal reflection*, φ being the phase difference between the reflected and the incident waves.

The refracted (transmitted) wave amplitude decreases exponentially as the reference point moves away from the boundary, i.e., this is an *inhomogeneous* wave:

$$p_2 = AW\exp[-k_2\sqrt{\sin^2\theta_1 - n^2}\,z + \mathrm{i}(\xi x - \omega t)]\,.$$

Since this wave does not carry energy away from the boundary, the reflection coefficient has the magnitude of unity. It is a complex number, however, $(\varphi \neq 0)$ at $\theta_2 > \theta_{cr}$. This fact causes distortion of the shape of a nonharmonic reflected plane wave as compared with the incident one. Let us consider an incident plane wave with an arbitrary dependence p on time and represent it as a Fourier integral over frequency. If $\theta_1 < \theta_{cr}$ [V is real, not depending on frequency, see (12.32)], we obtain for the reflected wave the same integral times V. If, however, V is complex, its constant phase being independent of frequency is equivalent to an additional increment of length φ/k in the medium, which is already frequency dependent. As a result, the shapes of the reflected as well as of the transmitted (W is complex, too) waves become distorted.

12.2.3 Energy and Symmetry Considerations

According to the *energy conservation law*, normal to the boundary and averaged over a period of the wave, energy flux in the incident wave must be equal to the sum of the corresponding fluxes in the reflected and refracted waves. We have, according to (12.25),

$$I_{1z}^{+} = \frac{|A|^2}{2\rho_1 c_1}\cos\theta_1\,, \quad I_{1z}^{-} = -\frac{|A|^2|V|^2}{2\rho_1 c_1}\cos\theta_1\,,$$
$$I_{2z} = \frac{|A|^2|W|^2}{2\rho_2 c_2}\cos\theta_2\,. \tag{12.39}$$

Hence, the following relation must hold:

$$\frac{\cos\theta_1}{\rho_1 c_1}(1-|V|^2) = \frac{\cos\theta_2}{\rho_2 c_2}|W|^2\,.$$

In fact, it can easily be verified that if we substitute for V and W from (12.33, 34) the last relation is identically satisfied.

It follows from (12.30) that the amplitude of the acoustic pressure in the refracted wave is $1 + V$ times that of the incident one. In the case of normal incidence of a sound wave upon the air-water interface from the air, for example, we have $\rho_1 c_1 = 42\ \mathrm{g\cdot cm^{-2}\,s^{-1}}$, $\rho_2 c_2 = 1.5 \times 10^5\ \mathrm{g\cdot cm^{-2}\,s^{-1}}$ and $V \simeq 1$, according to (12.32). Hence, the sound pressure in the water will be twice as large as that in the incident wave. On the other hand, if the wave is incident from the water, we obtain $V \simeq -1$ and $W \simeq 0$, i.e., the pressure in the transmitted wave is very small compared with that in the incident one. Hence, there is no symmetry in wave transmission through the boundary in opposite directions with regard to the sound pressure. The same can be shown with respect to the particle's velocities.

We will show, however, that such a symmetry does exist with respect to the energy fluxes normal to the boundary. Indeed, we have for fluxes the following ratio between the transmitted and incident waves, according to (12.39),

$$\frac{I_{2z}}{I_{1z}^{+}} = \frac{\cos\theta_2}{\cos\theta_1}\frac{\rho_1 c_1}{\rho_2 c_2}|W|^2 = \frac{\cos\theta_2}{\cos\theta_1}\frac{\rho_1 c_1}{\rho_2 c_2}|1+V|^2\,. \tag{12.40}$$

Let us first assume that the incidence is normal ($\theta_1 = \theta_2 = 0$). Then, using (12.32a), we find for the ratio above

$$\frac{I_2}{I_1^{+}} = 4\frac{\rho_1 c_1 \rho_2 c_2}{(\rho_1 c_1 + \rho_2 c_2)^2}\,,$$

which is invariant with respect to the interchanges $\rho_1 \rightleftarrows \rho_2$, $c_1 \rightleftarrows c_2$. Hence, at normal incidence the fraction of energy which penetrates the boundary is the same in both directions. The same result also holds in the case of oblique incidence. In fact, the relation (12.40) with (12.32) yields

$$\frac{I_{2z}}{I_{1z}^{+}} = 4\frac{\rho_1 c_1 \rho_2 c_2 \cos\theta_1 \cos\theta_2}{(\rho_1 c_1 \cos\theta_2 + \rho_2 c_2 \cos\theta_1)^2}\,,$$

which is a symmetric formula with respect to the interchanges

$$\rho_1 \rightleftarrows \rho_2\,, \quad c_1 \rightleftarrows c_2\,, \quad \theta_1 \rightleftarrows \theta_2\,.$$

Naturally, we have excluded the case of total internal reflection from this discussion where $I_{2z} = 0$ (the energy flux in the second media is parallel to the boundary).

Note also the symmetry of the relationship for those angles of incidence, for which the boundaries are completely transparent. When an incident wave is propagating from the first medium into the second one, we have the angle θ'_1 determined by (12.36). By changing $\theta'_1 \to \theta'_2$, $m \to 1/m$ and $n \to 1/n$, we also obtain θ'_2 from the same formula which is the angle of complete transparency when a wave is propagating from the second medium into the first one. It is easy to prove that these angles are related to one another by the refraction law $c_2 \sin\theta'_1 = c_1 \sin\theta'_2$, which is also symmetric with respect to the interchange $c_1 \rightleftarrows c_2$, $\theta'_1 \rightleftarrows \theta'_2$.

12.2.4 A Slowly-Varying Medium. Geometrical-Acoustics Approximation

The wave equation (12.3) holds when $c = c(\boldsymbol{r})$ is a continuous function of coordinates. We introduce the new quantity, i.e., the *refractive index of a medium* $\mu(\boldsymbol{r}) = c_0/c(\boldsymbol{r})$, where c_0 is the sound velocity at a fixed point. Now we obtain from (12.3) for harmonic waves the *Helmholtz equation* analogous to (12.8):

$$\Delta\psi + k_0^2\mu^2(\boldsymbol{r})\psi = 0\,, \quad k_0^2 = \omega^2/c_0^2\,. \tag{12.41}$$

It was shown above that in the case of a homogeneous medium, the plane wave of constant amplitude is the solution of the Helmholtz equation. For an inhomogeneous medium and high frequencies ($k_0 \to \infty$), we look for a solution in the form

$$\psi(\boldsymbol{r}) = A(\boldsymbol{r})\exp[ik_0 f(\boldsymbol{r})]\,, \tag{12.42}$$

where the amplitude $A(\boldsymbol{r})$ is a *slowly-varying* function of position and can be considered as constant in a small region (but large compared with $\lambda_0 = 2\pi/k_0$). $f(\boldsymbol{r})$ is also a *slowly-varying* function and in the same region can be represented as a linear function of coordinates by expanding it into a Maclaurin series. Thus, (12.42) can be considered a *locally plane wave.*

Substituting (12.42) into (12.41) and separating real and imaginary parts yields a system of two equations:

$$\begin{aligned} &\Delta A - k_0^2 A[(\nabla f)^2 - \mu^2] = 0\,, \\ &2\nabla A \cdot \nabla f + A\,\Delta f = 0\,. \end{aligned} \tag{12.43}$$

The term ΔA in the former can be neglected as compared with the others if $k_0 \to \infty$, and we obtain the so-called *eikonal equation:*

$$(\nabla f)^2 = \mu^2 . \tag{12.44}$$

Surfaces of constant phase $f(\boldsymbol{r}) = \text{const}$ are termed *fronts*. The curves orthogonal to fronts (tangents to these curves are collinear with ∇f at any point) are called *rays*. Let $\boldsymbol{r} = \boldsymbol{r}(s)$ be an equation of a ray where s is an arc length along the ray. The unit vector along the normal to the wave front is $\boldsymbol{n} = \nabla f/|\nabla f| = \nabla f/\mu$ on the one hand, and $d\boldsymbol{r}/ds$ on the other. Hence, we have the relation $\mu\, d\boldsymbol{r}/ds = \nabla f$. Differentiating it with respect to s and using the vector identity [see also (6.31)]

$$d(\nabla f)/ds = (\boldsymbol{n} \cdot \nabla)\nabla f = \mu^{-1}(\nabla f \cdot \nabla)\nabla f = \mu^{-1}\nabla[(\nabla f)^2/2] - \mu^{-1}\nabla f \times \operatorname{curl}(\nabla f),$$

we obtain, taking (12.44) into account and $\operatorname{curl}(\nabla f) = 0$, the *ray equation*

$$\frac{d}{ds}\left(\mu \frac{d\boldsymbol{r}}{ds}\right) = \nabla \mu . \tag{12.45}$$

With the ray we determine an *acoustic path length*

$$f = \int_0^s \mu\, ds \tag{12.46}$$

by integrating along the ray. It is assumed here that $f = f_0 = 0$ at $s = 0$. To determine the wave amplitude $A(\boldsymbol{r})$, we turn to the second equation of (12.43). With $\nabla f = \mu \boldsymbol{n}$ its first term is $2\mu\nabla A \cdot \boldsymbol{n} = 2\mu(dA/ds)$. To consider the second term we construct a *ray tube* composed of all rays leaving the wave front f_0 in a small area dS_0 (Fig. 12.2). We integrate $\Delta f \equiv \nabla(\nabla f)$ over a small volume V bounded by the walls of the tube and two cross sections of the latter a distance ds apart. Since V is small, we have

$$\int_V \Delta f\, dV = \Delta f \cdot V = \Delta f\, dS_1\, ds .$$

On the other hand, applying the Gauss divergence theorem we have

$$\int_V \Delta f\, dV = \int_{dS_1 + dS_2} \nabla f \cdot \boldsymbol{n}\, dS = \mu_2\, dS_2 - \mu_1\, dS_1 ,$$

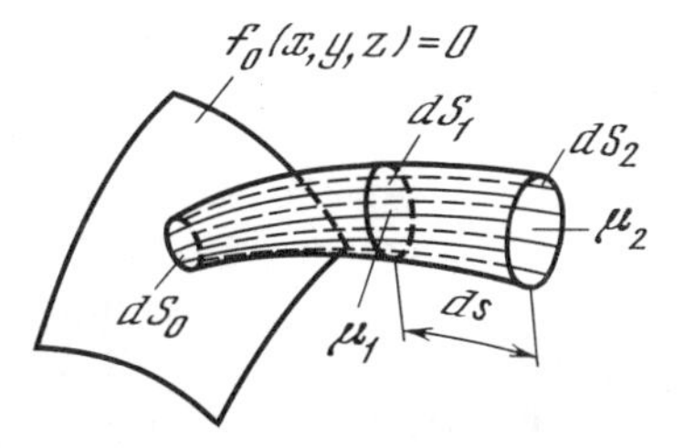

Fig. 12.2. Tube of rays emanating from the frontal surface $f_0(x, y, z) = 0$

since $\boldsymbol{n}$ is orthogonal to ∇f at lateral walls of the tube. We introduce the new quantity $\varkappa = \lim_{dS_0 \to 0} (dS/dS_0)$ called the *tube's divergence*. Equating the right-hand sides of the last two equations and dividing them by dS_0, we obtain in the limit $dS_0 \to 0$,

$$\Delta f = \varkappa^{-1} d(\mu\varkappa)/ds .$$

Now the second equation of (12.43) can be written as

$$2\varkappa\mu \frac{dA}{ds} + A \frac{d}{ds}(\mu\varkappa) = 0 \quad \text{or} \quad \frac{d}{ds}(A^2\mu\varkappa) = 0 . \tag{12.47}$$

Hence, the following relation holds for any ray:

$$A^2\mu\varkappa = \text{const} .$$

Assume $A = A_0$ at the initial front $f_0 = 0$ where $\mu = \mu_0$, $\varkappa = \varkappa_0 = 1$. Then, (12.42) for a locally plane wave becomes

$$\psi = A_0(\mu_0/\mu\varkappa)^{1/2} \exp(\mathrm{i}k_0 f) = A_0[c(\boldsymbol{r})/c_0\varkappa]^{1/2} \exp(\mathrm{i}k_0 f) . \tag{12.48}$$

We obtain the particle velocity using (12.11) and differentiating only the exponent because k_0 is supposed to be large, we get

$$\boldsymbol{v} = p\boldsymbol{n}/\rho_0 c(\boldsymbol{r}) , \tag{12.49}$$

as in a plane wave. The velocity vector is collinear with the tangent to the ray. The energy flux averaged over the period is, according to (12.22), $\boldsymbol{I} = (|p|^2/2\rho_0 c)\boldsymbol{n}$ and is collinear with the velocity vector. The average energy flux through the cross section of a ray tube dS, or

$$\bar{I}\, dS = (A^2/2\rho_0 c)\, dS = (dS_0/2\rho_0 c_0)A^2\mu\varkappa ,$$

remains constant along a fixed tube according to (12.47). This is the *energy conservation law*.

The theory developed above may be considerably simplified in the case of a *plane stratified medium*, i.e., when the sound velocity is a function of only one coordinate. In the atmosphere and in the ocean, this coordinate is vertical (z). When $c = c(z)$, $\nabla\mu$ in (12.45) has a component only in the z-direction and each ray is a curve in the plane. Assuming that this is in the (x, z)-plane, we obtain for the ray equation $z = z(x)$. Let $\boldsymbol{n} = d\boldsymbol{r}/ds$ be the unit vector tangent to the ray. Multiplying (12.45) by the unit vector $\boldsymbol{e}_x$ along the x-axis and taking into account that $\boldsymbol{e}_x \cdot \nabla\mu = 0$, we obtain the important relation $\mu\boldsymbol{n} \cdot \boldsymbol{e}_x = \text{const}$ for a fixed ray. Let χ be a *grazing angle* or an angle which a tangent to the ray makes

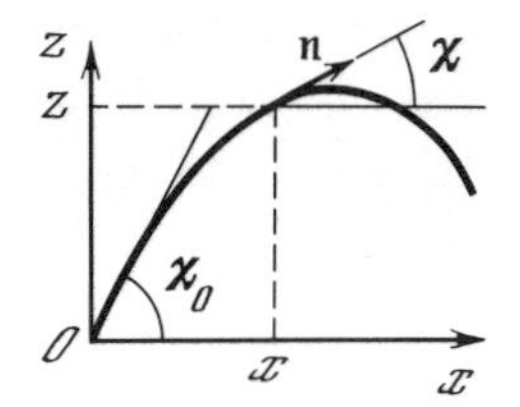

Fig. 12.3. Snell's law relating the grazing angle χ of the ray to χ_0

with the horizontal plane (Fig. 12.3). Then, $\boldsymbol{n} \cdot \boldsymbol{e}_x = \cos\chi$ and this relation assumes the form of the well-known *Snell's law*

$$\mu(z)\cos\chi(z) = \mu_0 \cos\chi_0\,, \tag{12.50}$$

where $\mu_0 \equiv \mu(z_0)$, $\chi_0 \equiv \chi(z_0)$ and z_0 is an arbitrary but fixed coordinate. Now we obtain

$$\frac{dz}{dx} = \tan\chi = \pm\frac{\sqrt{1-\cos^2\chi}}{\cos\chi} = \pm\frac{\sqrt{\mu^2(z)-\mu_0^2\cos^2\chi_0}}{\mu_0\cos\chi_0}\,,$$

or after integration,

$$x = \pm\mu_0\cos\chi_0 \int_{z_0}^{z} \frac{dz}{\sqrt{\mu^2(z)-\mu_0^2\cos^2\chi_0}}\,. \tag{12.51}$$

Analogously, (12.46) yields for the eikonal

$$f = \int_0^s \mu\,ds = \int_{x_0}^{x} \frac{\mu\,dx}{\cos\chi} = \int_{x_0}^{x} \frac{\mu^2\,dx}{\mu_0\cos\chi_0}$$

$$= (x-x_0)\mu_0\cos\chi_0 \pm \int_{z_0}^{z} \sqrt{\mu^2-\mu_0^2\cos^2\chi_0}\,dz\,.$$

The first term in (12.43) must be small compared with others, or

$$\Delta A/A \ll k_0^2 \tag{12.52}$$

for the geometrical (ray) acoustics to be valid. In other words, the change in the wave's amplitude at a distance of one wavelength must be small. From (12.48) we can see that the change in the medium's parameters at the distance of one wavelength must be small. These conditions are necessary but not sufficient, however. Indeed, the amplitude depends not only on μ but also on the tube's divergence $\varkappa$ according to (12.48). Hence, the condition of applicability of geometrical acoustics must also take the geometrical properties of rays into account.

The geometrical-acoustics approximation can be applied without any difficulty to (12.2a) where the medium's density $\rho_0(\boldsymbol{r})$ dependence on the coordinates is taken into account. An expression for the amplitude in (12.48) is now $A = A_0(\rho_0 c/\rho_{00} c_0 \varkappa)^{1/2}$ where $\rho_{00} = \rho_0(\boldsymbol{r}_0)$.

12.2.5 Acoustics Equations for Moving Media

Now we consider sound propagation in a moving medium which, e.g., currents in the ocean, winds in the atmosphere, etc. Bearing these important examples in mind, we assume that the current velocity is in the horizontal direction and depends on the vertical coordinate only: $\boldsymbol{v}_0 = \{u_0(z), 0, 0\}$. One can easily see that the terms $(\boldsymbol{v}_0 \cdot \nabla)\boldsymbol{v} = u_0(z)\partial\boldsymbol{v}/\partial x$ and $(\boldsymbol{v} \cdot \nabla)\boldsymbol{v}_0 = w\, d\boldsymbol{v}_0/dz$ must now be introduced into the linearized Euler equation (10.3) and the term $u_0\partial\rho/\partial x = (u_0/c^2)\partial p/\partial x$ into the continuity equation. Here, $\boldsymbol{v} = \{u, v, w\}$ is the acoustic particle velocity.

As a result, we obtain the following basic acoustic equations instead of system (12.2):

$$\frac{\partial \boldsymbol{v}}{\partial t} + u_0 \frac{\partial \boldsymbol{v}}{\partial x} + w \frac{d\boldsymbol{v}_0}{dz} = -\frac{\nabla p}{\rho_0}, \quad \frac{\partial p}{\partial t} + u_0 \frac{\partial p}{\partial x} + \rho_0 c^2 \nabla \boldsymbol{v} = 0. \tag{12.53}$$

A large variety of problems can be solved with the help of these equations. For example, the refraction of acoustic waves due to variation with height of the velocity of the moving medium can be found. We confine ourselves below to a simpler problem, however, i.e., the reflection of a sound wave at a horizontal interface between two fluids of the same density and sound velocity; the first of these fluids (from where the wave is incident) is at rest, whereas the second moves horizontally in the plane of the wave incidence (x, z-plane) at a constant velocity u_0 (Fig. 12.4).

In the first medium ($z < 0$), the pressure due to the incident and the reflected wave is

$$p_1 = A[\exp(\mathrm{i}k_{1z}z) + V\exp(-\mathrm{i}k_{1z}z)]\exp[\mathrm{i}(\xi x - \omega t)], \quad k_{1z} = \sqrt{(\omega/c)^2 - \xi^2}.$$

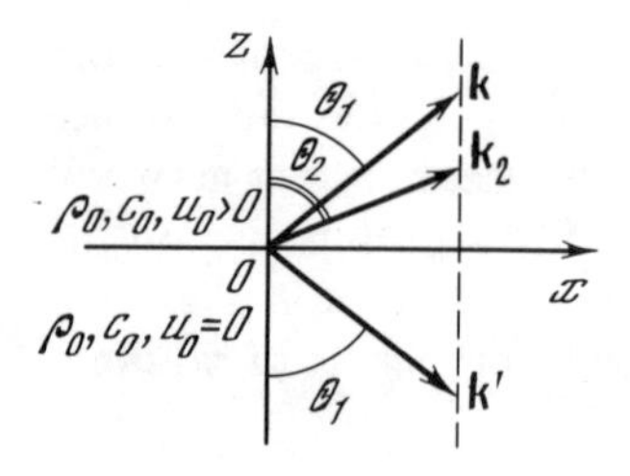

Fig. 12.4. Set of wave vector's in the case of sound reflection at the interface between a moving medium and a medium at rest

From (12.53) with $dv_0/dz = 0$, we obtain an equation for the pressure in the second, moving medium ($z > 0$):

$$\left(\frac{\partial}{\partial t} + u_0 \frac{\partial}{\partial x}\right)^2 p_2 - c^2 \Delta p_2 = 0 .$$

Its solution in the form of a plane wave leaving the interface is

$$p_2 = AW \exp[\mathrm{i}(k_{2x}x + k_{2z}z - \omega_2 t)] ,$$
$$k_{2z} = \sqrt{[(\omega_2 - u_0 k_{2x})/c]^2 - k_{2x}^2} .$$

Two conditions must be satisfied at the interface $z = 0$; namely the continuity of pressures ($p_1|_{z=0} = p_2|_{z=0}$) and the kinematic condition (10.4) where $\zeta(x, t)$ is the vertical displacement of the interface. We obtain from the latter by linearization:

$$w_1|_{z=0} - \frac{\partial \zeta}{\partial t} = 0 = w_2|_{z=0} - \frac{\partial \zeta}{\partial t} - u_0 \frac{\partial \zeta}{\partial x} .$$

Hence, it follows that $\partial\zeta/\partial t = -\mathrm{i}\omega\zeta = w_1|_{z=0}$ and $w_2|_{z=0} = (\partial\zeta/\partial t) + u_0\partial\zeta/\partial x = (1 - M\xi/k)w_1|_{z=0}$ where $k = \omega/c$ and $M = u_0/c$ is the Mach number. By also taking into account the relationship between w_j and p_j ($j = 1, 2$), which follows from the first equation in (12.53), we obtain as a result for $z = 0$,

$$(1 + V)\exp[\mathrm{i}(\xi x - \omega t)] = W \exp[\mathrm{i}(k_{2x}x - \omega_2 t)] ,$$
$$(1 - M\xi/k)(\omega\rho_0)^{-1} k_{1z}(1 - V)\exp[\mathrm{i}(\xi x - \omega t)]$$
$$= [(\omega_2 - k_{2x}u_0)\rho_0]^{-1} k_{2z} W \exp[\mathrm{i}(k_{2x}x - \omega_2 t)] .$$

These conditions can be satisfied for arbitrary x and t only if $\omega_2 = \omega$ and $k_{2x} = \xi$. Hence, the wave number $k_2 = \sqrt{k_{2x}^2 + k_{2z}^2} = k(1 - M\xi/k)$ in a moving medium is less than that of the incident wave $k = \omega/c$ if the wave travels along the current ($\xi > 0$ as in Fig. 12.4) and greater if the incident wave travels against the current ($\xi < 0$). The angle θ_2 is related to θ_1 by

$$\xi = k \sin\theta_1 = k_2 \sin\theta_2 \quad \text{or} \quad \sin\theta_2 = (1 - M\sin\theta_1)^{-1} \sin\theta_1 .$$

Total internal reflection occurs if

$$\sin\theta_1 > 1 - M\sin\theta_1 \quad \text{or} \quad \sin\theta_1 > (1 + M)^{-1} ,$$

i.e., when incidence is in the direction of the current ($\xi > 0$) and the angle of incidence is large enough. Introducing a new parameter $\alpha = 1 - M\sin\theta_1$, we can write the coefficients of reflection and transmission in the form

$$V = \frac{\alpha^2 \cos\theta_1 - \sqrt{\alpha^2 - \sin^2\theta_1}}{\alpha^2 \cos\theta_1 + \sqrt{\alpha^2 - \sin^2\theta_1}}, \qquad W = \frac{2\alpha^2 \cos\theta_1}{\alpha^2 \cos\theta_1 + \sqrt{\alpha^2 - \sin^2\theta_1}} .$$

12.2.6 Guided Propagation of Sound

We consider geophysical sound waveguides in horizontally stratified fluids. The simplest waveguide of this kind is the homogeneous ocean with a perfectly rigid bottom. At its surface $z = 0$, we have $p = 0$ and $V = -1$ according to Sect. 12.2.1. Hence, below the surface ($z < 0$) the sound pressure can be written as

$$p = A \sin k_z z \exp[i(\xi x - \omega t)], \qquad k_z = \sqrt{(\omega/c)^2 - \xi^2}\,. \tag{12.54}$$

At the bottom $z = -H$, the condition $w = 0$ must be satisfied, hence

$$\cos k_z H = 0\,, \qquad k_z H = (n - 1/2)\pi\,, \qquad n = 1, 2, \ldots. \tag{12.55}$$

Thus we obtain the *dispersion relation* for *normal modes:*

$$\begin{aligned} \xi_n(k) &= k[1 - (\pi/kH)^2(n - 1/2)^2]^{1/2}\,, \\ c_{\mathrm{Ph}}^{(n)} &= c[1 - (\pi/kH)^2(n - 1/2)^2]^{-1/2}\,. \end{aligned} \tag{12.56}$$

Here $k = \omega/c$ and $c_{\mathrm{Ph}}^{(n)} = \omega/\xi_n$ is the *mode's phase velocity*. Now the sound pressure for the nth mode can be written as

$$\begin{aligned} p_n &= A_n \varphi_n(z) \exp[i(\xi_n x - \omega t)]\,, \\ \varphi_n &= \sin[(n - 1/2)\pi z/H]\,. \end{aligned} \tag{12.57}$$

The group velocity is

$$c_{\mathrm{g}}^{(n)} = d\omega/d\xi_n = c[1 - (\pi/kH)^2(n - 1/2)^2]^{1/2}\,.$$

We can see from (12.56) that the horizontal wave number ξ_n is real only for a finite number of modes for fixed ω. Such modes are termed *propagating modes.* For higher n, the quantity ξ_n becomes purely imaginary and the sound pressure decreases exponentially with the horizontal distance. The frequency $\omega_{\mathrm{c}}^{(n)}$ at which the nth mode undergoes a transition from a propagating to a nonpropagating one is termed the *cut-off frequency* of the nth mode. We have, according to (12.56), $\omega_{\mathrm{c}}^{(n)} = (n - 1/2)\pi c/H$.

The model of an oceanic waveguide as a homogeneous layer with perfectly rigid bottom is a highly idealized one. The next approximation to reality is also a homogeneous layer but with a homogeneous liquid halfspace of density ρ_{b} and sound velocity c_{b} at $-\infty < z < -H$ as the bottom. If $c_{\mathrm{b}} > c$ (c being the sound velocity in the layer), a number of propagating modes of the type in (12.54) exist, their phase velocity being less than the sound velocity in the bottom ($\xi > \omega/c_{\mathrm{b}}$). The inhomogeneous wave

$$p_{\mathrm{b}} = B \exp[\sqrt{\xi^2 - (\omega/c_{\mathrm{b}})^2}(z + H) + i(\xi x - \omega t)] \tag{12.58}$$

arises in the halfspace $z < -H$, which carries no energy away from the boundary.

To determine the mode parameters, we equate the ratio p/w at $z = -H$ obtained from (12.54) (remember that $\mathrm{i}\omega\rho w = \partial p/\partial z$) to the impedance of the liquid homogeneous halfspace $Z_\mathrm{b} = (p_\mathrm{b}/w_\mathrm{b})_{z=-H}$ determined by (12.58):

$$-(\mathrm{i}\omega\rho/k_z)\tan k_z H = \mathrm{i}\omega\rho_\mathrm{b}[\xi^2 - (\omega/c_\mathrm{b})^2]^{-1/2},$$

whence introducing $\sigma = k_z H = H\sqrt{(\omega/c)^2 - \xi^2}$, we obtain a dispersion relation to determine $\xi_n(\omega)$:

$$\tan\sigma = (\rho_\mathrm{b}/\rho)\sigma[(kH)^2 - (k_\mathrm{b}H)^2 - \sigma^2]^{-1/2}, \qquad k_\mathrm{b} = \omega/c_\mathrm{b}. \tag{12.59}$$

At sufficiently high frequency, this equation has a finite number of roots $\sigma_n(\omega)$ such that $\pi(n - 1/2) < \sigma_n(\omega) < n\pi$, $n = 1, 2, \ldots$. The corresponding modes are given by (12.57) with $\varphi_n = \sin(\sigma_n z/H)$, $\xi_n = k[1 - (\sigma_n/kH)^2]^{1/2}$ and $c_\mathrm{Ph}^{(n)} = c[1 - (\sigma_n/kH)^2]^{-1/2}$. The cut-off frequency of the nth mode is determined from $\sigma_n = \pi(n - 1/2)$ and is, according to (12.59),

$$\omega_\mathrm{c}^{(n)} = (n - 1/2)(1 - \mu^2)^{-1/2}\pi c/H, \qquad \mu = c/c_\mathrm{b}.$$

The oceanic sound waveguide can be created not only by reflections at the surface and bottom, but also by sound refraction in the water layers with a variable $c(z)$. A typical dependence of the sound velocity on the vertical coordinate [$c(z)$—profile] in a deep ocean is plotted in Fig. 12.5a. The sound velocity attains its minimum value at $z = z_0$, the *axis of the waveguide* (*underwater-sound channel*). It increases at shallower depths due to rising temperature and at greater depths due to hydrostatic pressure rising with depth. It also turns out that the energy of a mode propagating with the phase velocity $c_\mathrm{Ph} = \omega/\xi = c(z_1) = c(z_2)$ (Fig. 12.5a) is localized mainly inside the layer $z_2 < z < z_1$. Outside this layer, inhomogeneous waves are created. The rays corresponding to this mode are shown in Fig. 12.5b. For a discussion of mode propagation, we look for the solution of (12.3) in the form

$$p = A\varphi(z)\exp[\mathrm{i}(\xi x - \omega t)]$$

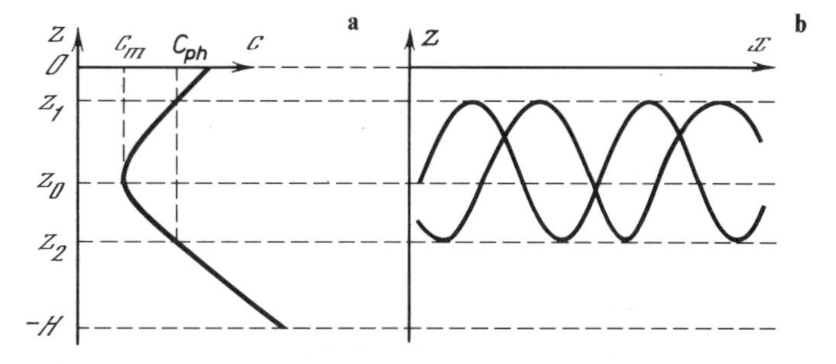

Fig. 12.5. (**a**) Typical dependence of sound velocity on the vertical coordinate [$c(z)$-profile] in a deep ocean. (**b**) Illustrating two rays propagating in the sound channel

which yields an equation for $\varphi(z)$, namely

$$\frac{d^2\varphi}{dz^2} + \omega^2[c^{-2}(z) - c_{\mathrm{Ph}}^{-2}]\varphi = 0. \tag{12.60}$$

This is analogous to (10.59) for internal gravitational waves. Adding the corresponding boundary conditions, for example, $\varphi(0) = 0$ at a free boundary and the continuity of p and w at the bottom (if it can be considered as a liquid one), we obtain a boundary-value problem, whose solution gives *eigenvalues* $c_{\mathrm{Ph}}^{(n)}$ and *eigenfunctions* $\varphi_n(z)$.

A qualitative discussion of this boundary-value problem is quite analogous to that for internal waves (Sect. 10.5). It appears that a finite number of propagating modes exist for sufficiently high ω, their phase velocities being less than the sound velocity at the bottom $[c_{\mathrm{Ph}}^{(n)} < c_{\mathrm{b}}]$. The amplitudes of these modes decrease exponentially if $z > z_1^{(n)}(\omega)$ or $z < z_2^{(n)}(\omega)$ where $c_{\mathrm{Ph}}^{(n)} = c(z_1^{(n)}) = c(z_2^{(n)})$. The modes with $c(-H) < c_{\mathrm{Ph}}^{(n)} < c_{\mathrm{b}}$ do not have small amplitudes through all the depths from the surface to the bottom like the modes in a homogeneous liquid layer above a liquid halfspace (see above). At the cut-off frequency $\omega_{\mathrm{c}}^{(n)}$, the phase velocity of the nth mode is equal to the sound velocity at the bottom.

12.3 Spherical Waves

12.3.1 Spherically-Symmetric Solution of the Wave Equation

A spherical wave generated by a pulsating sphere of small radius (*monopole*) is another simple solution to the wave equation (12.3). To obtain this solution we write (12.3) in spherical coordinates:

$$c^{-2}\frac{\partial^2 p}{\partial t^2} = r^{-2}\frac{\partial}{\partial r}\left(r^2\frac{\partial p}{\partial r}\right) = r^{-1}\frac{\partial^2(rp)}{\partial r^2}, \tag{12.61}$$

where $p = p(r, t)$ and $r = \sqrt{x^2 + y^2 + z^2}$. We see that the function rp satisfies the usual one-dimensional wave equation with a general solution analogous to (12.5) but with x replaced by r, i.e.,

$$p = \frac{f(r - ct)}{r} + \frac{g(r + ct)}{r}. \tag{12.62}$$

The first term represents a *diverging* wave and the second a *converging* wave. The pressure dependence on t does not change when the wave is propagated, as in the case of plane waves, but the pressure itself decreases as r increases.

Substituting (12.62) into (12.4) (assuming $t_0 = -\infty$), we find that the particle velocity is directed along $\boldsymbol{r}$, its magnitude being

$$v_r = v = \pm\frac{p}{\rho_0 c} + \frac{1}{\rho_0 r}\int_{-\infty}^{t} p\,dt\,. \tag{12.63}$$

The + and − signs correspond to the diverging and converging waves, respectively. The integral term here depending on the prior history of the wave becomes negligible at large r (*far field*) and we have at such distances $v \simeq p/\rho_0 c$, as in a plane wave. In the case of an incompressible fluid ($c \to \infty$), the first term in (12.63) vanishes whereas the second one corresponding to the spherical-symmetric solution of the Laplace equation for velocity potential, remains.

If the sound pulse has finite duration, i.e., p, $v \to 0$ as $t \to \pm\infty$, it follows from (12.63) that the total momentum of the pulse is zero: $\int_{-\infty}^{\infty} p\,dt = 0$. Hence, unlike the plane wave, the spherical wave cannot occupy only rarefactions ($p < 0$) or only compression ($p > 0$) regions.

If f and g are harmonic functions of time, using (12.62) we obtain the, *harmonic spherical-symmetric waves* of the type

$$\begin{aligned} p &= \frac{A}{r}\exp[\mathrm{i}(\pm kr - \omega t)]\,, \\ v &= \frac{A}{\rho_0 c r}\left(\pm 1 + \frac{\mathrm{i}}{kr}\right)\exp[\mathrm{i}(\pm kr - \omega t)]\,. \end{aligned} \tag{12.64}$$

Here v was obtained using (12.11). We can see that the distance $r_0 \sim k^{-1}$ divides the *far* or *wave zone* of the sound field (where $kr \gg 1$) from the *near zone* (where $kr \ll 1$). In the near zone, the second term in parentheses in the expression for v prevails.

Let the surface velocity of a pulsating sphere be specified by $v_r = v_0\exp(-\mathrm{i}\omega t)$. In the linear approximation (small v_0), we can neglect the change of the sphere's radius R_0, and equate v_r to v given by (12.64) at $r = R_0$, choosing the upper sign (diverging wave). As a result, we obtain the pressure amplitude

$$A = \frac{\rho_0 \omega R_0^2 v_0}{\mathrm{i} + kR_0}\exp(-\mathrm{i}kR_0)\,. \tag{12.65}$$

12.3.2 Volume Velocity or the Strength of the Source. Reaction of the Medium

If the radius of a pulsating sphere is small ($kR_0 \ll 1$), (12.65) becomes $A \simeq -\mathrm{i}\omega\rho_0 R_0^2 v_0$. The so-called *volume velocity*

$$V_0 = 4\pi R_0^2 v_0 \tag{12.66}$$

or the *strength* of the source is a useful quantity in this case; it is the volume displaced by the sphere per unit time. In terms of V_0, (12.64) reads

$$
\begin{aligned}
p &= -\frac{\mathrm{i}\omega\rho_0}{4\pi r} V_0 \exp[\mathrm{i}(kr - \omega t)], \\
v &= -\frac{\mathrm{i}\omega V_0}{4\pi c r}\left(1 + \frac{\mathrm{i}}{kr}\right)\exp[\mathrm{i}(kr - \omega t)].
\end{aligned} \tag{12.67}
$$

We will now show that any acoustic source (not only a pulsating sphere) generates a spherical wave (12.67) if it is small as compared with the wavelength, and its volume velocity V_0 is finite. We first consider sound generation by an arbitrary volume V assuming that a *surplus* or *defect of mass* can occur at any point of this volume. The mass change in the element of volume dV will be specified by the quantity $(\rho_0 + \rho)q(\boldsymbol{r}, t)\,dV$, where $q(\boldsymbol{r}, t)$ is a known function. Now, the mass conservation law (6.7) assumes the form

$$
\frac{d}{dt}\int_V (\rho_0 + \rho)\,dV = \int_V (\rho_0 + \rho)q\,dV
$$

which yields the *generalized form of the continuity equation* (6.9):

$$
\frac{\partial\rho}{\partial t} + \boldsymbol{v}\cdot\nabla\rho + (\rho_0 + \rho)\nabla\boldsymbol{v} = (\rho_0 + \rho)q\,.
$$

Linearizing this equation with respect to the acoustic quantities $\boldsymbol{v}$ and ρ and assuming q to be small, too, we obtain

$$
\frac{\partial\rho}{\partial t} + \rho_0\nabla\boldsymbol{v} = \rho_0 q \quad \text{or} \quad \frac{\partial p}{\partial t} + \rho_0 c^2\nabla\boldsymbol{v} = \rho_0 c^2 q\,. \tag{12.68}
$$

Now, if we take into account the linearized Euler equation (12.4), we obtain the *inhomogeneous wave equation*

$$
\varDelta p - c^{-2}\,\partial^2 p/\partial t^2 = -\rho_0\,\partial q/\partial t \tag{12.69}
$$

and the *inhomogeneous Helmholtz equation* for harmonic waves

$$
\varDelta p + k^2 p = \mathrm{i}\omega\rho_0 q\,, \qquad k = \omega/c\,. \tag{12.70}
$$

Equation (12.67) is the solution to this equation for the point source at $\boldsymbol{r} = 0$: $q(\boldsymbol{r}) = q_0\,\delta(\boldsymbol{r})$ where $V_0 = \iiint_{-\infty}^{\infty} q\,dV = q_0$ is the volume velocity.

We determine the *total energy flux* in a spherical wave. Assuming $V_0 = |V_0|\exp(\mathrm{i}\alpha)$, $\varphi = kr - \omega t + \alpha$ and using the real parts of (12.67), we obtain for a specific flux in the $\boldsymbol{r}$-direction,

$$
pv = \frac{\rho_0 c k^2}{16\pi^2}\frac{|V_0|^2}{r^2}\left(\sin^2\varphi + \frac{\sin 2\varphi}{2kr}\right). \tag{12.71}
$$

After averaging over the period of the wave and multiplying by $4\pi r^2$, this yields the total energy flux rate through a sphere of arbitrary radius r or *the power radiated*

$$I = 4\pi r^2 \overline{pv} = \rho_0 c k^2 |V_0|^2/8\pi . \tag{12.72}$$

The term $\sin(2\varphi)/2kr$ specifying the near field in (12.71) vanishes on averaging. This term is very important, however, for determining the medium's reaction to the source. We define the *impedance of a spherical wave* at an arbitrary distance r, i.e., the quantity $Z = p/v$ which, according to (12.67), is

$$Z = \mathrm{i}\rho_0 \omega r (ikr - 1)^{-1} . \tag{12.73}$$

At small distances and particularly at the surface of a pulsating sphere of small radius R_0 ($kR_0 \ll 1$), this impedance is mainly due to the near field term and is purely imaginary: $Z(R_0) \simeq -\mathrm{i}\omega\rho_0 R_0$. Hence, the pressure at the sphere's surface will be $p|_{r=R_0} = -\mathrm{i}\omega\rho_0 v|_{r=R_0} = \rho_0 R_0 a$, a denoting the sphere's surface acceleration. The total force on the sphere's surface is now $F = 4\pi\rho_0 R_0^3 a = Ma$ with $M = 4\pi\rho_0 R_0^3$. Thus, the fluid's reaction to the small source is specified in terms of the induced mass M in an incompressible fluid (Exercise 7.5.13).

12.3.3 Acoustic Dipole

Let us consider two monopoles at a distance $\boldsymbol{r}_0$ (Fig. 12.6). These monopoles are assumed to have the same strengths, operating with opposite phases. Their total sound field is, according to (12.67),

$$p = -\frac{\mathrm{i}\omega\rho_0}{4\pi} V_0 \left[\frac{\exp(\mathrm{i}k|\boldsymbol{r} + \boldsymbol{r}_0/2|)}{|\boldsymbol{r} + \boldsymbol{r}_0/2|} - \frac{\exp(\mathrm{i}k|\boldsymbol{r} - \boldsymbol{r}_0/2|)}{|\boldsymbol{r} - \boldsymbol{r}_0/2|} \right] . \tag{12.74}$$

At $r \gg r_0$ we have, according to Fig. 12.6, $|\boldsymbol{r} \pm \boldsymbol{r}_0/2| \simeq r[1 \pm (r_0/2r)\cos\theta]$. Retaining only the terms of order r_0/r in the exponents, we obtain

$$p \simeq \frac{\omega\rho_0 V_0}{2\pi r} \sin\left(\frac{1}{2} k r_0 \cos\theta\right) \exp(\mathrm{i}kr) . \tag{12.75}$$

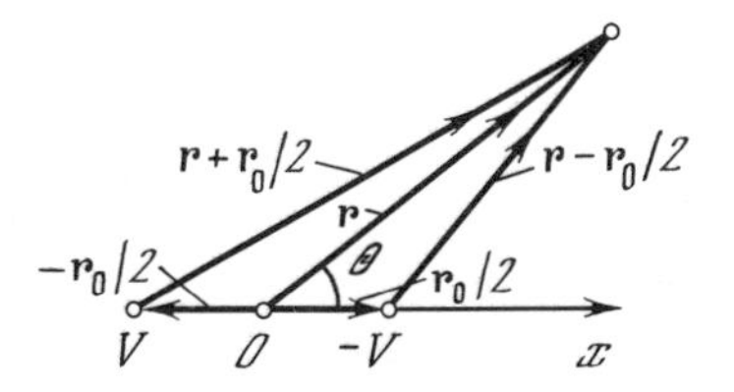

Fig. 12.6. Geometry for the calculation of the sound field of a dipole

Hence, we have the spherical wave $r^{-1}\exp(ikr)$ again but with a *direction factor* $\sin[(kr_0/2)\cos\theta]$. The radiation pattern has many lobes if $kr_0 \gg 1$, since the direction factor attains maxima a number of times in the interval $0 < \theta < \pi$. We obtain an *acoustic dipole* as in the other extreme case when the distance between monopoles is small compared with the wavelength ($kr_0 \ll 1$). Equation (12.74) then leads to

$$p \simeq \frac{\rho_0 \omega k}{4\pi r} V_0 r_0 \left(1 + \frac{i}{kr}\right) \cos\theta \exp(ikr), \tag{12.76}$$

i.e., the dipole's field has a *cosinusoidal* direction pattern. The line which connects the monopoles and the product $V_0 r_0$ are termed the *axis* and *strength* of the dipole, respectively. The sound field is maximum in the axial direction and zero perpendicular to it.

Note that (12.76) can also be written in the form

$$p = -\frac{i\omega\rho_0}{4\pi} V_0 \boldsymbol{r}_0 \cdot \nabla\left(\frac{\exp(ikr)}{r}\right), \tag{12.77}$$

i.e., the dipole field can be obtained by differentiating the monopole field in the direction of the dipole axis.

Whereas the monopole is a small pulsating sphere, it can be proved that the dipole field corresponds to sound generation by a small *oscillating* sphere, i.e., by a small rigid sphere with its center oscillating according to the law $\boldsymbol{v} = \boldsymbol{v}_0 \exp(-i\omega t)$. As in the case of a monopole, the force of the fluid's reaction to this sphere can be defined in terms of an induced mass of the accelerated sphere in an incompressible fluid, see (7.40).

12.4 Exercises

12.4.1. Find the solution $p(x,t)$ of the one-dimensional wave equation under the conditions that the pressure $p(x,0) \equiv p_0(x)$ and particle velocities $v(x,0) \equiv v_0(x)$ at $t = 0$ are given.

Solution: The general solution to the wave equation is $p(x,t) = f(x - ct) + g(x + ct)$. The corresponding relation for the velocity is $v(x,t) = (\rho_0 c)^{-1} \cdot [f(x - ct) - g(x + ct)]$. Initial conditions at $t = 0$ give

$$f(x) + g(x) = p_0(x), \qquad f(x) - g(x) = \rho_0 c v_0(x).$$

This yields immediately $f(x) = [p_0(x) + \rho_0 c v_0(x)]/2$, $g(x) = [p_0(x) - \rho_0 c v_0(x)]/2$. Hence,

$$p(x,t) = 2^{-1}[p_0(x - ct) + \rho_0 c v_0(x - ct) + p_0(x + ct) - \rho_0 c v_0(x + ct)].$$

12.4.2. Find the sound wave generated by the normal velocity distribution $v_z|_{z=0} = v_0 \exp[i(k_x x - \omega t)]$ over the plane $z = 0$.

Solution: Assume the sound wave in the halfspace $z > 0$ to be in the form

$$p = A \exp[i(k_x x + k_z z - \omega t)], \quad k_z = \sqrt{(\omega/c)^2 - k_x^2}.$$

The normal particle velocity in this wave at $z = 0$, namely

$$v_z|_{z=0} = (i\omega\rho_0)^{-1}(\partial p/\partial z)_{z=0} = Ak_z(\omega\rho_0)^{-1} \exp[i(k_x x - \omega t)],$$

must be equal to the plane's normal velocity prescribed by the problem's condition. This yields for the amplitude of the sound wave A:

$$A = \rho_0 \omega v_0/k_z = \rho_0 \omega[(\omega/c)^2 - k_x^2]^{-1/2} v_0 = \rho_0 c(\cos\theta)^{-1} v_0,$$

where θ is the angle between $\boldsymbol{k}$ and the z-axis.

12.4.3. A plane harmonic wave $p_+ = A \exp[i(kx - \omega t)]$ propagating along the tube is reflected at one end which is closed by a mobile piston with mass m per unit area. Determine the piston's impedance and the reflection coefficient.

Solution: The reflected wave is $p_- = AV \exp[i(kx + \omega t)]$. Under the action of the total pressure $p = (p_+ + p_-)_{x=0}$, the piston moves obeying Newton's second law $m(dv/dt) = p$ or $v = (-i\omega)^{-1}p$ for harmonic waves. For the piston's impedance we obtain $Z = p/v = -i\omega m$, hence the reflection coefficient is

$$V = (i\omega m + \rho_0 c)/(i\omega m - \rho_0 c).$$

12.4.4. The same as the previous exercise but for the case where a plug of thickness h ($kh \ll 1$) exists at the end of the tube and the latter is supported by a rigid bottom. The Young's modulus of the plug material is E.

Solution: The displacement u_0 at $x = 0$ is related to the pressure p_0 by Hooke's law $p_0 = Eu_0/h$. The corresponding velocity is $v_0 = -i\omega u_0$. Hence, the impedance at $x = 0$ and the reflection coefficient are

$$Z = p_0/v_0 = iE/\omega h, \quad V = (iE - \rho_0 c\omega h)/(iE + \rho_0 c\omega h).$$

12.4.5. For a sound wave with oblique incidence, find the input impedance Z_i of the fluid layer of thickness d on the fluid halfspace with the input impedance Z_1. The density and the sound velocity in the layer are ρ_0 and c_0. Consider the particular case where the halfspace is homogeneous with parameters ρ_1 and c_1.

Solution: The horizontal projections of the wave vectors as well as the frequencies must be the same in the layer and in the halfspace (Fig. 12.7) and equal to say, ξ and ω, respectively. Omitting the factor $\exp[i(\xi x - \omega t)]$ everywhere, we

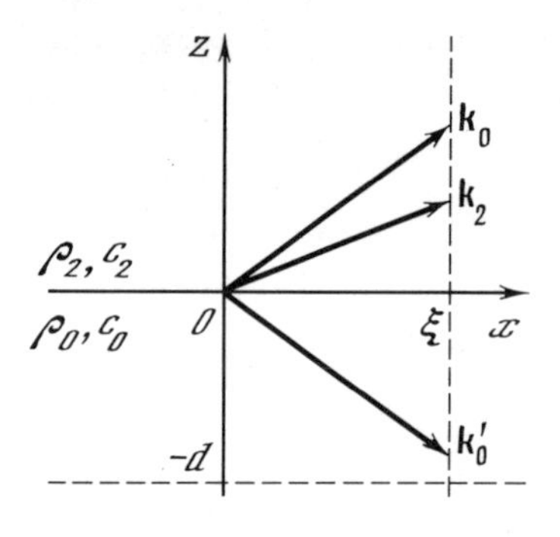

Fig. 12.7. Reflection of a plane wave at a fluid layer ($-d < z < 0$) in contact with a homogeneous fluid half-space ($0 < z < \infty$)

have for the pressure and the normal velocity in the layer

$$p = A_0 \sin k_{0z}z + B_0 \cos k_{0z}z ,$$
$$v_z = (\mathrm{i}\omega\rho_0)^{-1}k_{0z}(A_0 \cos k_{0z}z - B_0 \sin k_{0z}z),$$
$$k_{0z} = \sqrt{(\omega/c_0)^2 - \xi^2} .$$

Assuming here $z = 0$, we find $A_0 = \mathrm{i}\omega\rho_0 k_{0z}^{-1} v_z(0)$, $B_0 = p(0)$. Hence, we have for the layer

$$p(z) = \mathrm{i}\omega\rho_0 k_{0z}^{-1} v_z(0) \sin k_{0z}z + p(0) \cos k_{0z}z ,$$
$$v_z(z) = v_z(0) \cos k_{0z}z - k_{0z}(\mathrm{i}\omega\rho_0)^{-1} \sin k_{0z}z$$

so that the impedance is

$$Z(z) = \frac{p(z)}{v_z(z)} = \frac{Z_1 + \mathrm{i}Z_0 \tan k_{0z}z}{Z_0 + \mathrm{i}Z_1 \tan k_{0z}z} Z_0 , \tag{12.78}$$

where $Z_0 = \omega\rho_0 k_{0z}^{-1} = \rho_0 c_0 \cos^{-1}\theta$ is the normal impedance for a plane wave in the layer and $Z_1 = p(0)/v(0)$. The input impedance is

$$Z_\mathrm{i} = Z_0(-d) = \frac{Z_1 - \mathrm{i}Z_0 \tan k_{0z}d}{Z_0 - \mathrm{i}Z_1 \tan k_{0z}d} Z_0 . \tag{12.79}$$

If the halfspace $z > 0$ is homogeneous, only the outgoing wave $p_1 = A \exp[\mathrm{i}(\xi x + k_{1z}z - \omega t)]$, $k_{1z} = \sqrt{(\omega/c_1)^2 - \xi^2}$ can exist in it. Hence,

$$Z_1 = Z(0) = (p_1/v_{1z})_{z=0} = \omega\rho_1 k_{1z}^{-1} = \omega\rho_1[(\omega/c_1)^2 - \xi^2]^{-1/2} .$$

12.4.6. A homogeneous layer with parameters ρ_0, c_0 and d is placed between two halfspaces with parameters ρ_1, c_1 and ρ_2, c_2. Find the relationship between the parameters of a layer and the halfspaces which ensures a complete transmission through the layer of a wave with oblique incidence.

Solution: Under the assumption that the wave is incident from the halfspace 2, the reflection coefficient can be found using (12.33) and assuming $Z_1 \to Z_2 = \rho_2 c_2 \cos^{-1}\theta_2 = \omega\rho_2 k_{2z}^{-1}$, $k_{2z} = \sqrt{(\omega/c_2)^2 - \xi^2}$, $Z_2 \to Z_i$. Here, Z_i is the input impedance of the layer in contact with the homogeneous halfspace, which was found in the previous exercise. The condition $V = 0$ gives $Z_2 = Z_i$ or with (12.79),

$$Z_2 = \frac{Z_1 - iZ_0 \tan k_{0z}d}{Z_0 - iZ_1 \tan k_{0z}d} Z_0 .$$

This equation is satisfied in one of three cases:

i) $Z_0 = Z_1 = Z_2$. The normal impedances of all three media are equal;

ii) $k_{0z}d = n\pi$, $n = 1, 2, \ldots, Z_1 = Z_2$. If the incidence is normal, this means that $d = n\lambda/2$ and $\rho_1 c_1 = \rho_2 c_2$, i.e., the layer's thickness must be half the wavelength times an integral number, and the characteristic impedances of the halfspaces must be equal;

iii) $k_{0z}d = (n + 1/2)\pi$, $n = 0, 1, 2, \ldots, Z_0 = \sqrt{Z_1 Z_2}$, i.e., the layer's thickness must be a quarter of the wavelength times an odd integer, and the layer's normal impedance must be the geometric mean of those of the halfspaces (Exercise 2.7.6).

12.4.7. Find the reflection coefficient of an acoustic wave at the interface of a fluid with parameters ρ_0, c_0 and a solid halfspace with density ρ_1 and the velocities of longitudinal and transverse waves c_l and c_t, respectively. The wave is incident at the angle θ from the fluid.

Solution: Only a vertically polarized outgoing wave is generated at the interface. The projections of all wave vectors onto the boundary are the same, i.e., $k_{0x} = k_0 \sin\theta = \xi = k_x = \varkappa_x$ (Fig. 12.8) where $k_0 = \omega/c_0$, $k = \omega/c_l$ and $\varkappa = \omega/c_t$.

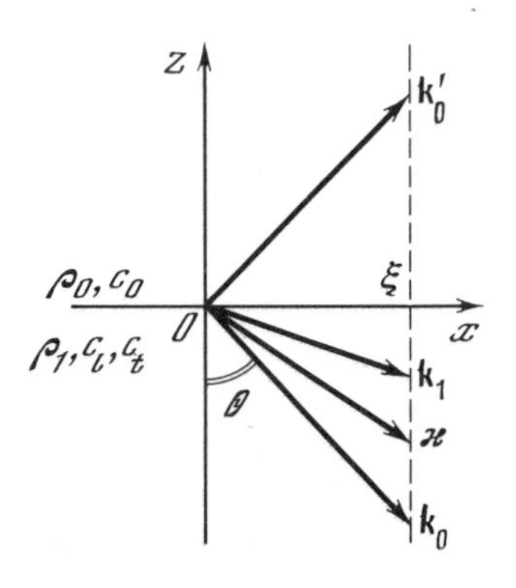

Fig. 12.8. Reflection of a sound plane wave at an interface between fluid ($0 < z < \infty$) and solid ($-\infty < z < 0$) halfspaces. $\boldsymbol{k}_0$, $\boldsymbol{k}'_0$, $\boldsymbol{k}_1$ and $\boldsymbol{\kappa}$ are the wave vectors of the incident, reflected, refracted longitudinal and refracted transverse waves, respectively

The wave potentials in the solid are

$$\varphi = W_l \exp[i(\xi x + k_z z - \omega t)], \quad z < 0,$$

$$\psi = W_t \exp[i(\xi x + \varkappa_z z - \omega t)], \quad z < 0,$$

where $k_z = \sqrt{(\omega/c_1)^2 - \xi^2}$ and $\varkappa_z = \sqrt{(\omega/c_t)^2 - \xi^2}$. According to (4.2), (3.6, 22), we have for the normal particle velocity and the components of the stress tensor at the boundary:

$$v_z|_{z=0} = -i\omega(ik_z W_l + i\xi W_t),$$

$$\sigma_{xz}|_{z=0} = -\rho_1 c_t^2[2k_z\xi W_l + (2\xi^2 - \varkappa^2)W_t],$$

$$\sigma_{zz}|_{z=0} = \rho_1 c_t^2[(2\xi^2 - \varkappa^2)W_l - 2\xi\varkappa_z W_t].$$

The factor $\exp[i(\xi x - \omega t)]$ common for all quantities is omitted here. The fluid does not support shear stress, hence $\sigma_{xz}|_{z=0} = 0$ or $W_t = 2k_z\xi(2\xi^2 - \varkappa^2)^{-1}W_l$. To obtain the input impedance of a solid, we note that the normal particle velocity must be continuous at the boundary and the pressure in the fluid equals the normal stress in the solid with opposite sign (by definition, the pressure gives the force which a given fluid element exerts on the surrounding fluid; on the other hand, the stress specifies the force exerted by the surrounding medium on a given volume). Hence,

$$Z_i = -(\sigma_{zz}/v_z)_{z=0} = -\rho_1\omega[(2\xi^2 - \varkappa^2)^2 + 4\xi^2 k_z \varkappa_z]/k_z\varkappa^4. \tag{12.80}$$

Now the reflection coefficient can be determined according to (12.33):

$$V = \frac{Z_i - Z_0}{Z_i + Z_0}, \quad Z_0 = -\frac{\rho_0\omega}{k_{0z}}, \quad k_{0z} = \sqrt{(\omega/c_0)^2 - \xi^2} = k_0\cos\theta. \tag{12.81}$$

12.4.8. A boundary (Stonely) wave can propagate along the interface between a fluid and a solid, analogous to the Rayleigh wave at the free boundary of the solid. Using the results of the previous exercise, derive an equation for velocity of this wave.

Solution: The boundary wave must be an inhomogeneous one and decrease exponentially going away from the boundary; hence,

$$k_z = -i\sqrt{\xi^2 - (\omega/c_l)^2} = -i\omega\sqrt{S - S_l},$$

$$\varkappa_z = -i\sqrt{\xi^2 - (\omega/c_t)^2} = -i\omega\sqrt{S - S_t},$$

$$k_{0z} = i\sqrt{\xi^2 - (\omega/c_0)^2} = i\omega\sqrt{S - S_0},$$

where $S_l = c_l^{-2}$, $S_t = c_t^{-2}$, $S_0 = c_0^{-2}$ and $S = \xi^2/\omega^2 = c_s^{-2}$ and c_s is the Stonely-wave velocity. We see that $c_s < c_0$, $c_s < c_t < c_l$. The boundary wave follows from the condition $V = 0$ (see the analogous reasoning for a Rayleigh wave, Sect. 4.3.1), hence according to (12.81), $Z_i = Z_0 = i\rho_0(S - S_0)^{-1/2}$ or, taking into account (12.80), $f_s(S) = f_r(S) - (\rho_0/\rho_1)S_t^2(S - S_l)^{1/2}(S - S_0)^{-1/2}$. It is the required equation for a Stonely wave where the function $f_r(S) = 4S[(S - S_l)(S - S_t)]^{1/2} - (2S - S_t)^2$ was introduced in the theory of a Rayleigh wave, see (4.33). The equation $f_s(S) = 0$ always has the root $S_s > \max\{S_0, S_t\}$. Indeed, as $S \to \infty$ we have $f_s(S) \to 2S(S_t - S_l) > 0$. On the other hand, we have $f_s(S_0) = -\infty$ if $S_0 > S_t$ and $f_s(S_t) = -S_t^2 - (\rho_0/\rho_1)S_t^2(S_t - S_l)^{1/2}(S_t - S_0)^{-1/2} < 0$ if $S_0 < S_t$. We see that the function $f_s(S)$ has opposite signs at the ends of the interval $(\max\{S_0, S_t\}, \infty)$, hence there must be $f_s(S_s) = 0$ at some point $S_s > \max\{S_0, S_t\}$. In the particular case where $S_t \ll S_0 < S_s$ ($c_s < c_0 \ll c_t$), it is easy to obtain

$$S_s \simeq S_0[1 + (\rho_0 c_0^2/2\rho_1 c_1^2)^2(1 - c_t^2/c_1^2)^{-2}],$$

$$c_s \simeq c_0[1 - 2^{-1}(\rho_0 c_0^2/2\rho_1 c_1^2)^2(1 - c_t^2/c_1^2)^{-2}].$$

The velocity of a Stonely wave is somewhat less than the sound velocity in the fluid of this problem. This implies, in turn, a small attenuation of the amplitude in the z-direction. Note that $f_s(S) = 0$ does not involve the frequency, i.e., the Stonely wave propagates without dispersion.

12.4.9. Consider a sound wave, taking gravity into account. Suppose that the sound velocity as well as the equilibrium entropy are constant throughout the fluid.

Solution: We have $p_0 = p_0(\rho_0)$ for the equilibrium state. The hydrostatic equation $-g\rho_0 = dp_0/dz = (dp_0/d\rho_0)\,d\rho_0/dz = c^2\,d\rho_0/dz$ yields on integration ($c^2 =$ const): $\rho_0(z) = \rho_0(0)\exp(-gz/c^2)$. The Väisälä frequency $N^2 = -g(\rho_0^{-1}\,d\rho_0/dz + g/c^2)$ is here zero. Now the system of linearized hydrodynamic equations (10.3) produces the following system of acoustic equations:

$$\frac{\partial \boldsymbol{v}}{\partial t} + \nabla\left(\frac{p}{\rho_0}\right) = 0, \quad \rho_0^{-1}\frac{\partial p}{\partial t} - gw + c^2\nabla\boldsymbol{v} = 0.$$

Introducing the new function $\mathscr{P} = p/\rho_0$, we obtain

$$\Delta\mathscr{P} - c^{-2}\frac{\partial^2\mathscr{P}}{\partial t^2} - gc^{-2}\frac{\partial\mathscr{P}}{\partial z} = 0.$$

We look for the solution in the form $\mathscr{P} = A\exp[i(\boldsymbol{\varkappa}\cdot\boldsymbol{r} - \omega t)]$, $\boldsymbol{\varkappa} = [\varkappa_x, \varkappa_y, \varkappa_z\}$. Substitution of this expression gives $\omega^2 = \varkappa^2 c^2 + ig\varkappa_z$. On the other hand, if we introduce the vector $\boldsymbol{k} = \{k_x, k_y, k_z\}$ so that $\varkappa_z = k_z + i\alpha$, $\varkappa_x = k_x$, $\varkappa_y = k_y$, we obtain, separating real and imaginary parts in the last equation, $\alpha = -g/2c^2$ and $\omega^2 = (k^2 + \alpha^2)c^2$. Hence, we obtain the sound wave

$$p = A\rho_0(0)\exp(-gz/2c^2)\exp[i(\boldsymbol{k}\cdot\boldsymbol{r} - \omega t)],$$

$$\boldsymbol{v} = (i\omega)^{-1}\nabla\mathscr{P} = (\boldsymbol{\varkappa}/\omega)A\exp(gz/2c^2)\exp[i(\boldsymbol{k}\cdot\boldsymbol{r} - \omega t)]$$

propagating along $\boldsymbol{k}$ with p decreasing and $\boldsymbol{v}$ increasing when z increases (the z-component of the energy flux averaged over a period is constant). The dispersion relation for this wave is

$$\omega = c\sqrt{k^2 + (g/2c^2)^2}, \qquad c_{\mathrm{Ph}} = \omega/k = c\sqrt{1 + (g/2c^2k)^2},$$

i.e., dispersion occurs at a very low frequency when $k \lesssim k_{\min} = g/2c^2$. The quantity $\omega_{\min} = g/2c$ is the minimum sound frequency in a gravitational field. In the case of water, $\omega_{\min} = 3 \times 10^{-3}\ \mathrm{c}^{-1}$. The wavelength $\lambda_{\max}$ corresponding to $k_{\min}$, when dispersion becomes significant, is $\lambda_{\max} = 2\pi/k_{\min} = 4\pi c^2/\mathrm{g} \simeq 3000$ km for water.

12.4.10. Obtain the dispersion relation for gravity-acoustic waves in an isothermal atmosphere.

Solution: Assuming that the atmosphere behaves as an ideal gas, we have for the equilibrium state $p_0(z) = [p_0(0)/\rho_0(0)]\rho_0(z)$. Waves propagate under isentropic conditions in the atmosphere, i.e., $p\rho^{-\gamma} = \mathrm{const}$ where $\gamma = c_p/c_v = 1.4$. Hence, the sound velocity is $c^2 = (\partial p/\partial \rho)_s = \gamma p_0/\rho_0 = \gamma p_0(0)/\rho_0(0) = \mathrm{const}$ and $p_0(z) = (c^2/\gamma)\rho_0(z)$. We have the hydrostatic equation $dp_0/dz = (c^2/\gamma)\,d\rho_0/dz = -g\rho_0(z)$ which yields after integration, $\rho_0(z) = \rho_0(0)\exp(-\gamma g z/c^2)$ and also, $N^2 = -g(\rho_0^{-1}\,d\rho_0/dz + g/c^2) = (\gamma - 1)g^2/c^2$. Now the linearized system of hydrodynamic equations becomes

$$\partial \boldsymbol{v}/\partial t + \rho_0^{-1}\nabla p + g\rho_0^{-1}\rho\nabla z = 0, \qquad \partial \rho/\partial t + (d\rho_0/dz)w$$
$$+ \rho_0 \nabla \boldsymbol{v} = 0, \qquad \partial \rho/\partial t = c^{-2}\,\partial p/\partial t + \rho_0 g^{-1} N^2 w.$$

Introducing the new function $\mathscr{P} = p/\rho_0$, we obtain in the case of harmonic waves with frequency ω,

$$-\mathrm{i}\omega\boldsymbol{v} + \nabla\mathscr{P} - (g^{-1}N^2\mathscr{P} - \mathrm{i}\omega^{-1}N^2 w)\nabla z = 0,$$
$$-\mathrm{i}\omega c^{-2}\mathscr{P} - g c^{-2} w + \nabla \boldsymbol{v} = 0.$$

We have two linear equations with constant coefficients so that it is natural to look for solutions $\mathscr{P}$ and $\boldsymbol{v}$ proportional to $\exp(\mathrm{i}\boldsymbol{\varkappa}\cdot\boldsymbol{r})$, $\boldsymbol{\varkappa} = \{\xi, \boldsymbol{\varkappa}_z\}$. Substitution yields the following relationship between $\boldsymbol{\varkappa}$ and ω:

$$\xi^2 - c^{-2}\omega^2 - (N/\omega)^2\xi^2 + \varkappa_z^2 + \mathrm{i}\varkappa_z g^{-1}(N^2 + g^2c^{-2}) = 0.$$

Substituting $\varkappa_z = k_z + \mathrm{i}\alpha$ and separating the real and imaginary parts, we obtain

$$\alpha = -(N^2 + g^2c^{-2})/2g = (2\rho_0)^{-1}d\rho_0/dz = -\gamma g/2c^2,$$
$$\xi^2 + k_z^2 - (\omega/c)^2 - (N/\omega)^2\xi^2 + \alpha^2 = 0.$$

Solving the last equation for ω^2 yields two branches of the dispersion relation

$$\omega_\pm^2 = \frac{k^2+\alpha^2}{2}c^2\left(1 \pm \sqrt{1 - \frac{4k^2N^2\sin^2\theta}{c^2(k^2+\alpha^2)^2}}\right),$$

where $k^2 = \xi^2 + k_z^2$, and θ is the angle which $\boldsymbol{k} = \{\xi, k_z\}$ makes with the vertical (z-axis). The second term under the square root sign attains its maximum at $k^2 = \alpha^2$, this maximum being $(c\alpha)^{-2}N^2\sin^2\theta < 4\gamma^{-2}(\gamma - 1) \simeq 0.8$. By a series expanding we obtain

$$\omega_\pm^2 \simeq \frac{k^2+\alpha^2}{2}c^2\left(1 \pm 1 \mp \frac{2k^2N^2\sin^2\theta}{c^2(k^2+\alpha^2)^2}\right).$$

The first branch

$$\omega_+^2 \simeq (k^2+\alpha^2)c^2\left(1 - \frac{k^2N^2\sin^2\theta}{c^2(k^2+\alpha^2)^2}\right)$$

includes an error of no more than 8% and corresponds to acoustic waves. These waves propagate without dispersion if $k^2 \gg \alpha^2$. Dispersion as well as a weak dependence on N^2 (not more than 10%) and also anisotropy of the wave propagation occur if $k^2 \lesssim \alpha^2$. The minimum frequency of the acoustic waves is $\omega_{\min} = -\alpha c = \gamma g/2c$. The second branch $\omega_-^2 = N^2k^2(k^2+\alpha^2)^{-1}\sin^2\theta$ corresponds to internal waves.

12.4.11. Describe the sound field of a point source (monopole) in the halfspace bounded by a rigid or a free plane boundary.

Solution: Let the boundary be at $z = 0$ and the source at the point $(0, 0, z_0/2)$. We need the solution of the Helmholtz equation (12.8) which behaves as (12.67) in the source vicinity $r \to 0$ $[r^2 = x^2 + y^2 + (z - z_0/2)^2]$ under the following boundary conditions:

a) $v_z|_{z=0} = 0$ if the boundary is rigid;
b) $p|_{z=0} = 0$ if the boundary is free.

We can easily see that the required solution is a superposition of the fields of our source and that of its "image" or the source at the point $(0, 0, -z_0/2)$ obtained by the mirror reflection of the "real" source at the boundary. The strength of the "image" is the same as that of a real source. It has the same phase as the real source in case (a) and the opposite phase in case (b). The sound field at $z > 0$ in the latter case is described by (12.74) where $\boldsymbol{r}_0 = -z_0\nabla z$. Hence, the monopole near the free boundary ($kz_0 \ll 1$) is equivalent to a dipole with a vertical axis.

12.4.12. Compare the total power radiated by a monopole in free space situated at a distance $z_0/2 \ll k^{-1}$ from the free boundary.

Solution: The power radiated by a monopole in free space is, according to (12.72), $I_{\mathrm{m}} = (\rho_0 c/8\pi)k^2|V_0|^2$. In the case of a dipole we can use (12.76) for the sound pressure with $r_0 = z_0$ (Exercise 12.4.11). Assuming $kr \gg 1$, we have for the particle velocity $v_r = (i\omega\rho_0)^{-1}\,\partial p/\partial r \simeq (z_0 k^2/4\pi r)V_0 \cos\theta \exp(ikr)$. Now using the real parts of p and v_r, we obtain for the specific energy-flux rate

$$I_{\mathrm{r}} = pv_r = \frac{\omega\rho_0}{16\pi^2}|V_0|^2 z_0^2 k^3 \cos^2\theta \,\frac{\cos^2(kr - \omega t)}{r^2}.$$

Time averaging and integrating over the halfspace of radius r gives the total radiated power:

$$I_{\mathrm{d}} = 2\pi \frac{\omega\rho_0}{32\pi^2}|V_0|^2 z_0^2 k^3 \int_0^{\pi/2} \cos^2\theta \sin\theta\, d\theta = \frac{\rho_0 c}{48\pi}|V_0|^2 (z_0 k)^2 k^2.$$

Hence, the presence of the free boundary decreases the radiated power considerably so that $I_{\mathrm{d}}/I_{\mathrm{m}} = (z_0 k)^2/6 \ll 1$.

12.4.13. Obtain the dispersion relation for the axially symmetric sound modes in a circular tube (radius R) with the fixed (rigid) wall.

12.4.14. Find the eigen spherically-symmetric vibrations of the fluid sphere (radius R). Consider the case of a free boundary (the drop) and fixed one.

13. Magnetohydrodynamics

When a conducting fluid moves in a magnetic field, additional forces acting on the fluid particles arise. On the other hand, electric currents showing up in a moving fluid change the external magnetic field. The interaction of a fluid velocity field and a magnetic field causes some pecularities of fluid motion and complicates a description of this motion considerably. In this chapter we turn to *magnetohydrodynamics*—the branch of science which provides us with such a description when the medium is assumed to be continuous.

The liquid core of the Earth, ionized gases (plasma) in the ionosphere in the interior of the sun and stars, interstellar space, etc., are examples in nature of conducting fluids in magnetic fields. Intensive development of magnetohydrodynamics and plasma physics was recently related to some applied problems such as magnetohydrodynamic generators, controlled thermonuclear fusion and so on. In some important cases, however, a description of the plasma's behaviour in a magnetic field is insufficient in the framework of the magnetohydrodynamics approximation considered below.

13.1 Basic Concepts of Magnetohydrodynamics

13.1.1 Fundamental Equations

The basic magnetohydrodynamic equations are those of hydrodynamics and electrodynamics in which the relationship between a particle's motion and a magnetic field is taken into account. This relationship can be expressed in terms of the force that the *magnetic field* $\boldsymbol{H}$ exerts on a *current* with density $\boldsymbol{J}$. As is known from physics, this force per unit volume is

$$\boldsymbol{f} = c^{-1}\boldsymbol{J} \times \boldsymbol{H}, \tag{13.1}$$

where c is the *velocity of light in vacuum*. Here and below we use the *Gauss* system of units (CGS) which seems to us to be most convenient.

The Navier-Stokes equation (8.8) with $a = \eta$ and $b = -2\eta/3$ (Exercise 8.4.1) can be written [including the force (13.1)] in the form

$$\rho\frac{\partial \boldsymbol{v}}{\partial t} + \rho(\boldsymbol{v} \cdot \nabla)\boldsymbol{v} = -\nabla p + \eta\,\Delta\boldsymbol{v} + \frac{\eta}{3}\nabla(\nabla\boldsymbol{v}) + \frac{1}{c}\boldsymbol{J} \times \boldsymbol{H}. \tag{13.2}$$

The continuity equation (6.9) still holds:

$$\frac{d\rho}{dt} = \frac{\partial \rho}{\partial t} + \boldsymbol{v} \cdot \nabla \rho = -\rho \nabla \boldsymbol{v} . \tag{13.3}$$

Neglecting entropy variations, we assume the state equation as

$$p = p(\rho) . \tag{13.4}$$

In the case of an incompressible fluid ($\rho = \text{const}$), we use $\operatorname{div} \boldsymbol{v} = 0$ instead of (13.3, 4).

Now we turn to the equations of electrodynamic. The current density is found from *Ohm's law:*

$$\boldsymbol{J} = \sigma \boldsymbol{E}_* = \sigma(\boldsymbol{E} + c^{-1} \boldsymbol{v} \times \boldsymbol{H}) , \tag{13.5}$$

where σ is the *electrical conductivity*, $\boldsymbol{E}$ the *electric field* and $\boldsymbol{E}_*$ is the field in the coordinate system moving together with the fluid. Nonmagnetic media (*magnetic permeability* $\mu = 1$) are usually considered in magnetohydrodynamics and the processes are assumed to be rather slow so that the *displacement current* can be disregarded. Then we have *Maxwell's equations* for $\boldsymbol{E}$ and $\boldsymbol{H}$:

$$\operatorname{curl} \boldsymbol{E} = -c^{-1} \frac{\partial \boldsymbol{H}}{\partial t} , \quad \operatorname{curl} \boldsymbol{H} = \frac{4\pi}{c} \boldsymbol{J} , \quad \operatorname{div} \boldsymbol{H} = 0 . \tag{13.6}$$

Equation (13.2–6) are sufficient to determine the unknown vector and scalar quantities ρ, p, $\boldsymbol{v}$, $\boldsymbol{J}$, $\boldsymbol{E}$, $\boldsymbol{H}$.

13.1.2 The Magnetic Pressure. Freezing of the Magnetic Field in a Fluid

Expressing $\boldsymbol{J}$ in terms of $\boldsymbol{H}$ by the second equation in (13.6), substituting it into (13.2) and transforming the term $\operatorname{curl} \boldsymbol{H} \times \boldsymbol{H}$ on the right-hand side using a known formula of vector analysis

$\operatorname{curl} \boldsymbol{H} \times \boldsymbol{H} = (\boldsymbol{H} \cdot \nabla)\boldsymbol{H} - \nabla H^2/2$ gives

$$\rho \frac{d\boldsymbol{v}}{dt} = -\nabla\left(p + \frac{H^2}{8\pi}\right) + \frac{(\boldsymbol{H} \cdot \nabla)\boldsymbol{H}}{4\pi} + \eta \, \Delta \boldsymbol{v} + \frac{\eta}{3} \nabla(\nabla \boldsymbol{v}) . \tag{13.7}$$

Assume for a moment that the magnetic-field gradient is perpendicular to the direction of the field itself. Then $(\boldsymbol{H} \cdot \nabla)\boldsymbol{H} = 0$ and the last equation yields the relation for the component $\boldsymbol{v}_\perp$ perpendicular to $\boldsymbol{H}$:

$$\rho \frac{d\boldsymbol{v}_\perp}{dt} = -\nabla_\perp\left(p + \frac{H^2}{8\pi}\right) + \boldsymbol{F}_\perp ,$$

where $\boldsymbol{F}_\perp$ is the viscous-force component perpendicular to $\boldsymbol{H}$. It follows from this equation that a conducting fluid moves in a direction normal to the magnetic field in such a way as if magnetic pressure $H^2/8\pi$ occurred in the fluid simultaneously with pressure p. Hence, one can exert a magnetic pressure on the conducting fluid, for example, by pushing a magnetic piston, or by confining the flow by a "magnetic walls".

Now turn to (13.5, 6) and eliminate $\boldsymbol{E}$ and $\boldsymbol{J}$ from them. Assuming $\sigma = \text{const}$, we obtain as a result

$$\frac{\partial \boldsymbol{H}}{\partial t} = \operatorname{curl}(\boldsymbol{v} \times \boldsymbol{H}) - \frac{c^2}{4\pi\sigma} \operatorname{curl}\operatorname{curl}\boldsymbol{H}.$$

This can be written in a somewhat different form if we use the vector-analysis formula $\operatorname{curl}\operatorname{curl} = \nabla \operatorname{div} - \Delta$ and take into account that $\operatorname{div}\boldsymbol{H} = 0$. Then we obtain

$$\frac{\partial \boldsymbol{H}}{\partial t} = \operatorname{curl}(\boldsymbol{v} \times \boldsymbol{H}) + \frac{c^2}{4\pi\sigma} \Delta\boldsymbol{H}. \tag{13.8}$$

Equation (3.8) allows us to find the magnetic field $\boldsymbol{H}$ if the fluid particle velocity field $\boldsymbol{v}$ is known. In particular, for the medium at rest ($\boldsymbol{v} = 0$), we have a diffusion equation for the field $\boldsymbol{H}$:

$$\partial\boldsymbol{H}/\partial t = D_{\mathrm{m}} \Delta\boldsymbol{H}, \tag{13.9}$$

where the quantity $D_{\mathrm{m}} = c^2/4\pi\sigma$ stands for the *diffusion coefficient*. Hence, the magnetic field diffuses through the substance in the course of time. The *diffusion (penetration) depth* at a given t can be estimated as $h \sim \sqrt{D_{\mathrm{m}} t} = c\sqrt{t/4\pi\sigma}$. In the case of a periodic process with frequency ω, one has $t \sim \omega^{-1}$ and $h_\omega \sim c/\sqrt{4\pi\sigma\omega}$. That is why alternating current occurs only in a thin surface layer of thickness h_ω in a conductor (*skin effect*). The greater the conductivity σ and the higher the frequency ω, the more pronounced the phenomenon, i.e., the less the penetration depth.

We now consider the extreme case of a perfectly conducting fluid ($\sigma \to \infty$), (13.8) being transformed into

$$\partial\boldsymbol{H}/\partial t = \operatorname{curl}(\boldsymbol{v} \times \boldsymbol{H}) \tag{13.10}$$

which is identical with that for vorticity in an ideal incompressible liquid [the latter can be obtained using the results of Exercise 8.4.2, setting $\eta = 0$, $p = p(\rho)$ and $\nabla\boldsymbol{v} = 0$]. Hence, in the same manner as vortex lines, the magnetic lines of force become captured by the fluid, i.e., they move together with it. It is conventionally said that the magnetic field is "*frozen*" in the fluid. This result could also be foreseen according to the following rather general consideration. Suppose for the moment that a perfectly conducting fluid crosses the magnetic lines of force while moving, then an electromotive force must arise in this fluid which, in turn, produces a current with infinite density, which is impossible.

An electric field in a coordinate system moving together with a fluid must vanish in a perfectly conducting fluid according to (13.5), i.e., $\boldsymbol{E}_* = \boldsymbol{E} + \boldsymbol{v} \times \boldsymbol{H}/c = 0$ whence $\boldsymbol{E} = -\boldsymbol{v} \times \boldsymbol{H}/c = -\boldsymbol{v}_\perp \times \boldsymbol{H}/c$. The last relation imposes certain constraints on the fluid velocity component across the magnetic field, whereas the velocity component $\boldsymbol{v}_{||}$ along the field $\boldsymbol{H}$ can be arbitrary. Taking the vector product of the last relation and $\boldsymbol{H}$, and treating the double vector product by a known formula of vector analysis, we obtain with $\boldsymbol{v}_\perp \cdot \boldsymbol{H} = 0$,

$$\boldsymbol{v}_\perp = c\boldsymbol{E} \times \boldsymbol{H}/H^2 = c\boldsymbol{E}_\perp \times \boldsymbol{H}/H^2, \quad v_\perp = |\boldsymbol{v}_\perp| = cE_\perp/H, \tag{13.11}$$

where $\boldsymbol{E}_\perp$ is the projection of the electric field onto the plane perpendicular to $\boldsymbol{H}$. Hence, in mutually orthogonal magnetic and electric fields the perfectly conducting fluid is bound to move across the magnetic lines of force at a speed specified by (13.11). This motion is called *electric drift* and the corresponding velocity $\boldsymbol{v}_\perp$ is the *electric drift velocity*. The latter is always perpendicular to the plane containing the vectors $\boldsymbol{E}$ and $\boldsymbol{H}$.

We have considered two extreme cases where the dominant term on the right-hand side of (13.8) is either the first term (convective) or the second one (diffusive). In general, the quantity analogous to the Reynolds number determines the relative importance of either term. This quantity is the ratio of vD (D being the characteristic scale) to the diffusion coefficient. In a viscous liquid, kinematic viscosity stands for the diffusion coefficient and an ordinary Reynolds number vD/ν shows to what extent convection prevails over vortex diffusion. In magnetohydrodynamics D_{m} is the diffusion coefficient (sometimes this quantity is called *magnetic viscosity*) and the relation $\mathrm{Re}_{\mathrm{m}} = vD/D_{\mathrm{m}} = 4\pi\sigma vDc^{-2}$ is called the *magnetic Reynolds number*. For $\mathrm{Re}_{\mathrm{m}} \gg 1$ convection prevails, and for the whole fluid (except the boundary layers) the perfect conductivity approximation holds. A more detailed discussion of dimensionless (similarity) parameters will follow in Exercise 13.3.1.

13.1.3 The Poiseuille (Hartmann) Flow

Let us consider the steady laminar Poiseuille flow of a conducting fluid between two plane-parallel plates at rest. We assume the external homogeneous magnetic field $\boldsymbol{H}_0$ to be perpendicular to the plates. The problem, approximately in this formulation, was first solved by Hartmann, therefore the flow is often referred to as the *Hartmann* flow.

We take the x-axis in the direction of the velocity vector of the flow, the y-axis in the direction of an external magnetic field $\boldsymbol{H}_0$ and the z-axis perpendicular to both. As for the usual Poiseuille flow, the velocity $\boldsymbol{v} = \{v_x, 0, 0\}$ can depend only on y $[v_x = v = v(y)]$; nothing depends on z by virtue of the problem's symmetry. The moving fluid "stretches" the magnetic lines of force in the direction of motion, hence the component of the magnetic field H_x appears. Moreover, an electric current arises in the fluid and its density can easily be found from Ohm's law (13.5). This current is in the z-direction, since the vector $\boldsymbol{v} \times \boldsymbol{H}/c$ has a nonzero component only in this direction. Also assuming the existence of a static electric field $\boldsymbol{E}_0$ along the z-axis for generality, we obtain for the electric current density

$$J = J_z = \sigma E^* = \sigma(E_0 + vH_0/c). \tag{13.12}$$

Substituting this into (13.2) gives equation for the flow velocity $v(y)$,

$$-\partial p/\partial x + \eta\, d^2v/dy^2 - \sigma(E_0 + vH_0/c)H_0/c = 0. \tag{13.13}$$

Here E_0 and H_0 are constants. The gradient $\partial p/\partial x$ is constant, which follows from differentiating (13.2) with respect to x and taking into account that $\boldsymbol{v}$, $\boldsymbol{J}$ and $\boldsymbol{H}$ do not depend on x: $\nabla(\partial p/\partial x) = 0$, $\partial p/\partial x = \text{const}$. As a result, we obtain an ordinary differential equation with constant coefficients for the velocity $v(y)$:

$$\frac{d^2v}{dy^2} - \frac{\sigma H_0^2}{\eta c} v = \left(\frac{\partial p}{\partial x} + \frac{\sigma}{c} E_0 H_0\right)\eta^{-1}. \tag{13.13a}$$

Its solution of this equation satisfying the boundary conditions $v(\pm d) = 0$ is

$$v = -\left(\frac{\partial p}{\partial x} + \frac{\sigma}{c} E_0 H_0\right)\frac{c^2}{\sigma H_0^2}\left[1 - \cosh\left(\frac{My}{d}\right)\cosh^{-1}M\right], \tag{13.14}$$

where the dimensionless quantity $M = H_0 d c^{-1}\sqrt{\sigma/\eta}$ is called the *Hartmann number*.[9] As a result, we have a complete solution of our problem. The component of the magnetic field H_x can be found from the second equation in (13.6):

$$dH_x/dy = -4\pi J_z c^{-1} = -4\pi\sigma c^{-1}(E_0 + H_0 v c^{-1}).$$

Relation (13.14) shows that Hartmann's flow (unlike an ordinary Poiseuille flow) can be affected not only by the pressure gradient change dp/dx but by a change in the extreme electric field $E_z = E_0$ as well. The velocity profile $v(y)$ depends essentially on the Hartmann number which is, as can easily be seen from (13.13), the ratio of magnetic $F_\mathrm{m} \sim \sigma H_0^2 c^{-1} v$ and viscous $F_\mathrm{v} \sim \eta\, d^{-2} v$ forces: $M \sim F_\mathrm{m}/F_\mathrm{v}$. For the case when viscous forces are dominant ($M \ll 1$), we find from (13.14)

$$v = -\left(\frac{\partial p}{\partial x} + \frac{\sigma}{c} E_0 H_0\right)\frac{d^2}{2\eta}\left(1 - \frac{y^2}{d^2}\right) \tag{13.14a}$$

[9] Do not confuse this with the Mach number introduced above

which coincides with (8.22) for Poiseuille flow (parabolic velocity profile) if we take $E_0 = 0$, $y \to y + d$, $2d = h$. When the magnetic forces ($M \gg 1$) are dominant the result is quite different:

$$v = -\left(\frac{\partial p}{\partial x} + \frac{\sigma}{c} E_0 H_0\right) \frac{c^2}{\sigma H_0^2} \left\{1 - \exp\left[M\left(1 - \frac{|y|}{d}\right)\right]\right\}. \tag{13.14b}$$

The velocity is practically constant and equal to $v_0 = -(\partial p/\partial x + \sigma E_0 H_0/c)$ $c^2/\sigma H_0^2$ over almost the entire fluid. Only in thin boundary layers of thickness $h \sim d/M$ near the plates does the velocity change rapidly from v_0 to 0.

13.2 Magnetohydrodynamic Waves

13.2.1 Alfvén Waves

The magnetic force $\boldsymbol{J} \times H/c$ in the equation of motion (13.2) for a conducting fluid introduces a new type of wave. Indeed, we can deduce from (13.7) that the magnetic field gives rise to "magnetic pressure" $H^2/8\pi$, and the force $(\boldsymbol{H} \cdot \nabla)\boldsymbol{H}/4\pi$ whose component along the i-axis is

$$\frac{1}{4\pi} H_k \frac{\partial H_i}{\partial x_k} = \frac{1}{4\pi} \frac{\partial}{\partial x_k}(H_i H_k) - \frac{1}{4\pi} H_i \frac{\partial H_k}{\partial x_k} = \frac{1}{4\pi} \frac{\partial}{\partial x_k}(H_i H_k).$$

Here, $\partial H_k/\partial x_k = \operatorname{div} \boldsymbol{H} = 0$ was taken into account. Hence, the effect of the force $(\boldsymbol{H} \cdot \nabla)\boldsymbol{H}/4\pi$ is equivalent to that of the stresses $\sigma_{ik} = H_i H_k/4\pi$. Taking the axis $x_1 = x$ along $\boldsymbol{H}$ at some arbitrary point, we have $\sigma_{11} = \sigma_{xx} = H^2/4\pi$, $\sigma_{12} = \sigma_{xy} = \sigma_{13} = \cdots = 0$ at this point. Hence, the force $(\boldsymbol{H} \cdot \nabla)\boldsymbol{H}/4\pi$ is equivalent to the tension $H^2/4\pi$ along the magnetic line of force (which was first pointed out by Faraday). We see now that the magnetic lines of force in a well-conducting medium experience tension on the one hand, and possess the inertia of the medium they are frozen in on the other hand. Hence, they become similar to a stretched string and can oscillate as a string. Such oscillations propagate in the form of waves which are called *Alfvén waves*.

We can obtain an equation for these waves by linearization of the original set of the magnetohydrodynamic equations, assuming the fluid to be non-viscous ($\eta = 0$), incompressible ($\nabla \boldsymbol{v} = 0$) and perfectly conducting ($\sigma \to \infty$) and the external magnetic field as homogeneous ($\boldsymbol{H}_0 = \text{const}$). Let $\boldsymbol{H}$ be the magnetic field disturbance caused by the fluid motion so that the total field is $\boldsymbol{H}_0 + \boldsymbol{H}$. Now we obtain from (13.7, 10), neglecting the terms containing the powers of $\boldsymbol{H}$ and $\boldsymbol{v}$ higher than the first,

$$\rho \frac{\partial \boldsymbol{v}}{\partial t} = -\nabla\left(p + \frac{\boldsymbol{H}_0 \cdot \boldsymbol{H}}{4\pi}\right) + \frac{(\boldsymbol{H}_0 \cdot \nabla)\boldsymbol{H}}{4\pi} = 0, \quad \frac{\partial \boldsymbol{H}}{\partial t} = \operatorname{curl}(\boldsymbol{v} \times \boldsymbol{H}_0)$$
$$= (\boldsymbol{H}_0 \cdot \nabla)\boldsymbol{v}. \tag{13.15}$$

Applying the operation ∇ to the first equation and taking into account that $\operatorname{div} \boldsymbol{H} = 0$, $\operatorname{div} \boldsymbol{v} = 0$, we obtain

$$\Delta(p + \boldsymbol{H}_0 \cdot \boldsymbol{H}/4\pi) = 0. \tag{13.16}$$

The solution of the Laplace equation which is bounded everywhere in space can only be constant, i.e., $p + \boldsymbol{H} \cdot \boldsymbol{H}/4\pi = \text{const}$. Taking this into account and choosing the x-axis along the external magnetic field $\boldsymbol{H}_0$, we obtain from (13.15),

$$\frac{\partial \boldsymbol{H}}{\partial x} = \frac{4\pi\rho}{H_0}\frac{\partial \boldsymbol{v}}{\partial t}, \quad \frac{\partial \boldsymbol{v}}{\partial x} = \frac{1}{H_0}\frac{\partial \boldsymbol{H}}{\partial t}, \tag{13.17}$$

whence the wave equation

$$\frac{\partial^2 \boldsymbol{H}}{\partial x^2} = c_{\mathrm{a}}^{-2}\frac{\partial^2 \boldsymbol{H}}{\partial t^2} \tag{13.18}$$

and an analogous equation for $\boldsymbol{v}$ follow. Here,

$$c_{\mathrm{a}} = H_0/\sqrt{4\pi\rho} \tag{13.19}$$

is the *velocity of the Alfvén waves.*

We look for a solution to (13.18) in the form of a plane wave:

$$\boldsymbol{H} = \boldsymbol{F}(x \pm c_{\mathrm{a}}t). \tag{13.20}$$

It follows from div $\boldsymbol{H} = 0$ that $F'_x = 0$, $F_x = \text{const}$ since $\partial/\partial y = \partial/\partial z = 0$. For the wavelike solution we have to set const $= 0$. Hence, the disturbance in a plane Alfvén wave is orthogonal to the direction of propagation. One can also obtain an expression analogous to (13.20) for the velocity field:

$$\boldsymbol{v} = \boldsymbol{G}\,(x \pm c_{\mathrm{a}}t), \tag{13.21}$$

v_x being zero as well. By virtue of (13.17), the quantities $\boldsymbol{v}$ and $\boldsymbol{H}$ in the plane wave are related to each other by

$$\boldsymbol{v} = \pm c_{\mathrm{a}}\boldsymbol{H}/H_0. \tag{13.22}$$

Note that (13.18) is also satisfied by a more general kind of wave

$$\boldsymbol{H} = \boldsymbol{F}(x \pm c_{\mathrm{a}}t, y, z) \tag{13.23}$$

for which (13.22) still holds. This fact implies a specific dispersion for the Alfvén waves. Indeed, let us consider the solution of (13.18) in the form of harmonic plane waves

$$\boldsymbol{H} = \boldsymbol{A}\exp[\mathrm{i}(\boldsymbol{k} \cdot \boldsymbol{R} - \omega t)], \tag{13.24}$$

where A is a constant vector, $\boldsymbol{k} = \{k_x, k_y, k_z\}$ and $\boldsymbol{R} = \{x, y, z\}$. Substituting (13.24) into (13.18) leads to the dispersion relation

$$\omega^2 = c_a^2 k_x^2, \quad \omega = c_a |k_x| \tag{13.25}$$

and the phase and the group velocities

$$\begin{aligned} c_{\text{Ph}} &= \omega/k = c_a k_x/k = c_a \cos\varphi, \\ \boldsymbol{c}_g &= \nabla_k \omega = c_a(\boldsymbol{H}_0/H_0)\,\text{sign}\,k_x. \end{aligned} \tag{13.26}$$

Thus, the phase velocity of the Alfvén waves depends on the propagation direction (i.e., on the angle φ which the wave vector $\boldsymbol{k}$ makes with the magnetic field $\boldsymbol{H}_0$), whereas the group velocity is equal to the Alfvén-wave velocity and is directed along $\boldsymbol{H}_0$. It follows from $\text{div}\,\boldsymbol{H} = 0$ that the magnetic field disturbance $\boldsymbol{H}$ and the wave vector $\boldsymbol{k}$ are mutually orthogonal. Since (13.22) holds for harmonic waves, we have $\boldsymbol{v} \perp \boldsymbol{k}$, too. Consequently, the Alfvén waves are *transverse* waves in an electromagnetic ($\boldsymbol{H} \perp \boldsymbol{k}$) as well as in a hydrodynamic ($\boldsymbol{v} \perp \boldsymbol{k}$) sense.

13.2.2 Magnetoacoustic Waves

We now consider wave motion in a compressible, perfectly conducting, inviscid fluid in the presence of a homogeneous magnetic field $\boldsymbol{H}_0$. In the linear approximation the fluid density ρ in the first equation of (13.15) can be assumed equal to its equilibrium value ρ_0. This equation together with the linearized continuity (13.3) and state (13.4) equations yields the system

$$\rho_0 \frac{\partial \boldsymbol{v}}{\partial t} = -\nabla\left(p + \frac{\boldsymbol{H}_0 \cdot \boldsymbol{H}}{4\pi}\right) + \frac{(\boldsymbol{H}_0 \cdot \nabla)\boldsymbol{H}}{4\pi},$$

$$\frac{\partial \rho}{\partial t} + \rho_0\,\text{div}\,\boldsymbol{v} = 0, \quad p = c_s^2 \rho,$$

where $c_s = (\partial p/\partial \rho)^{1/2}$ is the *adiabatic sound velocity* to which the index "s" is assigned in order not to confuse it with the velocity of light c. Differentiating the first equation with respect to t and taking into account the second and third equations, we obtain

$$\frac{\partial^2 \boldsymbol{v}}{\partial t^2} = \nabla\left(c_s^2\,\text{div}\,\boldsymbol{v} - \frac{\boldsymbol{H}_0}{4\pi\rho_0} \cdot \frac{\partial \boldsymbol{H}}{\partial t}\right) + \frac{\boldsymbol{H}_0 \cdot \nabla}{4\pi\rho_0} \frac{\partial \boldsymbol{H}}{\partial t}. \tag{13.27}$$

Using the relation $\text{curl}(\boldsymbol{v} \times \boldsymbol{H}_0) = (\boldsymbol{H}_0 \cdot \nabla)\boldsymbol{v} - \boldsymbol{H}_0\,\text{div}\,\boldsymbol{v}$, we rewrite the second equation of (13.15) in the form

$$\frac{\partial \boldsymbol{H}}{\partial t} = (\boldsymbol{H}_0 \cdot \nabla)\boldsymbol{v} - \boldsymbol{H}_0\,\text{div}\,\boldsymbol{v}. \tag{13.27a}$$

Two vector equations (13.27, 27a) for $\boldsymbol{v}$ and $\boldsymbol{H}$ constitute the system of *magnetoacoustical equations.*

We consider a plane wave, choosing the coordinate system in such a way that the x-axis is directed along $\boldsymbol{H}_0$ and the propagation direction of the wave (vector $\boldsymbol{k}$) lies in the (x, y)-plane. Nothing depends on the coordinate z in this case. We write (13.27a) in components

$$\frac{\partial H_x}{\partial t} = -H_0 \frac{\partial v_y}{\partial y}, \quad \frac{\partial H_y}{\partial t} = H_0 \frac{\partial v_y}{\partial x}, \quad \frac{\partial H_z}{\partial t} = H_0 \frac{\partial v_z}{\partial x}. \tag{13.28}$$

Taking these relations into account, we can easily obtain from (13.27) the following set of equations for the velocity components:

$$\begin{aligned} \frac{\partial^2 v_x}{\partial t^2} &= c_s^2 \left(\frac{\partial^2 v_x}{\partial x^2} + \frac{\partial^2 v_y}{\partial x \, \partial y} \right), \\ \frac{\partial^2 v_y}{\partial t^2} &= c_s^2 \left(\frac{\partial^2 v_x}{\partial x \, \partial y} + \frac{\partial^2 v_y}{\partial y^2} \right) + c_a^2 \left(\frac{\partial^2 v_y}{\partial x^2} + \frac{\partial^2 v_y}{\partial y^2} \right), \end{aligned} \tag{13.29}$$

$$\frac{\partial^2 v_z}{\partial t^2} = c_a^2 \frac{\partial^2 v_z}{\partial x^2}, \tag{13.30}$$

where c_a is the Alfvén velocity introduced above, see (13.19).

Note that the component v_z perpendicular both to the external magnetic field $\boldsymbol{H}_0$ and the propagation direction of the wave is not related to the other components of the velocity field and propagates independently with velocity c_a. Comparing the third equation of (13.28) with the second equation of (13.17) we can see that this is an ordinary Alfvén wave.

To discuss the equations for the two other components v_x and v_y, we consider some simple cases first.

i) A wave propagates in the x-direction ($\partial/\partial y = 0$). The system (13.29) decomposes into two independent equations:

$$\frac{\partial^2 v_x}{\partial t^2} = c_s^2 \frac{\partial^2 v_x}{\partial x^2}, \quad \frac{\partial^2 v_y}{\partial t^2} = c_a^2 \frac{\partial^2 v_y}{\partial x^2}. \tag{13.31}$$

The component v_x propagates at the sound velocity c_s. This is an ordinary sound wave not influenced by the magnetic field. The disturbance of the magnetic field is zero, according to (13.28). The component v_y propagates as an Alfvén wave.

ii) A wave propagates in the y-direction ($\partial/\partial x = 0$). From (13.29) we obtain

$$\frac{\partial^2 v_x}{\partial t^2} = 0, \quad \frac{\partial^2 v_y}{\partial t^2} = (c_s^2 + c_a^2) \frac{\partial^2 v_y}{\partial y^2}. \tag{13.32}$$

A longitudinal disturbance v_y propagates with velocity $\sqrt{c_s^2 + c_a^2}$ in this case, i.e., an acoustic-type wave occurs. However, by virtue of the first equation in (13.28), disturbance of the magnetic field ($H_x \neq 0$) occurs, too. If the external magnetic field is large ($c_a^2 \gg c_s^2$), this wave, called "*magnetic sound*", propagates at the Alfvén-wave velocity but the character of the particles' motion is different since the Alfvén waves are transverse, whereas the wave under consideration is longitudinal. Thus, in a conducting fluid, an acoustic-type wave can propagate in a direction perpendicular to the magnetic field; its velocity is determined by the elastic forces in the fluid and magnetic pressure as well.

13.2.3 Fast and Slow Magnetoacoustical Waves

Let us now discuss the general case $\boldsymbol{k} = \{k_x, k_y, 0\}$ and look for the solution of (13.29) in the form of a harmonic wave

$$\boldsymbol{v} = \boldsymbol{A} \exp[\mathrm{i}(\boldsymbol{k} \cdot \boldsymbol{r} - \omega t)], \tag{13.33}$$

where $\boldsymbol{A} = \{A_x, A_y\}$ is the constant vector and $\boldsymbol{r} = \{x, y\}$. Substitution of (13.33) into (13.29) leads to a homogeneous system of algebraic equations with respect to the quantities A_x and A_y:

$$\begin{aligned} &(\omega^2 - c_s^2 k_x^2) A_x - c_s^2 k_x k_y A_y = 0\,, \\ &c_s^2 k_x k_y A_x - (\omega^2 - c_s^2 k_y^2 - c_a^2 k^2) A_y = 0\,, \end{aligned} \tag{13.34}$$

whence it is seen that v_x and v_y are in phase, i.e., the particles' paths are straight lines. Equating the determinant of system (13.34) to zero, we obtain the *dispersion relation* for magnetoacoustic waves:

$$\omega^4 - (c_s^2 + c_a^2) k^2 \omega + c_s^2 c_a^2 k^4 \cos^2\theta = 0\,, \tag{13.35}$$

where $\theta = \arccos(k_x/k)$ is the angle which the vector $\boldsymbol{k}$ makes with the direction of the external magnetic field. Two solutions of (13.35)

$$\omega_\pm^2 = [(c_s^2 + c_a^2)/2 \pm \sqrt{(c_s^2 + c_a^2)^2/4 - c_s^2 c_a^2 \cos^2\theta}\,] k^2 \tag{13.36}$$

correspond to two types of waves propagating with the phase velocities $c_\pm = \omega_\pm/k$. Taking into account that $(c_s^2 + c_a^2)^2 - 4c_s^2 + c_a^2 \cos^2\theta \geq (c_s^2 - c_a^2)^2$, one can easily obtain relationships between c_s^2, c_a^2 and $c_\pm^2$:

$$c_-^2 \leq \min\{c_s^2, c_a^2\} \leq \max\{c_s^2, c_a^2\} \leq c_+^2\,. \tag{13.37}$$

Thus, the two types of waves can exist in a compressible conducting liquid in the general case. The phase velocity of one type exceeds neither the sound velocity nor the Alfvén-wave velocity. These waves are called *slow magnetoacoustic waves*. The other types of waves called *fast magnetoacoustic waves* have a greater velocity than both the sound and the Alfvén-wave velocities.

By virtue of $\operatorname{div} \boldsymbol{H} = 0$, magnetoacoustic waves are *transverse in an electromagnetic sense:* $\operatorname{div} \boldsymbol{H} = \mathrm{i}\boldsymbol{k} \cdot \boldsymbol{H} = 0$, $\boldsymbol{k} \perp \boldsymbol{H}$. But the angle which the particle velocity vector $\boldsymbol{v}$ makes with the wave vector $\boldsymbol{k}$ may vary in these waves. Two extreme cases $c_s^2 \gg c_a^2$ and $c_a^2 \gg c_s^2$ are considered in Exercise 13.3.4. In the first case the fast wave is almost longitudinal with $c_+ \simeq c_s$, whereas the slow one is almost transverse and propagates with velocity $c_- \simeq c_a \cos\theta$. In the second case, particles move parallel to the y-axis in a fast wave propagating with velocity $c_+ \simeq c_a$ and in the x-direction in a slow wave (velocity $c_- \simeq c_s \cos\theta$).

13.3 Exercises

13.3.1. Find the dimensionless similarity parameters for the conducting fluid motion in a magnetic field.

Solution: The basic dimensional parameters for the magnetohydrodynamic equations (13.2–6) are: characteristic velocity v_0, characteristic scale D, external magnetic field H_0. We have the orders of magnitude of different terms in (13.7): an inertial force $\boldsymbol{F}_i = \rho(\boldsymbol{v} \cdot \nabla)\boldsymbol{v} \sim \rho D^{-1} v_0^{-2}$, a viscous force $\boldsymbol{F}_v = \eta \Delta \boldsymbol{v} \sim \eta D^{-2} v_0$ and a magnetic force $\boldsymbol{F}_m \sim H_0^2/4\pi D$. Correspondingly, we can introduce two dimensionless parameters: an ordinary *Reynolds number*

$$\mathrm{Re} \sim F_i/F_v \sim \rho v_0 D/\eta$$

and the so-called *Alfvén number*

$$A \sim F_m/F_i \sim H_0^2/4\pi\rho v_0^2 \, .$$

Now consider two terms on the right-hand side of (13.8). The first one describing the "frozen-in" field has the order of magnitude $\operatorname{curl}(\boldsymbol{v} \times \boldsymbol{H}) \sim v_0 H_0 D^{-1}$. The second term corresponding to the diffusion has the order of magnitude $c^2 \Delta \boldsymbol{H}/4\pi\sigma \sim c^2 H_0^2/4\pi\sigma D^2$. The ratio of the former to the latter shows a degree of freezing of the field by the medium and is called the *magnetic Reynolds number:*

$$\mathrm{Re}_m = 4\pi\sigma c^{-2} v_0 D \, .$$

The field is "frozen-in" (complete freezing) for $\mathrm{Re}_m \gg 1$ and diffusive for $\mathrm{Re}_m \ll 1$.

The *Hartmann's number* $M = H_0 D c^{-1} \sqrt{\sigma/\eta}$ introduced in Sect. 13.1.3 can be expressed in terms of Re, A and Re_m as

$$M^2 = A \,\mathrm{Re}\,\mathrm{Re}_m \sim \mathrm{Re}_m F_m/F_v \, .$$

In some cases, the so-called *Stewart number* is introduced, too:

$$S = A\,\mathrm{Re}_m = \sigma H_0^2 D/\rho c^2 v_0 \, .$$

13.3.2. Examine the steady flow of a conducting fluid between two plane-parallel plates. One of the plates is at rest, the other moves in its plane with velocity v_0 (Couette flow). An external magnetic field $\boldsymbol{H}_0$ is perpendicular to the plates.

Solution: The problem again reduces to the solution of (13.13a) with $E_0 = 0$, $\partial p/\partial x = 0$ and with the boundary conditions $v(-d) = 0$, $v(d) = v_0$. The solution to this equation is

$$v(y) = v_0 \sinh[M(1 + y/d)]/\sinh 2M\,.$$

It becomes a linear function if $M \ll 1$ and the exponent

$$v(y) = v_0 \exp[-M(1 - y/d)] \quad \text{if} \quad M \gg 1.$$

13.3.3. Examine the Hartmann-Poiseuille flow (Sect. 13.1.3) assuming that the flow is bounded in the z-direction by perfectly conducting parallel walls at a distance large compared with the distance d between the plates which bound the flow in the y-direction.

Solution: The velocity $v(y)$ is again determined by (13.14), but the electric field now cannot be assumed arbitrary. Its value can be found from the condition that the total current perpendicular to the perfectly conducting walls vanishes $\int_{-d}^{d} J_z\, dy = 0$. Since

$$J_z = \sigma(E_0 + H_0 c^{-1} v) = \frac{c}{H_0}\frac{\partial p}{\partial x} + \left(\frac{c}{H_0}\frac{\partial p}{\partial x} + \sigma E_0\right)\frac{\cosh(My/d)}{\cosh M}\,,$$

we have

$$\int_{-d}^{d} J_z\, dy = \frac{2d}{M}\left[\sigma E_0 \tanh M - \frac{c}{H_0}\frac{\partial p}{\partial x}(M - \tanh M)\right].$$

Equating the latter to zero, we obtain

$$E_0 = \frac{c}{\sigma H_0}\frac{\partial p}{\partial x}\frac{M - \tanh M}{\tanh M}$$

and substituting this into (13.14), we find

$$v(y) = -\frac{c^2}{\sigma H_0^2}\frac{\partial p}{\partial x} M \frac{\cosh M - \cosh(My/d)}{\sinh M}\,.$$

13.3.4. Consider the motion of the fluid particles in magnetoacoustic waves for two extreme cases: a) $c_s \gg c_a$ and b) $c_a \gg c_s$.

Solution: If $\max\{c_\mathrm{s}, c_\mathrm{a}\} \gg \min\{c_\mathrm{s}, c_\mathrm{a}\}$, then the square root in (13.36) can be expanded in a series:

$$\sqrt{(c_\mathrm{s}^2 + c_\mathrm{a}^2)/4 - c_\mathrm{s}^2 c_\mathrm{a}^2 \cos^2\theta} \simeq \frac{c_\mathrm{s}^2 + c_\mathrm{a}^2}{2} + \frac{c_\mathrm{s}^2 c_\mathrm{a}^2}{c_\mathrm{s}^2 + c_\mathrm{a}^2} \cos^2\theta \,.$$

As a result, we obtain for the phase velocities

$$c_+^2 = c_\mathrm{s}^2 + c_\mathrm{a}^2 \simeq \max\{c_\mathrm{s}^2, c_\mathrm{a}^2\}\,,$$

$$c_-^2 = \frac{c_\mathrm{s}^2 c_\mathrm{a}^2}{c_\mathrm{s}^2 + c_\mathrm{a}^2} \cos^2\theta \simeq \min\{c_\mathrm{s}^2, c_\mathrm{a}^2\} \cos^2\theta \,.$$

From (13.34) we also have $v_y = (\omega^2 - c_\mathrm{s}^2 k_x^2)(c_\mathrm{s}^2 k_x k_y)^{-1} v_x$.

a) Let $c_\mathrm{s}^2 \gg c_\mathrm{a}^2$, then we have for the fast wave $c_+^2 \simeq c_\mathrm{s}^2$, $\omega_+^2 \simeq c_\mathrm{s}^2 k^2$ and $v_y \simeq (k_y/k_x) v_x$ or $v_x/v_y \simeq k_x/k_y$, i.e., the fast wave is longitudinal in this case. For the phase velocity of the slow wave we have $c_-^2 \simeq c_\mathrm{a}^2 \cos^2\theta$, so that $\omega_-^2 \simeq c_\mathrm{a}^2 k^2 \cos^2\theta = c_\mathrm{a}^2 k_x^2$ and $v_y \simeq -(k_x/k_y) v_x$, or $k_x v_x + k_y v_y \simeq 0$, i.e., $\boldsymbol{v} \perp \boldsymbol{k}$, meaning that the slow wave is transverse.

b) Analogously for $c_\mathrm{a}^2 \gg c_\mathrm{s}^2$, we have $c_+ \simeq c_\mathrm{a}$, $\omega_+^2 \simeq c_\mathrm{a}^2 k^2$ and

$$v_y \simeq \frac{c_\mathrm{a}^2}{c_\mathrm{s}^2} \frac{k^2}{k_x k_y} v_x = 2 \frac{c_\mathrm{a}^2}{c_\mathrm{s}^2} \frac{v_x}{\sin 2\theta} \gg v_x,$$

i.e., in the fast wave the fluid particles move practically in the y-direction. For the slow wave $c_-^2 \simeq c_\mathrm{s}^2 \cos^2\theta$, $\omega_-^2 \simeq c_\mathrm{s}^2 k^2 \cos^2\theta = c_\mathrm{s}^2 k_x^2$ and $v_y \simeq 0$, i.e., the motion is parallel to the x-axis.

14. Nonlinear Effects in Wave Propagation

Linear theory for different kinds of waves in fluids has so far been developed. The superposition principle holds in this theory, i.e., waves of different kinds as well as different modes of waves of the same kind propagate without interaction with each other. One must bear in mind, however, that linear equations are only approximate ones and the original hydrodynamic equations are substantially nonlinear. It is important to clarify the conditions under which a linear approximation is adequate and to consider new effects caused by nonlinear terms in the equations. To this very question is this chapter devoted.

First we will discuss the limitations imposed on the waves' amplitudes by the linear approximation. Then we will consider some typical problems of nonlinear wave theory such as the generation of higher harmonics, change of wave form, propagation in dispersive media, resonance interaction of waves and so on.

Our discussion of principal nonlinear effects will be based on simple model equations without specifying the nature of the wave. This approach may appear somewhat abstract but it has the advantage that it avoids a lot of mathematics which sometimes obscures the main features of the phenomena under consideration. Besides, this approach reveals the common nature of nonlinear phenomena for waves of any kind. Applications of the general theory to specific types of waves are given in the main text as well as in the exercises at the end of the chapter.

14.1 One-Dimensional Nonlinear Waves

14.1.1 The Nonlinearity Parameter

One of the nonlinear terms in the equation of motion (8.1) is an inertial one $(\boldsymbol{v} \cdot \nabla)\boldsymbol{v}$. We compare its order of magnitude with that of the linear term $\partial \boldsymbol{v}/\partial t$. If v_0 is the particle velocity amplitude, ω and k are the wave frequency and wave number, respectively, then $\partial \boldsymbol{v}/\partial t \sim \omega v_0$, $(\boldsymbol{v} \cdot \nabla)\boldsymbol{v} \sim k v_0^2$. The ratio of these terms is $\varepsilon = k v_0^2/\omega v_0 = v_0/c_{\mathrm{Ph}}$ where $c_{\mathrm{Ph}} = \omega/k$ is the phase velocity of the wave.

The linear approximation is valid if $\varepsilon \ll 1$, i.e., if the particle velocity is small compared with the phase velocity. The quantity ε is called the *nonlinear parameter*. For acoustic waves, for example, it is the Mach number $\varepsilon = v_0/c = M$.

In the case of surface waves in deep water, we have $\varepsilon = v_0/c_{\mathrm{Ph}} = ka$ where $a = v_0/\omega$ is the displacement amplitude of the surface. Hence, the requirement $\varepsilon \ll 1$ means a *small slope* or what is the same, a small surface displacement a compared with the wavelength $\lambda = 2\pi/k$. In the case of shallow water ($kH \ll 1$, H is the water depth), the nonlinear parameter must be defined in another way. The horizontal component v_x of the particle velocity is now $(kH)^{-1}$ times greater than the vertical one: $v_x \sim (kH)^{-1} v_z \sim (kH)^{-1}\omega a = (a/H)c_{\mathrm{Ph}}$. We have to use this very component for estimating ε, i.e., $\varepsilon = v_x/c_{\mathrm{Ph}} = a/H$. Hence, the vertical displacement must be small compared with the depth of the water. Naturally we also have $ka = (a/H)kH = \varepsilon kH \ll 1$ as before. Analogous requirements can also be obtained for other types of waves. For internal waves, for example, we have $\varepsilon = v_0/c_{\mathrm{Ph}} = ka \ll 1$ again, a being the amplitude of particle displacement. For waves in the thermocline of thickness d, the condition $\varepsilon = a/d \ll 1$ appears, too.

A small nonlinearity parameter ε is a necessary but not sufficient condition for the validity of the linear approximation in the general case. The effects of small terms in the equations can accumulate in the course of time or when a wave propagates over a long distance. Such an accumulation strongly depends on the rate of wave energy dissipation and on the wave dispersion. The latter causes the pulse to spread and a decrease in the time and space intervals of the effective wave interaction.

14.1.2 Model Equation. Generation of Second Harmonics

In order to discuss the main features of nonlinear phenomena independently of the wave's nature, we will consider a simple one-dimensional model equation

$$\frac{\partial u}{\partial t} + Lu = -\varepsilon u \frac{\partial u}{\partial x}. \tag{14.1}$$

The term $u\,\partial u/\partial x$ here is analogous to the term $(\boldsymbol{v} \cdot \nabla)\boldsymbol{v}$ in the case of the hydrodynamical motion equation; ε is a nonlinear parameter and a linear operator L specifies the dispersion law for linear waves. If $L = c_0\,\partial/\partial x$, for example, the linear solution of (14.1) (with $\varepsilon = 0$) represents an acoustic wave

$$u = u_0 \exp[\mathrm{i}(kx - \omega t)], \quad \omega = \omega(k) = c_0 k \tag{14.2}$$

propagating without dispersion.

The case where $L = c_0\,\partial/\partial x + \beta\,\partial^3/\partial x^3$ is very important in nonlinear wave theory. Relation (14.1) is called the *Korteweg-de Vries* (KdV) *equation* in this case. It was first obtained in the theory of shallow-water waves. The dispersion relation for linear waves in this case will be

$$\omega = c_0 k - \beta k^3. \tag{14.3}$$

It is shown in Exercise 14.3.1 that this dispersion law differs from that for waves in shallow water only by terms of order higher than $(kH)^4$.

Relation (14.1) becomes the so-called *Burgers equation* if $L = c_0\,\partial/\partial x - \alpha\,\partial^2/\partial x^2$. Its linear version describes the wave in the fluid with energy dissipation

$$u(x,t) = u_0 \exp[-\alpha k^2 t + \mathrm{i}k(x - c_0 t)]\,.$$

It is easy to obtain from (14.1) that in the general case of arbitrary L, the dispersion relation is

$$\omega = \omega(k) = \omega' - \mathrm{i}\omega'' = \mathrm{i}\exp(-\mathrm{i}kx)L\exp(\mathrm{i}kx)\,. \tag{14.4}$$

The real part of this relation describes the wave dispersion and the imaginary part corresponds to the wave attenuation, so that

$$u(x,t) = u_0 \exp[-\omega'' t + \mathrm{i}(kx - \omega' t)]\,.$$

We now consider a nonlinear process governed by the complete equation (14.1). We suppose that the dispersion law for linear waves is known; for convenience, we write it as $\omega(k) = \omega_k$. Let the disturbance $u(x,0) = \varepsilon a \exp(\mathrm{i}kx) + \text{c.c.}$ be given at $t = 0$. The symbol c.c. here and below means the complex conjugate quantity to the first (written out) term. Thus, $u(x,t)$ is automatically real, which is important in nonlinear problems. We look for the solution of (14.1) as a series in powers of ε: $u = \varepsilon u_1 + \varepsilon^2 u_2 + \varepsilon^3 u_3 + \cdots$. Its substitution into (14.1) gives, after equating the terms of the first power in ε, the linear version

$$\frac{\partial u_1}{\partial t} + Lu_1 = 0\,. \tag{14.5}$$

Its solution under the assumption that at $t = 0$ $u_1(x,0) = a\exp(\mathrm{i}kx) + \text{c.c.}$, is

$$u_1(x,t) = a\exp[\mathrm{i}(kx - \omega_k t)] + \text{c.c.}$$

Now equating the terms with ε^2, we obtain an inhomogeneous linear equation for u_2 under the initial condition $u_2(x,0) = 0$:

$$\frac{\partial u_2}{\partial t} + Lu_2 = -u_1\frac{\partial u_1}{\partial x} = -\mathrm{i}ka^2\exp[2\mathrm{i}(kx - \omega_k t)] + \text{c.c.} \tag{14.6}$$

If the right-hand side of (14.6) is not the solution of the homogeneous version of the same equation, or what is the same, $2\omega_k \neq \omega_{2k} \equiv \omega(2k)$, then the solution $u_2(x,t)$ is

$$u_2(x,t) = \frac{ka^2}{2\omega_k - \omega_{2k}}\{\exp[2\mathrm{i}(kx - \omega_k t)] - \exp[\mathrm{i}(2kx - \omega_{2k} t)]\} + \text{c.c.} \tag{14.7}$$

It can also be written in the form of a wave with twice the wave number and amplitude modulation

$$u_2 = -2ika^2 \frac{\sin(\Delta\omega t/2)}{\Delta\omega} \exp\left[i\left(2kx - \frac{\omega_{2k} + 2\omega_k}{2} t\right)\right] + \text{c.c.} \tag{14.7a}$$

If the difference $\Delta\omega = 2\omega_k - \omega_{2k}$ (called *detuning*) is large, then the second harmonic u_2 remains small at any time. The time interval $\tau_i = \pi/\Delta\omega = \pi/(2\omega_k - \omega_{2k})$ required for the amplitude of the second harmonic to reach its maximum value (beginning from zero) is referred to as the *characteristic time of interaction.* The greater the detuning $\Delta\omega$ (i.e., the more pronounced the dispersion), the less the interaction time τ_i.

If $\Delta\omega = 0$, for example, for nondispersive waves when $\omega_k = c_0 k$, it is not difficult to obtain from (14.7a), in the limit, a solution with the secular term:

$$u_2(x, t) = -ika^2 t \exp[i(2kx - \omega_{2k} t)] + \text{c.c.} \tag{14.8}$$

This case corresponds to the *resonance generation* of the second harmonic when the "exterior" force [right-hand side in (14.6)] is the solution of a homogeneous equation. The equality $2\omega_k = \omega_{2k}$ here is a particular case of the so-called *conditions of synchronism.* The second harmonic increases linearly with time in this case so that for any small ε, the linear approximation fails at some sufficiently large characteristic time τ_1. An estimate of this time can be obtained by equating εu_1 and $\varepsilon^2 u_2$ which gives $\varepsilon a = \varepsilon^2 k a^2 \tau_1$ or $\tau_1 = (\varepsilon k a)^{-1}$. In the case of acoustic waves when $\varepsilon a = u_0$ is the particle velocity amplitude, we have $\tau_1 = (k u_0)^{-1} \sim T \mathrm{M}^{-1}$ where $T = 2\pi/\omega$ is the wave's period and M the Mach number.

When the second harmonic increases, its interaction with the first one leads to the generation of the third harmonic in the case of nondispersive waves. Then the 4th harmonic comes into play in the same manner and so on. As time passes, the front of the wave becomes steeper as the higher harmonics are generated.

14.1.3 The Riemann Solution. Shock Waves

The exact solution of the one-dimensional nonlinear equation for sound waves was first obtained by Riemann in the last century. We consider an analogous solution to our model equation (14.1) without dispersion or dissipation. Assuming $L = c_0\, \partial/\partial x$ and omitting the parameter ε, we obtain

$$\frac{\partial u}{\partial t} + c_0 \frac{\partial u}{\partial x} = -u \frac{\partial u}{\partial x}. \tag{14.9}$$

Let $u(x, 0) \equiv f(x)$ at $t = 0$. We show that the function

$$u(x, t) = f[x - (c_0 + u)t] \tag{14.10}$$

describing the velocity $u(x, t)$ in an implicit form is the solution to (14.9). Indeed, we have

$$\frac{\partial u}{\partial t} = -\left(c_0 + u + \frac{\partial u}{\partial t} t\right) f', \quad \frac{\partial u}{\partial x} = \left(1 - \frac{\partial u}{\partial x} t\right) f',$$

whence we find

$$\frac{\partial u}{\partial t} = -\frac{c_0 + u}{1 + f't} f' \quad \text{and} \quad \frac{\partial u}{\partial x} = \frac{f'}{1 + f't}.$$

Equation (14.9) becomes an identity when these expressions are substituted.

The solution (14.10) of (19.9) has a simple physical meaning, namely the disturbance in the medium corresponding to some fixed u travels at a constant speed $c_0 + u$. Stronger disturbances are travelling faster, eventually overtaking the weaker ones. This causes a change in the wave's profile, in particular a steepening of the leading edge of the pulse if $u > 0$.

A change in wave profile can be illustrated graphically. Introducing the new variable $\xi = x - c_0 t$, we will fix the observer in the coordinate system moving at a velocity c_0. In this system the solution (14.10) becomes

$$u(\xi, t) = f(\xi - ut). \tag{14.10a}$$

The wave form $\xi = g(u)$ at an initial moment $t = 0$ is plotted in Fig. 14.1 (dotted line), where $g(u)$ is the inverse function of $u = f(\xi)$. At any time $t > 0$, we have from (14.10a), $\xi = g(u) + ut$. Hence, to plot $u(\xi, t)$, we just have to add the linear function ut (steepening when t increases) to the fixed profile $g(u)$. In Fig. 14.1, the wave profiles are, in addition, plotted for two consecutive moments t_1 and $t_2 > t_1$ (full lines). Two points ξ_1 and ξ_2 corresponding to zero disturbance remain in the same place in this representation.

The upper half of the curve $\xi(u)$ in Fig. 14.1 becomes steeper and steeper in the course of time t and eventually at $t = t_m$ the tangent becomes perpendicular to the ξ-axis at some point of the curve $\xi(u)$. Because of the relation

$$\partial \xi / \partial u = t + dg/du,$$

we have $t_m = \min(-dg/du)$. The multi-valued solution $u(\xi)$ for $t > t_m$ must be discarded as having no physical meaning. Instead discontinuity arises; intensive energy dissipation takes place in this region. The wave profile becomes triangular and so-called *shock waves* arise. Profiles of the pulse at four consecutive moments are shown in Fig 14.2 for the case $u > 0$. The strength of the discontinuity in the shock wave decreases with time whereas the duration of the wave (the pulse length) increases. The time dependence of these quantities can be found from the momentum conservation law $\mathscr{P} = \int_{-\infty}^{\infty} u \, d\xi =$ const (Exercise 14.3.3) which follows from (14.9).

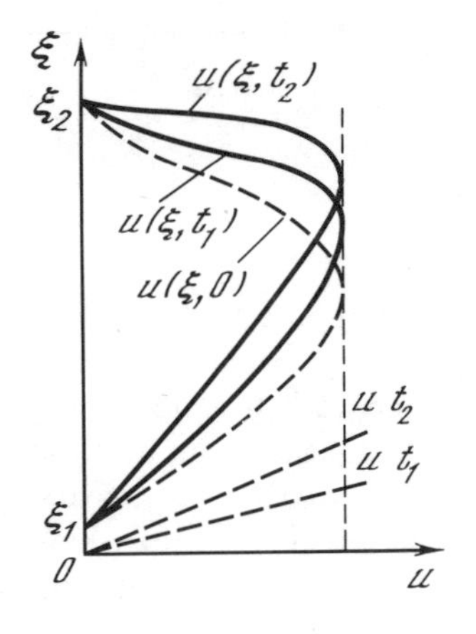

Fig. 14.1.

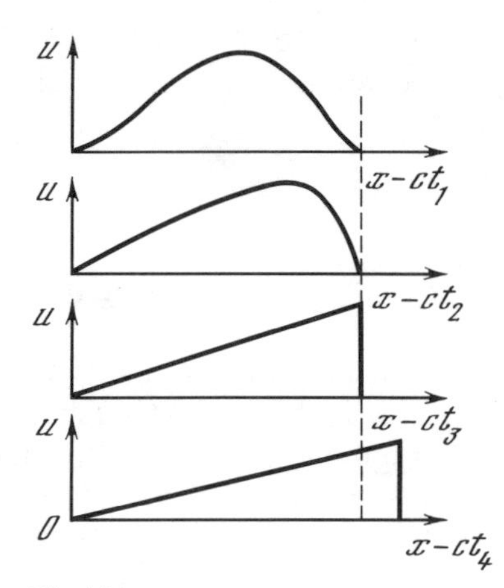

Fig. 14.2.

Fig. 14.1. Steepening of the profile of a nonlinear wave. Here, u is the particle velocity, $\xi = x - c_0 t$ and $u(\xi, 0)$ is the initial profile of the wave

Fig. 14.2. Profiles of a wave at four consecutive moment t

Disturbances with $u < 0$ or with u changing sign can be considered in an analogous way. It turns out, in particular, that any periodic disturbance (including a harmonic wave) eventually takes on a sawlike shape.

14.1.4 Dispersive Media. Solitons

Even a weak dispersion, for example, of waves in shallow water or long internal waves prevents shock front formation. The tendency to shock formation due to the nonlinearity of an equation is compensated by the opposite tendency for the shock to smear out due to dispersion. Interplay of these two factors leads to the formation of a *stationary nonlinear wave* traveling at a constant speed without changing shape.

We discuss this process for the solution of the KdV equation (Sect. 14.1.2):

$$\frac{\partial u}{\partial t} + c_0 \frac{\partial u}{\partial x} + \beta \frac{\partial^3 u}{\partial x^3} = -u \frac{\partial u}{\partial x}. \tag{14.11}$$

Introducing a new coordinate system $\xi = x - c_0 t$ moving at velocity c_0, we rewrite the (14.11) as

$$\frac{\partial u}{\partial t} = -\beta \frac{\partial^3 u}{\partial \xi^3} - u \frac{\partial u}{\partial \xi}. \tag{14.11a}$$

The first term on the right-hand side describes dispersion, the second one nonlinearity. Let us discuss qualitatively the interplay of these terms when a wave is propagating. Let l and A be the characteristic scale and the amplitude of a disturbance at $t = 0$, respectively. The orders of magnitude of dispersive and nonlinear terms will are then

$$\beta \frac{\partial^3 u}{\partial \xi^3} \sim \frac{\beta A}{l^3} \quad \text{and} \quad u \frac{\partial u}{\partial \xi} \sim \frac{A^2}{l},$$

the ratio of the second to the first being $q \sim Al^2\beta^{-1}$. Suppose that $q \gg 1$ *at* $t = 0$, i.e., we have a sufficiently long and strong disturbance. Hence, the nonlinear term is be dominant during the initial period which leads to a steepening of the wave, decreasing l. The parameter q decreases which leads to an increase in the role of dispersion, hence a tendency for our disturbance to spread appears. This reasoning suggests the possibility of such a solution to (14.11a) which corresponds to a stationary disturbance, propagating without change of form with constant $q \sim 1$. Hence, we look for a solution of the kind $u = f(\xi - ct)$, where c is a new constant, the *propagation velocity of a stationary wave.* Substitution of u into (14.11a) gives an ordinary differential equation for the function $f(\eta)$, where $\eta = \xi - ct$:

$$-cf' + \beta f''' + ff' = (-cf + \beta f'' + f^2/2)' = 0.$$

Integrating it once we obtain

$$\beta f'' - cf + f^2/2 = \text{const}. \tag{14.12}$$

The integration constant is assumed zero here. In fact, using the substitution $f = \tilde{f} + f_0$, $c = \tilde{c} + f_0$, we can rewrite (14.12) as

$$\beta\tilde{f}'' - \tilde{c}\tilde{f} + \tilde{f}^2/2 = f_0(\tilde{c} + f_0/2) + \text{const},$$

where the right-hand side vanishes with a proper choice of f_0. We must bear in mind, therefore, that an arbitrary constant can be added to any solution of (14.12) if the same constant is added to c which is equivalent to a transfer to the moving coordinate system.

Note that (14.12) is equivalent to the equation of motion for a particle of a mass β in a field of force with potential $U(f)$:

$$\beta f'' = -\frac{\partial U}{\partial f}, \quad U = \frac{f^3}{6} - \frac{cf^2}{2}. \tag{14.13}$$

It is well known that a bounded solution $f(\eta)$ exists only if the total energy $E = (\beta/2)(f')^2 + U$ is that for the particle inside the potential hole. The dependence $U(f)$ at $\beta > 0$ and $c > 0$ is plotted in Fig. 14.3a. A bound solution exists if $E \leq 0$ in this case.

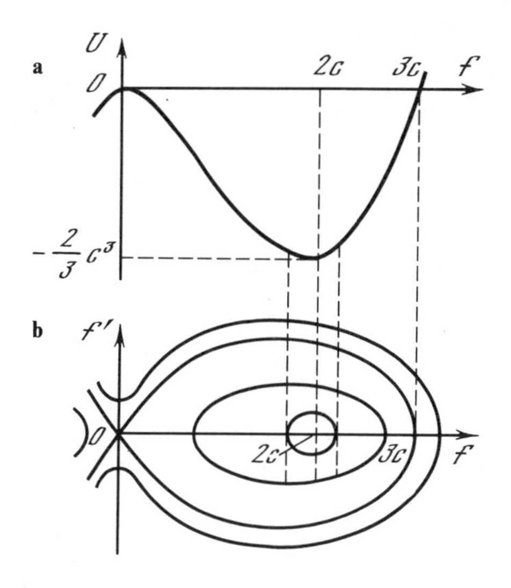

Fig. 14.3. (**a**) Equivalent potential $U(f)$ for the Korteveg-de Vries equation (**b**) Phase trajectories in the (f, f')-plane for this equation

Using the substitution $P = f'$, $f'' = dp/d\eta = p\,dp/df$, we write (14.13) as

$$\frac{d}{df}\left(\frac{\beta p^2}{2} + U\right) = 0\,,$$

which yields after integration

$$P \equiv \frac{df}{d\eta} = \pm\sqrt{2(E-U)/\beta}\,, \tag{14.14}$$

where the integration constant E describes the total energy. Using (14.13, 14), we can find the corresponding p for each value of f. The corresponding curves on the p, f-plane at different but fixed E are called *phase trajectories.* For the case under consideration, these trajectories are shown in Fig. 14.3b. They intersect the f-axis at points which can be found from the equation $E = U(f)$. The most interesting trajectory corresponds to $E = 0$ in which case this equation has a double root $f_{1,2} = 0$ and a simple root $f_3 = A = 3c$. The trajectory separates that part of the p, f plane where the trajectories are closed (*periodic motion*) from the part where trajectories go to infinity (*non-periodic motion*). That is why this trajectory is called *separating* one. It can be proved easily by substitution that the corresponding solution of (14.14) ($E = 0$) is

$$\begin{aligned} f(\eta) &= A\cosh^{-2}[(\eta + \eta_0)/\Delta]\,, \\ \Delta &= \sqrt{4\beta/c} = \sqrt{12\beta/A}\,, \end{aligned} \tag{14.15}$$

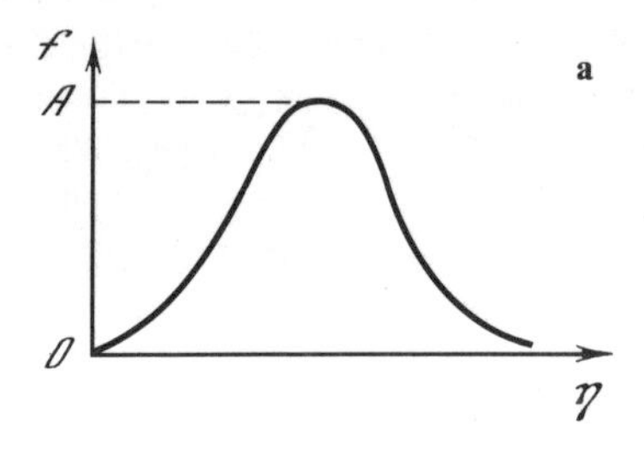

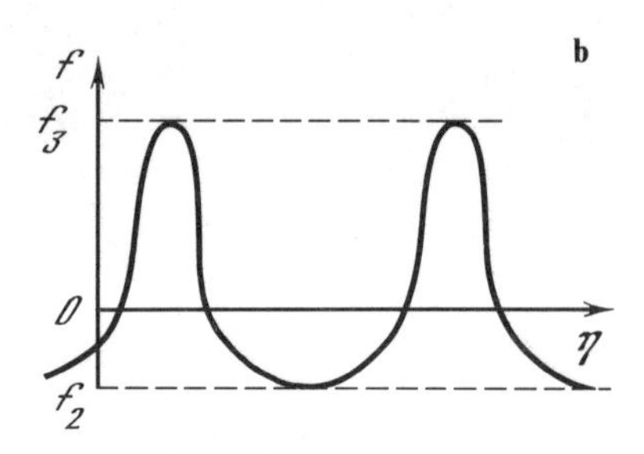

Fig. 14.4a,b. Two simple solutions of the Korteveg-de Vries equation: (**a**) solution, (**b**) cnoidal wave

called a *soliton.* Its velocity is determined in terms of its amplitude $c = A/3$. The quantity Δ is the *length scale* of the soliton; the integration constant η_0 determines the soliton's position in space. In the laboratory coordinate system, the soliton's velocity $v = c_0 + c$ always exceeds the phase velocity $c_{\mathrm{Ph}} = c_0 - \beta k^2$ of the harmonic waves. The soliton's form $f(\eta)$ is shown in Fig. 14.4a.

If $E < 0$, the solution $f(\eta)$ will be in the interval $f_2 < f < f_3$ where f_2 and f_3 are the 2nd and 3rd roots of $E = U(f)$ $(f_1 < f_2 < f_3)$. The root f_2 can even be negative since an arbitrary constant can always be added to it. Periodic solutions for $E < 0$ are called *cnoidal waves.* A typical form of this wave is shown in Fig. 14.4b. As $E \to 0$, the phase trajectory approaches the separating one. A representative point moving along this trajectory spends most time near f_2 and the solution becomes a periodic sequence of solitons.

If $E \simeq U_{\min} = U(2c) = -2c^3/3$, we can expand the right-hand side of (14.13) into a power series $(f - 2c)$ and obtain an equation for the harmonic process $\beta f'' = c(f - 2c)$. Its solution is a harmonic wave superimposed on the constant disturbance $2c$, i.e., $f = 2c + u_0 \sin k\eta$, $k = \sqrt{c/\beta}$, u_0 being an arbitrary amplitude. Thus, in a coordinate system moving at velocity $2c$, we obtain the usual linear solution of the KdV equation. Taking into account that $\eta = \xi - ct + 2ct = x - (c_0 - c)t$ in this system and also that $c = \beta k^2$, we obtain

$$u = u_0 \sin(kx - \omega t), \qquad \omega = k(c_0 - c) = kc_0 - \beta k^3 .$$

An exact solution of (14.11) confirms the existence of solitons and proves their stability. Moreover, an interesting new fact follows from this solution. Suppose we have an initial pulse which is negative if $\beta < 0$ and positive if $\beta > 0$ with $q = A\Delta^2/\beta > 12$ ($q = 12$ for a soliton). It turns out then that while propagating, this pulse breaks into a sequence of solitons arranged in order of their strength. The soliton with the largest amplitude (and the greatest speed) is first.

There is one more interesting fact about the multisoliton solution of the KdV equations. Suppose we have two solitons and that which goes first has a smaller amplitude. Then the other soliton which was initially behind will overtake the first in the course of time. We have some wave disturbance which is not just the sum of two solitons during some time. Later, however, this disturbance again breaks into two separate solitons but now that with the greater amplitude is in front. Hence, solitons behave as noninteracting linear waves in some sense. It turns out that the soliton's positions after interaction are somewhat different as they would have been if interaction were not taken into account.

14.2 Resonance Wave Interaction

14.2.1 Conditions of Synchronism

It was shown in Sect. 14.1.2 how resonance occurs in the process of wave interaction when dispersion is absent. The secular terms increasing linearly with time appear in the second approximation of the perturbation method in this case. It is very instructive to note that the phenomenon discussed there can be regarded as a *three-wave interaction* process. Indeed, the resonance condition $\omega(2k) = 2\omega(k)$ used there can also be written $k_3 = k_1 + k_2$, $\omega_3 = \omega_1 + \omega_2$, where $k_1 = k_2 = k$, $k_3 = 2k$, $\omega_j = \omega(k_j)$, $j = 1, 2, 3$. Other types of hydrodynamic waves reveal analogous but more general three-wave interactions, also due to the quadratic nonlinearity of the basic equations. We will first consider the general mechanism of such an interaction. Suppose we require a solution of some nonlinear partial differential equation:

$$L_1 w = \varepsilon L_2(w^2) + \cdots \tag{14.16}$$

where L_1 is a linear operator, and $L_1 w_1 = 0$ can be satisfied by a harmonic-wave solution of the kind

$$w_1 = a \exp[\mathrm{i}(\boldsymbol{k} \cdot \boldsymbol{r} - \omega t)] + \text{c.c.}, \qquad \omega = \omega(\boldsymbol{k}). \tag{14.17}$$

The term $\varepsilon L_2(w^2)$ on the right-hand side of (14.16) symbolically represents nonlinear quadratic terms, ε being the nonlinearity parameter. We look for a solution of (14.16) in the form $w = \varepsilon w_1 + \varepsilon^2 w_2 + \cdots$. Suppose we have two waves of the type (14.17) with a_j, $\boldsymbol{k}_j$ and $\omega_j = \omega(\boldsymbol{k}_j)$, $j = 1, 2$ as initial conditions at $t = 0$. For the function w_2 of a second approximation, we obtain an inhomogeneous linear equation $L_1 w_2 = L_2(w_1^2)$. The right-hand side of the latter is the sum of terms of the type $\exp[\mathrm{i}(\boldsymbol{k}_3 \cdot \boldsymbol{r} - \omega_3 t)]$, where the pairs $\boldsymbol{k}_3, \omega_3$ assume one of the following possible values: $\pm(\boldsymbol{k}_m \pm \boldsymbol{k}_n)$, $\pm(\omega_m \pm \omega_n)$, $m, n = 1, 2$. Resonance will take place if some of the terms on the right-hand side of the equation for w_2, for example, the term $\exp\{\mathrm{i}[(\boldsymbol{k}_1 + \boldsymbol{k}_2) \cdot \boldsymbol{r} - (\omega_1 + \omega_2)t]\}$, will satisfy a homogeneous linear equation, i.e., constitute a harmonic wave (14.17). The secular term appears in the solution w_2 in this case. Obviously, such a resonance arises only if the so-called *synchronism conditions*

$$\boldsymbol{k}_3 = \boldsymbol{k}_1 + \boldsymbol{k}_2, \qquad \omega_3 = \omega(\boldsymbol{k}_3) = \omega_1 + \omega_2 = \omega(\boldsymbol{k}_1) + \omega(\boldsymbol{k}_2) \tag{14.18}$$

take place. Note that the synchronism conditions for other combinations $\boldsymbol{k}_1$ and $\boldsymbol{k}_2$, ω_1 and ω_2 can also be written in the form of (14.18), using proper enumeration of waves in a given triad and assuming the wave frequencies to be positive. For example, if the conditions $\boldsymbol{k}_3 = \boldsymbol{k}_1 - \boldsymbol{k}_2$, $\omega_3 = \omega_1 - \omega_2$ are satisfied, they can be transformed into (14.18) by the index interchange $1 \rightleftarrows 3$.

It is shown in Exercise 14.3.5 for isotropic waves (when only the magnitude k is involved in the dispersion relation) that $\boldsymbol{k}_1$ and $\boldsymbol{k}_2$ can always be found satisfying (14.18) if the dispersion curve $\omega = \omega(k)$ increases monotonically ($\omega' > 0$) and is convex downward ($\omega'' > 0$). On the contrary, no such $\boldsymbol{k}_1$ and $\boldsymbol{k}_2$ exist if the dispersion curve is convex upward. This means, according to Figs. 10.3, 9, that resonance interactions are possible in the second approximation for surface capillary waves and are forbidden for surface gravity waves as well as for given modes of internal gravity waves. The synchronism conditions (14.18) can be satisfied almost always if one considers the interaction between different kinds of waves or between waves of the same kind but with different modes. For example, two surface gravity waves with close frequencies ($\omega_2 \simeq \omega_1$) and of almost opposite directions ($\boldsymbol{k}_2 \simeq -\boldsymbol{k}_1$) generate a sound wave of frequency $\omega_3 = \omega_1 + \omega_2 \simeq 2\omega_1$, with a horizontal component of the wave vector equal to $\boldsymbol{k}_3 = \boldsymbol{k}_1 + \boldsymbol{k}_2$ ($k_3 \ll k_1$) (Exercise 14.3.9). Analogously, the conditions (14.18) are satisfied in the wave triad consisting of two surface gravity waves (ω_2 and $\omega_3 \simeq \omega_2$) and one internal wave ($\omega_1 \ll \omega_2$). Hence, an internal wave can be generated in the interaction process of two surface waves.

14.2.2 The Method of Slowly-Varying Amplitudes

The method of successive approximations discussed above is valid only at the first stage of a nonlinear process when the amplitude of a newly generated wave is small. The time in which this amplitude becomes comparable with those of the initial waves inversely proportional to the nonlinearity parameter ε and is large compared with the wave period if $\varepsilon \ll 1$. This means that the amplitude of each interacting wave in a triad can be assumed to be a *slowly-varying* function of time. Hence, the wave field of the resonance triad can be written in the form

$$w = \sum_{j=1}^{3} a_j(\varepsilon t) \exp[\mathrm{i}(\boldsymbol{k}_j \cdot \boldsymbol{r} - \omega_j t)] + \text{c.c.} \tag{14.19}$$

Here, the amplitudes depend only on $\tau \equiv \varepsilon t$ which is called *slow time*.

Substituting (14.19) into (14.16), the terms $a_j L_1 \exp[\mathrm{i}(\boldsymbol{k}_j \cdot \boldsymbol{r} - \omega_j t)]$ vanish since each wave satisfies the linearized equation, but the additional terms $\varepsilon\, da_j/d\tau$, $\varepsilon^2\, d^2 a_j/d\tau^2$ arise. The amplitudes $a_j(\varepsilon t)$ do not need differentiation on the right-hand side of (14.16) since we have the factor ε on this side already. The terms on the right-hand side which are linear in ε, are double products of linear waves as in Sect. 14.2.1. Now, equating the terms with the first power in (14.16), we obtain the system of so-called *truncated equations:*

$$\dot{a}_1 = V_1 a_2^* a_3, \quad \dot{a}_2 = V_2 a_1^* a_3, \quad \dot{a}_3 = V_3 a_1 a_2. \tag{14.20}$$

Here the dot and the asterisk mean differentiation with respect to slow time and the complex conjugate value, respectively. The subscript $j = 3$ is ascribed to the wave of highest frequency. We assume a normalization of the wave's amplitudes such that the energy of the jth wave is proportional to $|a_j|^2$ with the coefficient of proportionality independent of the number j. Then it turns out that $V_j = \mathrm{i}\omega_j V$, where V is a real quantity specifying the wave's interaction. Equation (14.20) now become

$$\dot{a}_1 = \mathrm{i}\omega_1 V a_2^* a_3, \quad \dot{a}_2 = \mathrm{i}\omega_2 V a_1^* a_3, \quad \dot{a}_3 = \mathrm{i}\omega_3 V a_1 a_2. \tag{14.20a}$$

It is shown in Exercise 14.3.6 that the law of energy conservation

$$E = |a_1|^2 + |a_2|^2 + |a_3|^2 = \text{const} \tag{14.21}$$

as well as three more conservation laws

$$\begin{aligned} \omega_1^{-1}|a_1|^2 - \omega_2^{-1}|a_2|^2 &= H_{12}, \\ \omega_1^{-1}|a_1|^2 + \omega_3^{-1}|a_3|^2 &= H_{13}, \\ \omega_2^{-1}|a_2|^2 + \omega_3^{-1}|a_3|^2 &= H_{23} \end{aligned} \tag{14.22}$$

hold for a given triad of waves, where H_{12}, H_{13}, H_{23} are constants. Each of two relations in (14.22) can be deduced from the third one together with (14.21).

It follows from (14.22) that the energy of low-frequency waves (frequencies ω_1 and ω_2) varies in phase (rise or decrease simultaneously), whereas the energy of the high-frequency wave (frequency ω_3) varies with opposite phase. Hence, the high-frequency wave can acquire energy from the other two waves or, on the other hand, supply them with energy. The second and third relations of (14.22) can be written in the form:

$$|a_j|^2 = |a_j(0)|^2 + (\omega_j/\omega_3)[|a_3(0)|^2 - |a_3(\tau)|^2], \quad j = 1, 2$$

whence it follows that the energy obtained from the third wave is distributed between the first two in proportion to their frequencies.

Using conservation laws, one can obtain a general solution of system (14.20a) in quadratures (Exercise 14.3.6). Here we confine ourselves to the case where an amplitude of the high-frequency wave is kept constant with the help of an external source of energy. Then, differentiating one of the first two equations in (14.20a) with respect to τ, we obtain, taking the other into account,

$$\ddot{a}_j = \omega_1 \omega_2 V^2 |a_3|^2 a_j, \quad j = 1, 2.$$

It follows from this equation that the amplitudes of low-frequency waves increase exponentially at the expense of those of high-frequency, i.e.,

$$a_j \sim \exp(\alpha\tau), \qquad \alpha = (\omega_1\omega_2)^{1/2} V|a_3|.$$

This phenomenon, called *decay instability* of high-frequency waves, is often used for the "parametric" generation of low-frequency waves.

14.2.3 Multiwave Interaction

Very often the resonance triad cannot be considered separately from other waves. This is the case, for example, when one of the waves in the triad together with any two other waves, makes up a new resonance triad. A good example of this kind is the generation of higher harmonics in a medium without dispersion ($\omega = c_0 k$). Elementary resonance interaction here is of the type $k_1 + k_1 = k_2$, $\omega_1 + \omega_1 = \omega_2$. The theory of this process, but in a medium with strong dispersion, is given in Exercise 14.3.7. We have complete energy transition into the second harmonic as $t \to \infty$. Such a solution, however, is obviously inconsistent in the case of nondispersive waves because in this case a harmonic wave must eventually transform itself into a periodic shock wave (Riemann's solution). Inconsistency arises because of the disregard of resonant interactions of higher harmonics: $k_3 = k_1 + k_2$, $k_4 = 2k_2, \ldots$

There is the possibility, in principle, to take all multiwave interactions into account. We will discuss the corresponding method with an application to the one-dimensional equation (14.1). We look for the solution of this equation in the form of a Fourier integral

$$u(x,t) = \int_{-\infty}^{\infty} A_k(t) \exp(ikx)\, dk, \tag{14.23}$$

where the notation $A_k(t) \equiv A(k,t)$ is used and $A_{-k} = A_k^*$ since $u(x,t)$ is real. We assume further that $A_k = \alpha_k \exp(-i\omega_k t)$, where $\omega_k = \omega(k)$ is found from the linear problem and $\alpha_{-k} = \alpha_k^*$, $\omega_{-k} = -\omega_k$. Substituting (14.23) into (14.1) and taking $\partial/\partial x = ik$ and $L = i\omega_k$ into account yields a system of equations for amplitudes α_k:

$$\dot{\alpha}_k = -i\frac{\varepsilon k}{2} \int_{-\infty}^{\infty} \alpha_q \alpha_s \exp(-i\Delta_{qsk} t)\, dq, \qquad s = k - q, \quad \Delta_{qsk} = \omega_q + \omega_s - \omega_k. \tag{14.24}$$

If we have a harmonic wave $u(x,0) = a_1(0)\exp(ik_1 x) + \text{c.c.}$ at $t = 0$, i.e., $\alpha_k(0) = a_1(0)\,\delta(k - k_1) + a_1^*(0)\,\delta(k + k_1)$, the spectrum of $u(x,t)$ remains discrete at any time; the only nonzero components are $k_n = nk_1$, $\omega_n = \omega(k_n)$, $n = \pm 1, \pm 2, \ldots$ and

$$\alpha_k(t) = \sum_n a_n(t)\,\delta(k - k_n).$$

Equation (14.24) becomes the infinite system of ordinary differential equations

$$\dot{a}_n = -\frac{\mathrm{i}\varepsilon k_n}{2} \sum_m a_m a_l \exp(-\mathrm{i}\Delta_{mln} t), \qquad l = n - m, \tag{14.25}$$

where $\Delta_{mln} = \omega_m + \omega_l - \omega_n$, $a_{-n} = a_n^*$ and $a_0 = 0$. In the case of strong dispersion and when the resonance between the first and the second harmonics takes place as in Exercise 14.3.7, only Δ_{112} is zero. Then retaining in (14.25) only those waves with indices 1 and 2, we obtain the system of equations of Exercise 14.3.7.

If, however, dispersion is absent, as in the case of an acoustic wave, or is small at some frequencies as in the case of surface and internal gravity waves, more equations must be retained in the system (14.25). Numerical integration of this system allows us to describe the initial phase of the shock-wave formation as well as the breaking of a pulse into a sequence of solitons.

14.2.4 Nonlinear Dispersion

We will now consider harmonic wave propagation in a medium with strong dispersion. The second harmonic $2k_1$ will not be in resonance with the first one k_1. From (14.25) we have for the amplitudes of the first and the second harmonics

$$\begin{aligned} \dot{a}_1 &= -\mathrm{i}\varepsilon k_1 a_1^* a_2 \exp(-\mathrm{i}\,\Delta t), \\ \dot{a}_2 &= -\mathrm{i}\varepsilon k_1 a_1^2 \exp(\mathrm{i}\,\Delta t). \end{aligned} \tag{14.26}$$

Since the difference $\Delta = \omega_2 - 2\omega_1$ is not small, the amplitude a_2 remains small at any t and the solution to the second equation in (14.26) can be found under the assumption that $a_1 = \text{const}$. Hence,

$$a_2(t) = -\varepsilon k_1 \Delta^{-1} a_1^2(t) \exp(\mathrm{i}\,\Delta t). \tag{14.27}$$

Substituting this into the first equation of (14.26), we obtain for $a_1(t)$ the time derivative

$$\dot{a}_1(t) = \mathrm{i}\varepsilon^2 k_1^2 \Delta^{-1} |a_1|^2 a_1. \tag{14.28}$$

The solution to (14.28) is

$$a_1 = a_1(0) \exp(-\mathrm{i}\,\delta\omega t), \quad \text{where} \qquad \delta\omega = -\varepsilon^2 k_1^2 \Delta^{-1} |a_1|^2.$$

We see that a harmonic wave remains harmonic and only its frequency changes by $\delta\omega$, which is proportional to its amplitude squared. This phenomenon is called *nonlinear dispersion.*

Note that we have just obtained the frequency change in the ideal case of an infinite harmonic wave. In the more realistic case of a quasiharmonic wave of finite duration with a finite frequency band spectrum, the interaction of the constituents of this spectrum must be taken into account. It appears (Exercise 14.3.11) that this interaction also gives a frequency shift proportional to $\varepsilon^2 k_1^2 |a_1|^2$.

Note also that the right-hand side of (14.28) already corresponds to the nonlinear interaction of the third order, the synchronism condition for which are $k_3 + k_4 = k_1 + k_2$, $\omega_3 + \omega_4 = \omega_1 + \omega_2$. The case under consideration corresponds to $k_1 = k_2 = k_3 = k_4$. The terms of the same kind as on the right-hand side in (14.28) could be obtained in a direct way by adding terms of the 3rd order with respect to u in (14.1).

The synchronism conditions are satisfied for arbitrary dispersion if $k_1 = k_2 = k_3 = k_4$. Hence, nonlinear corrections to the frequency will always occur in waves with dispersion provided that the right-hand side in (14.28) is not zero. It can be shown, for example, that for surface gravity waves in the case of deep water (dispersion law: $\omega^2 = gk$), one has $\delta\omega/\omega = (ka)^2/2$, a being the amplitude of the vertical surface displacement.

14.3 Exercises

14.3.1. Show that the dispersion law for gravity waves in shallow water is that for linear waves of the KdV equation if small quantities of the order of $(kH)^4$ and higher are neglected.

Solution: Expanding the dispersion relation for gravity waves $\omega^2 = gk \tanh kH$ in a power series with $kH \ll 1$ gives

$$\omega = \sqrt{gk}\sqrt{kH - (kH)^3/3} \simeq \sqrt{gH}k[1 - (kH)^2/6] = c_0 k - \beta k^3$$

which coincides with (14.3) when $c_0 = \sqrt{gH}$ and $\beta = c_0 H^2/6$.

14.3.2. Obtain the Riemann solution for the one-dimensional nonlinear acoustic equations.

Solution: Suppose that the equilibrium density and entropy are constant throughout the fluid. Then the equation of state is $p = p(\rho)$ and $\partial p/\partial x = (\partial p/\partial \rho)\,\partial \rho/\partial x = c^2(\rho)\,\partial \rho/\partial x$. Taking this into account, we write the continuity equation (6.9) and the Euler equation (6.4) for the one-dimensional case as

$$\frac{\partial \rho}{\partial t} + \rho\frac{\partial u}{\partial x} + u\frac{\partial \rho}{\partial x} = 0, \qquad \frac{\partial u}{\partial t} + u\frac{\partial u}{\partial x} + \frac{c^2}{\rho}\frac{\partial \rho}{\partial x} = 0.$$

Introduce the Riemann function

$$R(\rho) = \int_{\rho_0}^{\rho} c(\rho)\frac{d\rho}{\rho}, \qquad \frac{\partial R}{\partial t} = \frac{c}{\rho}\frac{\partial \rho}{\partial t}, \qquad \frac{\partial R}{\partial x} = \frac{c}{\rho}\frac{\partial \rho}{\partial x}$$

and the two new functions $f = (R + u)/2$ and $g = (R - u)/2$, so that

$$\frac{\partial f}{\partial t} + (c + u)\frac{\partial f}{\partial x} = 0, \quad \frac{\partial g}{\partial t} - (c - u)\frac{\partial g}{\partial x} = 0.$$

Each of these equations is analogous to (14.9). The first of them describes the propagation of the disturbance f with velocity $c + u$; the second, the propagation of the disturbance g with velocity $u - c$. Since the velocity $u = f - g$ is changing, the propagation velocities for f and g are also changing, because, for example, f passes through points with different g. If, however, $g = 0$ at $t = 0$, it remains zero at any time. Then $R \equiv u$, $f \equiv u$ and $\rho = \rho(u)$ and we obtain $(\partial u/\partial t) + (c + u)\,\partial u/\partial x = 0$. Solving this equation under the initial condition $u(x, 0) = F(x)$, we obtain $u = F[x - (c + u)t]$ which is Riemann's wave traveling in the positive x-direction. Analogously, if $f = 0$ and $u(x, 0) = G(x)$, we obtain Riemann's wave $u = G[x + (c - u)t]$ traveling in the opposite direction (if $u < c$). In the general case, the disturbances F and G propagate independently provided that they do not overlap. Otherwise they interact with each other but begin to propagate independently again after the overlapping ceases.

14.3.3. Using the momentum conservation law find the dependence of $l(t)$, the shock wave's extent, and $u_m(t)$ its maximum amplitude on time, assuming a triangular shape of the wave at any time (Fig. 14.5).

Solution: In the moving frame of reference $\xi = x - c_0 t$, we have

$$u(\xi, t) = u_m(t)\xi/l(t) \quad \text{if} \quad 0 < \xi < l(t)$$

and $u(\xi, t) = 0$ otherwise.

Such a disturbance is shown in Fig. 14.5 for $t = 0$ and arbitrary. The total momentum of the wave is

$$\mathcal{P} = \int_{-\infty}^{\infty} u(\xi, t)\,d\xi = l(t)u_m(t)/2 = l_0 u_0/2,$$

where $l_0 = l(0)$, $u_0 = u_m(0)$. We can see from Fig. 14.5 that

$$l(t) = \xi(0, u_m) + u_m t = \frac{u_m}{u_0} l_0 + u_m t = \frac{l_0}{u_0}\left(1 + \frac{u_0 t}{l_0}\right)u_m$$

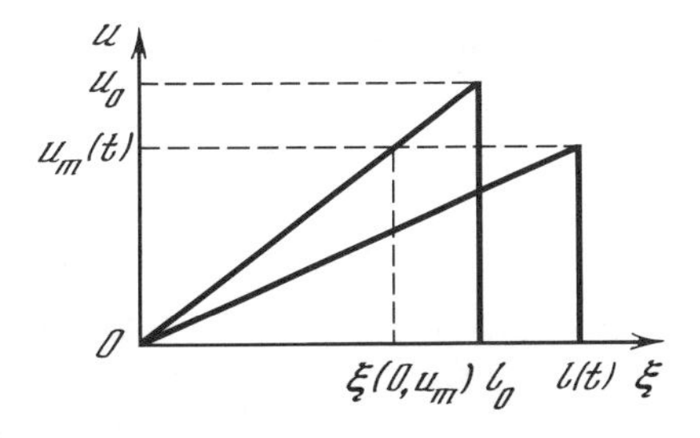

Fig. 14.5. Profile of a triangular shock wave at the initial moment $t = 0$ and at an arbitrary moment t

and obtain

$$u_{\mathrm{m}}(t) = u_0(1 + u_0 t/l_0)^{-1/2}, \qquad l(t) = l_0(1 + u_0 t/l_0)^{1/2}.$$

14.3.4. Find the stationary solution of the KdV equation in implicit form and the length (spatial period) of the cnoidal wave.

Solution: The expression under the radical on the right-hand side of (14.14) can be written as

$$\begin{aligned} 2\beta^{-1}(E - U) &= 2\beta^{-1}[E - (f^3/6) - (cf^2/2)] \\ &= (3\beta)^{-1}(f - f_1)(f - f_2)(f - f_3), \end{aligned}$$

where $f_1 < f_2 < f_3$ are the roots of $E - U(f) = 0$. After that, integration of (14.14) gives the required solution:

$$(3\beta)^{1/2}(\eta - \eta_0) = \pm \int_{f_2}^{f} [(\xi - f_1)(\xi - f_2)(\xi - f_3)]^{-1/2}\, d\xi .$$

Hence, the function $f(\eta)$ $[f_2 = f(\eta_0)]$ is expressed in terms of the elliptic integral of the first kind. The spatial period of the cnoidal wave is then

$$\begin{aligned} l &= 2(3\beta)^{1/2} \int_{f_2}^{f_3} [(\xi - f_1)(\xi - f_2)(\xi - f_3)]^{-1/2}\, d\xi \\ &= 4[3\beta/(f_3 - f_1)]^{1/2} \mathrm{K}[(f_3 - f_2)^{1/2} (f_3 - f_1)^{-1/2}], \end{aligned}$$

where $\mathrm{K}(s)$ is the complete elliptic integral.

14.3.5. Show that the synchronism conditions (14.18) can always be satisfied for isotropic waves $[\omega = \omega(k)]$ with $\omega(0) = 0$ if $\omega'(k) > 0$, $\omega''(k) > 0$ and these conditions cannot be satisfied if $\omega' > 0$, $\omega'' < 0$.

Solution: This follows from the following obvious reasoning. The dispersion curve $\omega(k)$ is convex downwards if $\omega'' > 0$, hence $\omega(|k_2 - k_1|) < \omega(k_1) + \omega(k_2) < \omega(k_1 + k_2)$. Introduce $\boldsymbol{k}_3 = \boldsymbol{k}_1 + \boldsymbol{k}_2$ so that when the angle between the vectors $\boldsymbol{k}_1$ and $\boldsymbol{k}_2$ changes between 0 and π, one has $|k_2 - k_1| \le k_3 \le k_1 + k_2$. Hence, the frequency $\omega_3 = \omega_3(k_3)$ will change continuously between $\omega(k_1 + k_2)$ and $\omega(|k_2 - k_1|)$ and will coincide with $\omega(k_1) + \omega(\boldsymbol{k}_2)$ at some point. If $\omega'' < 0$, however, we have $\omega(k_1) + \omega(k_2) > \omega(k_1 + k_2) \ge \omega(k_3)$ and such a coincidence is impossible.

14.3.6. Reduce the system of truncated equations (14.20a) to one ordinary differential equation.

Solution: Set $a_j = A_j \exp(i\varphi_j)$, $j = 1, 2, 3$ where $A_j = |a_j|$ is the amplitude and φ_j the phase of the jth wave. Now rewrite, for example, the first equation in (14.20a) as

$$\dot{A}_1 + iA_1\dot{\varphi}_1 = i\omega_1 V A_2 A_3 \exp[i(\varphi_3 - \varphi_1 - \varphi_2)].$$

The other two equations are written in an analogous way. Now introducing $\varphi \equiv \varphi_3 - \varphi_1 - \varphi_2$ and separating the real and imaginary parts, we obtain

$$\dot{A}_1 = -\omega_1 V A_2 A_3 \sin\psi, \quad \dot{\varphi}_1 = \omega_1 V A_1^{-1} A_2 A_3 \cos\psi = -A_1^{-1}\dot{A}_1 \cot\psi,$$
$$\dot{A}_2 = -\omega_2 V A_1 A_3 \sin\psi, \quad \dot{\varphi}_2 = \omega_2 V A_1 A_2^{-1} A_3 \cos\psi = -A_2^{-1}\dot{A}_2 \cot\psi,$$
$$\dot{A}_3 = \omega_3 V A_1 A_2 \sin\psi, \quad \dot{\varphi}_3 = \omega_3 V A_1 A_2 A_3^{-1} \cos\psi = A_3^{-1}\dot{A}_3 \cot\psi$$

and

$$\dot{\psi} = \dot{\varphi}_3 - \dot{\varphi}_1 - \dot{\varphi}_2 = \frac{\dot{A}_1 A_2 A_3 + \dot{A}_2 A_1 A_3 + \dot{A}_3 A_1 A_2}{A_1 A_2 A_3} \cot\psi$$
$$= \frac{d(A_1 A_2 A_3)/d\tau}{A_1 A_2 A_3} \frac{\cos\psi}{\sin\psi}.$$

Introducing $E_j \equiv A_j^2 = |a_j|^2$, we obtain a system of four equations with real quantities:

$$\dot{E}_1 = -2\omega_1 V (E_1 E_2 E_3)^{1/2} \sin\psi, \quad \dot{E}_2 = -2\omega_2 V (E_1 E_2 E_3)^{1/2} \sin\psi,$$
$$\dot{E}_3 = 2\omega_3 V (E_1 E_2 E_3)^{1/2} \sin\psi, \quad d[(E_1 E_2 E_3)^{1/2} \cos\psi]/d\tau = 0.$$

The conservation laws (14.21, 22) as well as the relation

$$(E_1 E_2 E_3)^{1/2} \cos\psi = \theta = \text{const} \quad \text{or} \quad \sin\psi = \pm[(E_1 E_2 E_3 - \theta^2)/E_1 E_2 E_3]^{1/2}$$

follow from these equations if the synchronism conditions (14.18) are also taken into account. Now express E_1 and E_2 in terms of E_3 from the second and the third equations in (14.32) and after that derive the required equation for E_3:

$$\dot{E}_3 = \pm 2\omega_1\omega_2 V [E_3(\omega_3 H_{13} - E_3)(\omega_3 H_{23} - E_3) - \omega_3^2\theta^2/\omega_1\omega_2]^{1/2}.$$

The latter can be solved in terms of the elliptic integrals (see also Exercise 14.3.4).

14.3.7 It was shown in Exercise 10.6.5 that the dispersion relation for gravity-capillary waves $\omega = \sqrt{gk + \gamma k^3}$ admits the resonance

$$\omega(2k_r) = 2\omega(k_r) \equiv 2\omega_r, \quad \text{where} \quad k_r = (g/2\gamma)^{1/2}.$$

Assuming dispersion of this kind for (14.1), integrate the corresponding truncated equations under the initial conditions

$$a_{k_r}(0) = a_1(0) = a_0\,, \qquad a_{2k_r}(0) = a_2(0) = 0\,.$$

Solution: Assume the solution of (14.1) in the form

$$u = a_1(\varepsilon t)\exp[\mathrm{i}(k_r x - \omega_r t)] + a_2(\varepsilon t)\exp[2\mathrm{i}(k_r x - \omega_r t)] + \text{c.c.}$$

We obtain, after substitution into (14.1),

$$\dot{a}_1 = -\mathrm{i}k_r a_1^* a_2\,, \qquad \dot{a}_2 = -\mathrm{i}k_r a_1^2\,.$$

In the same way as in the previous exercise we find

$$\dot{A}_1 = k_r A_1 A_2 \sin\psi\,, \qquad \dot{A}_2 = -k_r A_1^2 \sin\psi\,, \qquad d(A_1^2 A_2 \cos\psi)/d\tau = 0\,,$$

where

$$a_j = A_j \exp(\mathrm{i}\varphi_j)\,, \qquad \psi = \varphi_2 - 2\varphi_1\,, \qquad A_1(0) = A_0\,, \qquad A_2(0) = 0\,.$$

Now we obtain

$$A_1^2 + A_2^2 = A_0^2\,, \qquad A_1^2 A_2 \cos\psi = 0\,, \qquad \psi = -\pi/2\,, \qquad \sin\psi = -1\,,$$

which yields for A_2,

$$\dot{A}_2 = k_r(A_0^2 - A_2^2)\,, \qquad A_1(0) = 0$$

and after integration,

$$A_2 = A_0 \tanh(k_r A_0 t)\,, \qquad A_1 = A_0 \cosh^{-1}(k_r A_0 t)\,.$$

We see that the energy of the surface gravity waves can transfer into capillary waves.

14.3.8. Find the solution of the truncated equations (14.20a) under the condition that the amplitude of the low-frequency wave ω_1 is kept constant.

Solution: Differentiating the second or the third equation in (14.20a), we obtain taking into account the other equation $\ddot{a}_j = -\omega_2\omega_3 V|a_1|^2 a_j$, $j = 2, 3$, whence it follows $a_j = \alpha_j \cos\delta t + \beta_j \sin\delta t$, $\delta = (\omega_2\omega_3)^{1/2} V|a_1|$. The constants α_j, β_j can be determined from the initial conditions

$$a_j(0) = a_{j0}\,, \qquad \dot{a}_{20} = \mathrm{i}\omega_2 V a_1^* a_{30}\,, \qquad \dot{a}_3(0) = \mathrm{i}\omega_3 V a_1 a_{20}\,,$$

which gives $\alpha_2 = a_{20}$, $\alpha_3 = a_{30}$ and

$$\beta_2 = \mathrm{i}\omega_2 V \delta^{-1} a_1^* a_{30}\,, \qquad \beta_3 = \mathrm{i}\omega_3 V \delta^{-1} a_1 a_{20}\,.$$

If the amplitudes a_{20} and a_{30} are small at $t = 0$, they remain small at any time. This shows the stability of the low-frequency wave in the resonant triad.

14.3.9. Derive conditions under which two interacting surface gravity waves generate acoustic or internal gravity waves. Determine the directions of the generated waves.

Solution: Let $\boldsymbol{k}_1$, $\boldsymbol{k}_2$ and ω_1, ω_2 be the wave vectors and the frequencies of surface waves, respectively. While interacting they generate waves with $\boldsymbol{k} = \boldsymbol{k}_1 \pm \boldsymbol{k}_2$ and $\omega = \omega_1 \pm \omega_2$. The frequency of an internal wave ω_i must be much smaller than those of surface waves. This is possible if $\omega_\mathrm{i} = \omega_1 - \omega_2$, $\omega_2 \simeq \omega_1$, $k_2 \simeq k_1$. The internal wave travels in the direction which makes an angle θ_i with the vertical, so that $\sin\theta_\mathrm{i} = (\omega_1 - \omega_2)/N$ where N is the Väisälä frequency assumed constant.

The wavelength of a sound wave is much longer than that of a surface wave. Hence, the horizontal projection of the wave vector of the sound wave $\boldsymbol{k}_\mathrm{s}$ must be much smaller compared with $\boldsymbol{k}_1$. This is only possible for the combination of $\boldsymbol{k}_\mathrm{s} = \boldsymbol{k}_1 + \boldsymbol{k}_2$ with $\boldsymbol{k}_2 \simeq -\boldsymbol{k}_1$, so that $k_\mathrm{s} \ll k_1$. In this case, $\omega_2 \simeq \omega_1$ and the sound frequency $\omega_\mathrm{s} = \omega_1 + \omega_2 \simeq 2\omega_1$ is approximately twice the frequency of the surface waves. For the vertical component of the sound-wave vector we have $k_{\mathrm{s}z} = \sqrt{(\omega_\mathrm{s}/c)^2 - k_\mathrm{s}^2}$, where c is the sound velocity. The angle θ_s which the sound-wave vector makes with the vertical can be found from the relation $\sin\theta_\mathrm{s} = ck_{\mathrm{s}z}/2\omega_1$.

14.3.10. Find the amplitude of the sound wave generated by two surface gravity waves.

Solution: The total vertical displacement of the surface due to surface waves is

$$\zeta = \sum_{j=1}^{2} a_j \exp[\mathrm{i}(\boldsymbol{k}_j \cdot \boldsymbol{r} - \omega_j t)], \qquad \omega_j^2 = gk_j .$$

According to the previous exercise $\boldsymbol{k}_2 \simeq \boldsymbol{k}_1$, $\omega_2 \simeq \omega_1$; the horizontal component of the sound-wave vector is $\boldsymbol{k}_\mathrm{s} = \boldsymbol{k}_1 + \boldsymbol{k}_2$ and the sound frequency is $\omega_\mathrm{s} = \omega_1 + \omega_2$. We obtain the amplitude of the sound pressure using the perturbation method and substituting $\boldsymbol{v}_\mathrm{g}$ and p_g (the particle velocity and pressure of the gravity surface waves) into the nonlinear terms of the acoustic equations and boundary conditions. Since surface waves do not influence the medium density, the only nonlinear term in the acoustic equations (12.2) is $(\boldsymbol{v}_\mathrm{g} \cdot \nabla)\boldsymbol{v}_\mathrm{g}$. Taking into account that for a potential field $(\boldsymbol{v}_\mathrm{g} \cdot \nabla)\boldsymbol{v}_\mathrm{g} = \nabla(v_\mathrm{g}^2/2)$, we may write an equation for the sound pressure analogous to (12.3) as

$$\Delta p - c^{-2}\,\partial^2 p/\partial t^2 = -\rho_0\,\Delta(v_\mathrm{g}^2/2) \equiv F .$$

The dynamical boundary condition (10.5) at $\sigma = 0$ may be written in the second approximation (the gravity force can be neglected) as $p|_{z=0} = -\zeta(\partial p_g/\partial z)_{z=0} \equiv G$. The quantities p_g and $\boldsymbol{v}_g = \{\boldsymbol{u}_g, w_g\}$ are taken from (10.32) where $b_j = -\mathrm{i}\omega_j a_j$, $j = 1, 2$. Omitting the factor $\exp[\mathrm{i}(\boldsymbol{k}_j \cdot \boldsymbol{r} - \omega_j t)]$, we obtain

$$p_g = \rho_0 \sum_{j=1}^{2} \omega_j k_j^{-1} a_j \exp(k_j z), \qquad w_g = -\mathrm{i} \sum_{j=1}^{2} \omega_j a_j \exp(k_j z),$$

$$\boldsymbol{u}_g = -\mathrm{i} \sum_{j=1}^{2} \omega_j (\boldsymbol{k}_j/k_j) a_j \exp(k_j z).$$

To calculate F and G, we take only the resonance term $\exp[\mathrm{i}(\boldsymbol{k}_s \cdot \boldsymbol{r} - \omega_s t)]$ into account and neglect terms of the order $(k_s/k_j) \ll 1$, obtaining

$$F = \rho_0 |a_1|\,|a_2| 4k_1^2 \exp[(k_1 + k_2)z + \mathrm{i}(\boldsymbol{k}_s \cdot \boldsymbol{r} - \omega_s t)],$$
$$G = -\rho_0 \omega_1^2 |a_1|\,|a_2| \exp[\mathrm{i}(\boldsymbol{k}_s \cdot \boldsymbol{r} - \omega_s t)].$$

We look for the solution for p in the form

$$p = \{A \exp[(k_1 + k_2)z] + B \exp(\mathrm{i}k_{sz} z)\} \exp[\mathrm{i}(\boldsymbol{k}_s \cdot \boldsymbol{r} - \omega_s t)],$$
$$k_{sz}^2 = (\omega_s/c)^2 - k_s^2.$$

Substituting p into the equation and the boundary condition yields for the amplitude of the decaying term $A = \rho_0 |a_1|\,|a_2| \omega_1^2$ and that for the sound waves $B = -2\rho_0 |a_1|\,|a_2| \omega_1^2$.

14.3.11. Find the nonlinear frequency shift for a quasiharmonic wave train of finite extension in a medium with strong dispersion.

Solution: Let k_1 and $\Delta k \ll k_1$ be the wave number of the principal harmonic of the train and the width of the spectral band, respectively. The system (14.26) must now be extended, namely the interactions of spectral components of the wave train with long-wavelength waves ($\xi \leq \Delta k$) must be taken into account. For the corresponding spectral amplitude α_ξ, we obtain from (14.24)

$$\dot{\alpha}_\xi = -\frac{\mathrm{i}\varepsilon\xi}{2} \int_{\Delta k} [\alpha_{k_1+\eta}\alpha_{\xi-k_1-\eta} \exp(\mathrm{i}\Delta_+ t) + \alpha_{-k_1+\eta}\alpha_{\xi+k_1+\eta} \exp(\mathrm{i}\Delta_- t)]\, d\eta,$$

where

$$\Delta_\pm = \omega_\xi \mp \omega_{k_1+\eta} \pm \omega_{k_1+\eta\mp\xi} \simeq \omega_\xi \mp \omega_{k_1} \pm (d\omega/dk)_{k_1}\eta$$
$$\pm \omega_{k_1} \pm (d\omega/dk)_{k_1}(\eta - \xi) = \omega_\xi - c_g(k_1)\xi = [c_{\mathrm{Ph}}(0) - c_g(k_1)]\xi = \Delta_\xi.$$

One also has $\alpha_{k_1+\eta} \simeq \alpha_{k_1+\xi+\eta} \simeq \alpha_{k_1}$, $\alpha_{-k_1+\xi-\eta} \simeq \alpha_{-k_1+\eta} \simeq \alpha_{k_1}^*$, hence

$$\dot{\alpha}_\xi \simeq -\mathrm{i}\varepsilon\xi |\alpha_{k_1}|^2 \exp(\mathrm{i}\Delta_\xi t)\, \Delta k.$$

Assuming temporarily $\alpha_{k_1} = \text{const}$, we find

$$\alpha_\xi(t) \simeq -\frac{\varepsilon\xi}{\Delta_\xi}|\alpha_{k_1}|^2 \exp(\mathrm{i}\,\Delta_\xi t)\,\Delta k = -\frac{\varepsilon|\alpha_{k_1}|^2 \Delta k}{c_{\mathrm{Ph}}(0) - c_{\mathrm{g}}(k_1)} \exp(\mathrm{i}\,\Delta_\xi t)\,.$$

The following equation for α_k corresponds to such an interaction:

$$\alpha_k = -\mathrm{i}\varepsilon k \int\limits_{\Delta k} \alpha_\xi \alpha_{k-\xi} \exp(-\mathrm{i}\,\Delta_\xi t)\,d\xi \simeq \frac{\mathrm{i}\varepsilon^2 k_1 |\alpha_{k_1}|^2 \alpha_k (\Delta k)^2}{c_{\mathrm{Ph}}(0) - c_{\mathrm{g}}(k_1)}\,,$$

where $|k - k_1| \leq \Delta k$. Let $a_1 = \alpha_k\,\Delta k$ stand for the amplitude of the principal harmonic in the train, then taking into account the term corresponding to the interaction with a second harmonic, we obtain instead of (14.28),

$$\dot{a}_1 = \mathrm{i}\varepsilon^2 k_1^2 |a_1|^2 \{\Delta^{-1} + k_1^{-1}[c_{\mathrm{Ph}}(0) - c_{\mathrm{g}}(k_1)]^{-1}\} a_1\,.$$

Its solution $a_1 = a_1(0) \exp(-\mathrm{i}\,\delta\omega t)$, corresponds to harmonic waves with the frequency shift

$$\delta\omega = -\varepsilon^2 k_1^2 |a_1|^2 \{\Delta^{-1} + k_1^{-1}[c_{\mathrm{Ph}}(0) - c_{\mathrm{g}}(k_1)]^{-1}\}\,,$$

Hence, the nonlinear frequency shift has an additional term as compared with that obtained at the end of Sect. 14.2.4.

Appendix: Tensors

Different quantities are encountered in physics. The simplest of them, *scalars*, (mass, density, temperature, etc.) are pure numbers and have an absolute meaning independent of the coordinate system. Other quantities can be defined only relative to a certain frame of reference. The simplest of the latter are *vectors* which involve a magnitude and a direction; in three-dimensional space, they can be specified by three numbers—*components* along the axes of an orthogonal coordinate system, for example. Of course, the vector *per se* does not change if the coordinate system changes, but its components transform according to certain rules. These rules can easily be established for the case of a position vector $\boldsymbol{r}$ or a vector with components x_1, x_2, x_3 which are the coordinates of some point M. We shall confine ourselves to the three-dimensional case, an orthogonal coordinate system, and consider a rotation of this system. It is known that the "new" coordinates x'_1, x'_2, x'_3 (coordinates in the rotated system) of some point M are expressed in terms of initial ones x_1, x_2, x_3 (coordinates in the original system) by the relation

$$x'_i = \sum_{k=1}^{3} a_{ik}x_k \equiv a_{ik}x_k . \tag{A.1}$$

Here and below we use the convention that whenever the same index appears twice in the same term, this term must be automatically summed over this (the so-called *dummy*) index.

In (A.1), the coefficient a_{ik} is a projection of the unit vector $\boldsymbol{x}'_i$ of the $\boldsymbol{x}_i$-axes on the axes x_k (of the original system): $a_{ik} = \cos(\widehat{\boldsymbol{x}'_i, \boldsymbol{x}_k})$. The relations (A.1) also form the transformation law for the components of the position vector $\boldsymbol{r}$.

Now we can define an arbitrary vector in three dimensions: the set of quantities B_i $(i = 1, 2, 3)$ transformed (when the coordinate system is rotated) according to the rule

$$B'_i = a_{ik}B_k . \tag{A.2}$$

Analogously, a *tensor* is defined. A set of nine quantities g_{ik} transformed according to the rule

$$g'_{ik} = a_{il}a_{km}g_{lm} \tag{A.3}$$

is called a *second-order tensor*. A third-order tensor is a set of 27 quantities g_{ijk} transformed according to the rule $g'_{ijk} = a_{il}a_{js}a_{km}g_{lsm}$, etc. In all these transformations, a_{ik} are the same as in (A.1). Note that the vector and scalar can also be called tensors of first and zero order, respectively.

We now consider some relationships between the transformation coefficients a_{ik}. It is well known that rotation of the coordinate system can be specified by 3 quantities (Euler's angles). There are nine a_{ik}, however, which means that six relations must exist between a_{ik}. To establish these relations, we take into account that a_{ik} is a projection of the unit vector $\boldsymbol{x}'_i$ of the new coordinate system upon the axis x_k of the old one, i.e.,

$$\boldsymbol{x}'_i = a_{i1}\boldsymbol{x}_1 + a_{i2}\boldsymbol{x}_2 + a_{i3}\boldsymbol{x}_3 = a_{ik}\boldsymbol{x}_k . \tag{A.4}$$

Analogously, $\boldsymbol{x}'_l = a_{lm}\boldsymbol{x}_m$. We find the scalar product of the unit vectors

$$\boldsymbol{x}'_i \cdot \boldsymbol{x}'_l = a_{ik}a_{lm}\boldsymbol{x}_k \cdot \boldsymbol{x}_m . \tag{A.5}$$

Due to the orthogonality of the unit vectors one has

$$\boldsymbol{x}_k \cdot \boldsymbol{x}_m = \delta_{km} , \qquad \boldsymbol{x}'_i \cdot \boldsymbol{x}'_l = \delta_{il} , \tag{A.6}$$

where the Kronecker symbol δ_{km} is 1 for $k = m$ and 0 for $k \neq m$. Hence, on the right-hand side of (A.5), only terms with $k = m$ differ from zero so that

$$a_{ik}a_{lk} = \delta_{il} . \tag{A.7}$$

These are the required six relations (not nine by virtue of the symmetry relations $a_{ik}a_{lk} = a_{lk}a_{ik}$).

Relation (A.7) can also be written in another form. Consider an initial coordinate system x'_i and a new one x_k, taking into account also that the set a_{ki} specifies the inverse transform $\boldsymbol{x}_i = \cos(\widehat{\boldsymbol{x}_i, \boldsymbol{x}'_k})\boldsymbol{x}'_k = a_{ki}\boldsymbol{x}'_k$. Determining $\boldsymbol{x}_i \cdot x_l$ and a reasoning as above we find

$$a_{ki}a_{kl} = \delta_{il} . \tag{A.7a}$$

Obviously enough, all definitions and results considered above can be formulated in the space of n dimensions: $x_1, x_2, \ldots, x_n$.

Let us now consider some properties of tensors and prove some theorems.

Theorem A.1. If A_k is a vector and g_{ik} a second-order tensor, then $B_i = g_{ik}A_k$ are components of a vector.

To prove this we write the last relation in the new coordinate system $B'_i = g'_{ik}A'_k$ and use the transformation laws for g'_{ik} and A'_k: $g'_{ik} = a_{il}a_{km}g_{lm}$, $A'_k = a_{ks}A_s$. Then $B'_i = a_{il}a_{km}a_{ks}g_{lm}A_s$, but $a_{km}a_{ks} = \delta_{ms}$, $\delta_{ms}A_s = A_m$, therefore $B'_i = a_{il}g_{lm}A_m = a_{il}B_l$. Hence, the transformation law for B_i is that for the vector.

Theorem A.1a (inverse). If the set g_{ik} transforms an arbitrary vector A_k into the vector B_i by the formula $B_i = g_{ik}A_k$, then g_{ik} is a second-order tensor.

To prove it we note that in the new coordinate system we have $B'_i = g'_{ik}A'_k$. Taking into account that for vectors $B'_i = a_{il}B_l$, $A'_k = a_{km}A_m$, we obtain from the last relation $a_{il}B_l = g'_{ik}a_{km}A_m$. Multiplying this equation by a_{is} then summing over i, taking into account that $a_{is}a_{il} = \delta_{sl}$, $\delta_{sl}B_l = B_s$, we find $B_s = g'_{ik}a_{km}a_{is}A_m$.

Comparison of this relation with the initial one $B_i = g_{ik}A_k$ gives $g_{sm} = a_{km}a_{is}g'_{ik}$. Multiplying the last relation once more by $a_{ns}a_{pm}$ with subsequent summation over s, m and taking into account that $a_{km}a_{pm} = \delta_{kp}$, $a_{is}a_{ns} = \delta_{in}$, $\delta_{kp}\delta_{in}g'_{ik} = g'_{np}$, we finally obtain $g'_{np} = a_{ns}a_{pm}g_{sm}$ which is the tensor transformation.

Quite analogously, two other theorems can be proved, too!

Theorem A.2. If A_i and B_k are arbitrary vectors and g_{ik} the second-order tensor, then $M = g_{ik}A_iB_k$ is a scalar.

Theorem A.2a (inverse). If $M = g_{ik}A_iB_k$ is a scalar and A_i and B_k are arbitrary vectors, then g_{ik} is a second-order tensor.

Theorem A.3. The sum of the diagonal elements of a tensor is a scalar, i.e., *invariant* with respect to a rotation of the coordinate system. Indeed, we have from (A.3)

$$g'_{ii} = a_{il}a_{im}g_{lm} = \delta_{lm}g_{lm} = g_{mm}$$

which proves the theorem.

In vector calculus, the so-called *Gauss divergence theorem* or the relation

$$\int_V (\partial A_k/\partial x_k)\,dV = \int_S A_k n_k\,dS \tag{A.8}$$

is proved where V is a volume bounded by a surface S with the unit vector of external normal, $\boldsymbol{n}$. Now we are going to prove an analogous theorem for a tensor.

Theorem A.4. For a second-order tensor, the relation

$$\int_V (\partial g_{ik}/\partial x_k)\,dV = \int_S g_{ik}n_k\,dS \tag{A.8a}$$

holds which we can call the Gauss divergence theorem for a tensor. To prove it, we apply (A.8) to the vector $A_k = g_{ik}B_i$ where B_i is an arbitrary constant vector. Then (A.8a) can be rewritten as

$$B_i\left[\int_V (\partial g_{ik}/\partial x_k)\,dV - \int_S g_{ik}n_k\,dS\right] = 0\,.$$

Here, the brackets must be zero due to the arbitrariness of B_i and we obtain (A.8a) as a result.

Definition: A second-order tensor is *symmetric* (*antisymmetric*) if $g_{ki} = g_{ik}$ ($g_{ki} = -g_{ik}$). Obviously, a symmetric tensor has six independent components

$g_{11}, g_{22}, g_{33}, g_{12} = g_{21}, g_{13} = g_{31}, g_{23} = g_{32}$, and an antisymmetric one only three: $g_{12} = -g_{21}, g_{13} = -g_{31}, g_{23} = -g_{32}, g_{11} = g_{22} = g_{33} = 0$.

Theorem A.5. The symmetry or antisymmetry of a tensor is a property which remains invariant with respect to the rotation of the coordinate system.

To prove this we write

$$g'_{ki} = a_{kl}a_{im}g_{lm} = \pm a_{kl}a_{im}g_{ml} = \pm g'_{ik}.$$

Here, $+(-)$ corresponds to the symmetry (antisymmetry) of the tensor. The last relation proves the theorem.

Definition: A function of the tensor components is called an *invariant* of this tensor if this function does not change under a transformation of the coordinate system.

Theorem A.3 gives one such invariant (trace of the tensor) $g_{ii} = \text{const}$. To obtain other invariants, we first consider the concept of *principal values* and *principal axes* of a symmetric tensor.

Let us have a symmetric tensor g_{ik} and the vectors $\boldsymbol{A} = \boldsymbol{B} = \boldsymbol{r}$ ($A_i = B_i = x_i$). Then, by virtue of Theorem A.2, $M = g_{ik}x_ix_k$ is scalar. Now let $\boldsymbol{r}$, i.e., x_i vary but in such a manner that M remains constant and equal to unity. Then,

$$g_{ik}x_ix_k = g_{11}x_1^2 + g_{22}x_2^2 + g_{33}x_3^2 + 2g_{12}x_1x_2 + 2g_{13}x_1x_3 + 2g_{23}x_2x_3 = 1$$

describes a second-order surface called a *tensor surface*. In particular, when all $g_{ik} > 0$, this surface is an ellipse. It is known that by a suitable choice of axes x'_i, a quadratic function can be reduced to a canonical form so that the equation for a tensor surface $g'_{ik}x'_ix'_k$ becomes

$$\lambda_1(x'_1)^2 + \lambda_2(x'_2)^2 + \lambda_3(x'_3)^2 = 1 \tag{A.9}$$

or $g'_{11} = \lambda_1$, $g'_{22} = \lambda_2$, $g'_{33} = \lambda_3$, $g'_{ik} = 0$ if $k \neq i$. In other words, the tensor becomes a diagonal one. Such coordinates axes are called the *principal axes of a tensor* g_{ik} and the quantities λ_i its *principal values*.

Now consider a vector $\boldsymbol{A}$ which is parallel to one of the principal axes, for example, to the first. Then for the vector $B_i = g_{ik}A_k$, we have in a coordinate system whose axes coincide with the principal ones, $B'_1 = \lambda_1 A_1$, $B'_2 = B'_3 = 0$ or $\boldsymbol{B}' = \lambda_1 \boldsymbol{A}'$, i.e., the transformation $g'_{ik}A'_k$ just elongates, by a factor of λ_1, the vector $\boldsymbol{A}$. This fact cannot depend on the choice of a coordinate system, hence in any system we have $B_i = \lambda_1 A_i$ or $\boldsymbol{B} = \lambda_1 \boldsymbol{A}$ if vector $\boldsymbol{A}$ is parallel to the first principal axes of the tensor g_{ik} considered. This suggests a way for determining principal axes and principal values of second-order tensor. We have to find such values of λ for which the homogeneous system of equations

$$\begin{aligned} &g_{ik}A_k = \lambda A_i \quad \text{or} \\ &(g_{ik} - \lambda\delta_{ik})A_k = 0 \end{aligned} \tag{A.10}$$

has nonzero solutions. From algebra it is known that this is the case if the determinant of the system is zero, i.e.,

$$\begin{vmatrix} g_{11}-\lambda & g_{12} & g_{13} \\ g_{21} & g_{22}-\lambda & g_{23} \\ g_{31} & g_{32} & g_{33}-\lambda \end{vmatrix} = 0. \tag{A.11}$$

Equation (A.11) is one of 3rd degree in λ. Its 3 roots are the principal values of the tensor. For each λ_i ($i = 1, 2, 3$), we have a triad $A_k^{(i)}$ ($k = 1, 2, 3$) as a solution of (A.10). This triad determine three components of the vector A which is parallel to the ith principal axis.The principal axes are orthogonal to each other. To prove it consider, for example, the vectors $A_k^{(1)}$ and $A_k^{(2)}$ parallel to the principal axes 1 and 2. These vectors are solutions of (A.10), i.e., $g_{ik}A_k^{(1)} = \lambda_1 A_i^{(1)}$ and $g_{ik}A_k^{(2)} = \lambda_2 A_i^{(2)}$. Multiplying the first of these equations by $A_i^{(2)}$, the second by $A_i^{(1)}$ (summation over i is implied) and subtracting the second from the first, we obtain (taking into account that tensor g_{ik} is a symmetric one)

$$(\lambda_1 - \lambda_2)A_i^{(1)}A_i^{(2)} = 0.$$

We assume that all λ_i are different. Then, since $\lambda_1 - \lambda_2 \neq 0$, we have $A_i^{(1)}A_i^{(2)} = 0$ or $\boldsymbol{A}^{(1)} \cdot \boldsymbol{A}^{(2)} = 0$ which is the orthogonality condition for the vectors $\boldsymbol{A}^{(1)}$ and $\boldsymbol{A}^{(2)}$. If two or all three of the principal values coincide, the vectors $\boldsymbol{A}^{(i)}$ can be chosen mutually orthogonal due to the arbitrariness of the choice of the solutions to (A.10).

Now we can obtain more invariants of a second-order tensor. Note that the principal values, i.e., solutions to (A.11), must not depend on the choice of the coordinate system. Then, the coefficients of this equation must also be invariants with respect to a rotation of the axes. Evaluating the determinant (A.11), we write the equation for λ as $-\lambda^3 + I_1\lambda^2 - I_2'\lambda + D = 0$, where the coefficients $I_1 = g_{ii}$, $I_2' = g_{12}^2 + g_{13}^2 + g_{23}^2 - g_{11}g_{22} - g_{11}g_{33} - g_{22}g_{33}$,

$$D = \begin{vmatrix} g_{11} & g_{12} & g_{13} \\ g_{21} & g_{22} & g_{23} \\ g_{31} & g_{32} & g_{33} \end{vmatrix} \tag{A.12}$$

are invariants of the tensor g_{ik}. Invariance of I_1 was proved above independently (Theorem A.3). Instead of I_2', however, the invariant I_2 is usually used:

$$I_2 = I_1^2 2I_2' = g_{11}^2 + g_{22}^2 + g_{33}^2 + 2g_{12}^2 + 2g_{13}^2 + 2g_{23}^2 \tag{A.13}$$

which is always a positive quadratic function of the tensor elements.

Bibliographical Sketch

Part I

In writing this text the authors tried to present the material in such a way that readers familiar with university courses on physics and mathematics can successfully study this book, not being required to consult other literature. It is hoped that reading this book provides the reader with a good basis of knowledge about physical processes and phenomena which take place in continuous media and, in particularly, the wave motion. For the further study we recommend some additional literature.

The physics and general aspects of the theory of elasticity, presented in Chap. 1, 3 and Sect. 5.3., are treated in detail in:

Feynman, R., Leighton, R., Sands, M.: *The Feynman lectures on Physics,* Vol. 2 (Addison-Wesley, Reading, MA 1964)
Landau, L. D., Lifshitz, E. M.: *Teoriya uprugosti* (Theory of Elasticity) (Nauka, Moscow 1965)
Mase, G. E.: *Theory and Problems of Continuum Mechanics* (McGraw-Hill, New York 1970)
Sommerfeld, A.: *Mechanics of Deformable Bodies* (Academic, New York 1950)
Southwell, R. V.: *Introduction to the Theory of Elasticity* (Oxford U. Press, London 1966)
Timoshenko, S., Goodier, J. N.: *Theory of Elasticity,* 2nd. ed. (McGraw-Hill, New York 1951)

The theory of wave motion in rods and plates (Chaps. 2, 4, 5) as well as that of the propagation of longitudinal and transverse waves in elastic media are well covered in:

Brekhovskikh, L. M.: *Waves in Layered Media,* 2nd ed. (Academic, New York 1980)
Ewing, W. M., Yardetzky, W. S., Press, F.: *Elastic Waves in Layered Media* (McGraw-Hill, New York 1957)
Tolstoy, I.: *Wave Propagation* (McGraw-Hill, New York 1973)

Wave propagation in the media with dispersion (Sect. 2.6) is the subject of the monograph:

Karpman, V. I.: *Nonlinear Waves in Dispersive Media* (Pergamon, Oxford 1975)

and the collection of papers:

Leibovich, S., Seebas, A. R. (eds.): *Nonlinear Waves* (Cornell U. Press, Ithaca, NY 1974) (Chaps. 1 and 2)

Part II

There is a great amount of literature on principles of fluid dynamics to which Chaps. 6–8 are devoted. Some of the most useful ones are:

Batchelor, G. K.: *An Introduction to Fluid Mechanics* (Cambridge U. Press, London 1967)
Feynman, R., Leighton, R., Sands, M.: *The Feynman Lectures on Physics,* Vol. 2 (Addison-Wesley, Reading MA 1964)
Lamb, H.: *Hydrodynamics* (Dover, New York 1945)
Landau, L. D., Lifshitz, E. M.: *Fluid Mechanics* (Addison-Wesley, Reading MA 1959)
Milne-Thomson, L. M.: *Theoretical Hydrodynamics* (St. Martin's Press, New York 1960)

The similarity method discussed in Sect. 9.1.3 is also outlined in this literature.

In Sects. 8.3, 9.2.3 we have discussed the theory of boundary layers. This theory was also developed in:

Rosenhead, L. (ed.): *Laminar Boundary Layers* (Oxford U. Press, London 1964)
Schlichting, H.: *Grenzschicht-Theorie* (Braun, Karlsruhe 1951)

Aspects of the theory of turbulence (Chap. 9) can be found in:

Batchelor, G. K.: *The Theory of Homogeneous Turbulence* (Cambridge U. Press, London 1953)
Bradshaw, P. (ed.): *Turbulence*, 2nd. ed., Topics Appl. Phys., Vol. 12 (Springer, Berlin, Heidelberg 1978)
Monin, A. S., Yaglom, A. M.: *Statistical fluid Mechanics*, Vols. 1, 2 (MIT Press, Cambridge 1971)

Chapters 10 – 14 are devoted to the wave motion in fluids. More details on wave dynamics one can find in:

Gossard, E., Hooke, W.: *Waves in the Atmosphere* (Elsevier, New York 1975)
Kamenkovich, V. M.: *Fundamentals of Ocean Dynamics* (Elsevier, New York 1977)
LeBlond, P. H., Mysak, L. A.: *Waves in the Ocean* (Elsevier, New York 1978)
Lighthill, J.: *Waves in Fluids* (Cambridge U. Press, London 1978)
Tolstoy, I.: *Wave Propagation* (McGraw-Hill, New York 1973)
Whitham, G. M.: *Linear and Nonlinear Waves,* (Wiley, New York 1974)

In addition, a range of literature on specific types of waves in fluids is available. For example, for the study of gravitational and capillar surface waves (Chap. 10) we recommend:

Phillips, O. M.: *The Dynamics of the Upper Ocean,* 2nd ed. (Cambridge U. Press, New York 1977)
Stoker, J. J.: *Water Waves* (Interscience, New York 1953)

The reader interested in the theory of internal gravitational waves in the ocean and the atmosphere, including the effect of the earth rotation, will find the following works to be useful:

Eckart, C.: *Hydrodynamics of Ocean and Atmosphere* (Pergamon, New York 1969)
Krauss, W.: *Interne Wellen* (Gebrüder Borntraeger, Berlin-Nicolasee 1966)
Miroposlkii, Yu. Z.: *Dinamica vnutrennikh gravitatzionnikh voln v okeane* (The Dynamics of the Internal Gravity Waves in the Ocean) (Gidrometeoizdat, Leningrad 1981)
Phillips, O. M.: *The Dynamics of the Upper Ocean,* 2nd ed. (Cambridge U. Press, New York 1977)

A detailed analysis of the Rossby waaves (Chap. 2) was given in:

Kamenkovich, V. M.: *Fundamentals of Ocean Dynamics* (Elsevier, New York 1977)
Longuet-Higgins, M. S.: *Planetary Waves on a Rotating Sphere,* Proc. Roy. Soc. (London) A, **279**, 446 – 473 (1964)

We discussed the theory of acoustic waves in Chap 12. Among the numerous works we would like to call attention first of all to the fundamental work of

Rayleigh, J. W. S.: *The Theory of Sound*, Vols. 1, 2 (Dover, New York 1945)

and the monograph by

Skudrzyk, E.: *The Foundations of Acoustics* (Springer, Wien, New York 1971)

The propagation of acoustic waves in real inhomogeneous media (the ocean is such a medium) were studied in the literature. We can recommend the following books:

Brekhovskikh, L. M.: *Waves in Layered Media,* 2nd. ed. (Academic, New York 1980)
Brekhovskikh, L. M., Lysanov, Yu. P.: *Theoretical Fundamentals of Ocean Acoustics,* Springer Ser. Electrophys., Vol. 8 (Springer, Berlin, Heidelberg 1982)
DeSanto, J. A. (ed.): *Ocean Acoustics*, Topics Current Phys., Vol. 8 (Springer, Berlin, Heidelberg 1979)
Keller, J. B., Papadakis, J. S. (eds.): *Wave Propagation and Underwater Acoustics,* Lecture Notes Phys., Vol. 70 (Springer, Berlin, Heidelberg 1977)
Tolstoy, I., Clay, C. S.: *Ocean Acoustics* (McGraw-Hill, New York 1966)

A detailed investigation of sound propagation in inhomogeneous and moving media was given in Brekhovskikh L. M., Godin O. A.: *Acoustics of Layered Media I, II,* Springer Ser. Wave Phenom., vols. 5, 10 (Springer, Berlin, Heidelberg 1990, 1992)

In Chap. 13 we discussed some aspects of magnetohydrodynamics. Good exposees on the fundamentals of magnetohydrodynamics and on its applications to geophysics and astrophysics are available. Note the following of them:

Alexandrov, A. F., Bogdankevich, L. S., Rukhadze, A. A.: *Principles of Plasma Electrodynamics,* Springer Ser. Electrophys., Vol. 9 (Springer, Berlin, Heidelberg 1984)
Cowling, T. G.: *Magnetohydrodynamics* (Interscience, New York 1957)
Sherdiff, J. A.: *A Textbook of Magnetohydrodynamics* (Pergamon, New York 1965)
Stix, T. H.: *The Theory of Plasma Waves* (McGraw-Hill, New York 1962)

Problems of nonlinear wave theory (Chap. 14) are well presented in:

Karpman, V. I.: *Nonlinear Waves in Dispersive Media* (Pergamon, Oxford 1975)
Leibovich, S., Seebas, A. R. (eds.): *Nonlinear Waves* (Cornell U. Press, Ithaca, NY 1974)
Whitham, G. B.: *Linear and Nonlinear Waves* (Wiley, New York 1974)

A number of problems of nonlinear wave theory were considered at a discussion meeting on the nonlinear theory of wave propagation in dispersive systems (organized by M. J. Lighthill) whose proceedings have been published in the journal Proc. Roy. Soc. (London) A, **299** (1967).

The reader should bear in mind that the bibliography given here is rather incomplete and concerns mainly with the general problems of the mechanics of continua. However, those who are interested in literature on some specific problems should consult the references given in the books mentioned above.

When preparing this text the authors tried to use mathematical formalism familiar to physicists. Consequently, the following mathematical textbooks can be recommended:

Courant, R., Hilbert, D.: *Methods of Mathematical Physics* (Interscience, New York 1962)
McConnel, A. J.: *Application of Tensor Analysis* (Dover, New York 1957)
Morse, P. M., Feshbach, H.: *Methods of Theoretical Physics,* Vols. 1, 2 (McGraw-Hill, New York 1953)
Simmonds, J. G.: *A Brief on Tensor Analysis,* Undergraduate Texts in Mathematics (Springer, New York 1982)
Sommerfeld, A.: *Partial Differential Equations in Physics* (Interscience, New York 1953)
Thirring, W.: *A Course in Mathematical Physics,* Vols. 1 – 4 (Springer, Wien 1978 – 1983)

Subject Index

Springer-Verlag and the Environment

We at Springer-Verlag firmly believe that an international science publisher has a special obligation to the environment, and our corporate policies consistently reflect this conviction.

We also expect our business partners – paper mills, printers, packaging manufacturers, etc. – to commit themselves to using environmentally friendly materials and production processes.

The paper in this book is made from low- or no-chlorine pulp and is acid free, in conformance with international standards for paper permanency.